Green Analytical Chemistry in India

Vidhya Bhusari · Minal Ghante
Editors

Green Analytical Chemistry in India

Innovations for a Sustainable Future

Springer

Editors
Vidhya Bhusari
Department of Pharmaceutical Chemistry
Kalyan Jadhav's Educational Institute
Trinity College of Pharmacy
Pune, Maharashtra, India

Minal Ghante
Department of Pharmaceutical Quality Assurance Techniques, Sinhgad Technical Education Society's, Smt. Kashibai Navale College of Pharmacy
Savitribai Phule Pune University
Pune, Maharashtra, India

ISBN 978-3-032-24563-2 ISBN 978-3-032-24564-9 (eBook)
https://doi.org/10.1007/978-3-032-24564-9

This Springer imprint is published by the registered company Springer Nature Switzerland AG
The registered company address is: Gewerbestrasse 11, 6330 Cham, Switzerland

Contents

Principles and Practices of Green Analytical Chemistry

Vidhya K. Bhusari, Moreshwar P. Mahajan, and Vasundhara N. Sawant

Abstract Green Analytical Chemistry (GAC) represents a transformative approach to analytical science that integrates the principles of environmental sustainability into every stage of analysis—from sampling to detection. By extending the 12 Principles of Green Chemistry to analytical methodologies, GAC seeks to minimize the use of hazardous reagents, reduce waste generation, improve energy efficiency, and ensure operator safety without compromising analytical accuracy or precision. This chapter outlines the foundational concepts of GAC, including waste prevention, solvent substitution, miniaturization, and real-time monitoring, and emphasizes the role of renewable feedstocks and safer chemicals. Key sustainability metrics such as the E-factor, atom economy, Process Mass Intensity (PMI), Analytical Eco-Scale, and Analytical GREEnness Evaluator (AGREE) are discussed as quantitative tools for assessing and improving method greenness. The chapter also explores life cycle assessment (LCA) as a comprehensive framework for evaluating environmental impact, along with the use of green solvents—ionic liquids, deep eutectic solvents, and water-based systems. Finally, strategies for designing eco-friendly analytical methods and digital tools for predicting greenness are presented. Together, these approaches demonstrate how analytical chemistry can evolve toward a sustainable and responsible discipline aligned with global environmental goals.

Keywords Green analytical chemistry · Sustainability · Eco-scale · AGREE · Life cycle assessment · Green solvents · Deep eutectic solvents · Ionic liquids · Miniaturization

V. K. Bhusari (✉)
Department of Pharmaceutical Chemistry, KJ's Educational Institute, Trinity College of Pharmacy, Pune, Maharashtra, India
e-mail: vidhyabhusari@gmail.com

M. P. Mahajan
Department of Pharmaceutical Chemistry, Sinhgad Technical Education Society's, Sinhgad Institute of Pharmacy, Narhe, Pune, Maharashtra, India

V. N. Sawant
Department of Pharmaceutical Chemistry, Sinhgad Technical Education Society's, Smt. Kashibai Navale College of Pharmacy, Pune, Maharashtra, India

V. Bhusari and M. Ghante (eds.), *Green Analytical Chemistry in India*,
https://doi.org/10.1007/978-3-032-24564-9_1

1 Green Analytical Chemistry (GAC) and the 12 Principles of Green Chemistry

The 12 Principles of Green Chemistry [1], which were first created to direct chemical synthesis, are directly expanded upon and modified for use in analytical procedures by Green Analytical Chemistry (GAC). Analytical chemistry is concerned with identifying, measuring, and characterizing substances, as opposed to synthetic chemistry, which is more concerned with producing molecules. However, these procedures may generate toxic wastes, require a lot of energy, and use a lot of hazardous chemicals. Thus, in order to reduce environmental impact while maintaining analytical accuracy, dependability, and safety, GAC reinterprets the 12 principles [2]. All phases of analysis are affected by the adaptation, which turns analytical practice into a more sustainable science. These phases include sampling, sample preparation, separation, detection, and data processing (Fig. 1).

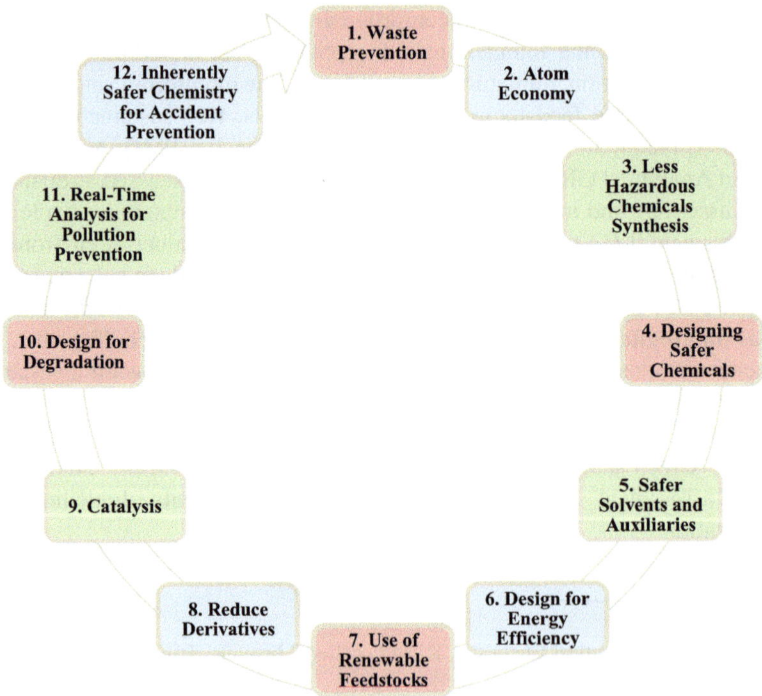

Fig. 1 Twelve principles of green chemistry

1. **Waste Prevention**
   Reducing waste at the source instead of managing it after production is the most straight forward way to apply this idea. Large amounts of solvent waste are frequently produced by conventional analytical procedures such as large-volume chromatography or Soxhlet extraction. Rather, GAC encourages miniaturization by using methods like liquid-phase microextraction (LPME), solid-phase microextraction (SPME), and microfluidic "lab-on-a-chip" systems. These significantly reduce the production of hazardous waste in addition to consuming microliter or nanoliter volumes of solvent. In order to avoid expensive and environmentally damaging waste treatment processes, preventive thinking also promotes the use of techniques that are naturally less wasteful [3].
2. **Atom Economy**
   Atom economy is mainly used in chemical synthesis, but it can be applied to analytical chemistry by reducing or avoiding derivatization reactions. Derivatization adds additional reagents, produces byproducts, and complicates waste streams. It is frequently used to increase analyte volatility or detectability. GAC promotes direct analysis techniques, such as headspace gas chromatography without derivatization or sophisticated detectors that can handle complex matrices without requiring chemical modification, such as Fourier-transform infrared spectroscopy (FTIR) and mass spectrometry (MS). This increases efficiency and lowers the need for chemicals by guaranteeing that the majority of the analyte being studied is measured directly [4].
3. **Less Hazardous Chemical Syntheses**
   This idea relates to selecting safer reagents and solvents in analytical settings. Conventional techniques frequently use liquid–liquid extraction with chlorinated or extremely hazardous organic solvents (such as dichloromethane or chloroform). Greener substitutes for these include water-based systems, ionic liquids (ILs), deep eutectic solvents (DESs), and natural deep eutectic solvents (NADES). In addition to being less toxic, these alternatives are frequently recyclable and biodegradable. Subcritical water extraction, for example, totally eliminates dangerous organic solvents by using only heated water under high pressure [5].
4. **Designing Safer Chemicals**
   Although they are not primarily in charge of creating molecules, analytical chemists are in charge of choosing the reagents used in detection procedures. Risks to human health and the environment can be considerably decreased by using non-toxic dyes, environmentally friendly colorimetric reagents, enzyme-based biosensors, or antibody-based immunoassays. This idea is supported, for instance, by the substitution of enzyme–substrate systems for spectrophotometric detection in place of carcinogenic diazonium salts [6].
5. **Safer Solvents and Auxiliaries**
   Solvent substitution is essential to GAC since solvents make up the majority of chemical use and waste in analytical workflows. Methods consist of: supercritical CO_2 extraction (instead of solvents derived from petroleum). Subcritical

water extraction, which uses water that is between 100 and 374 °C under pressure as a green solvent. Micellar solutions are chromatographic separation systems based on aqueous surfactants. In addition to reducing toxicity, these methods frequently allow for quicker analysis times and better recoveries [4, 5].

6. **Design for Energy Efficiency**
   Pressurized liquid extraction (PLE), ultrasound-assisted extraction (UAE), and microwave-assisted extraction (MAE) have supplanted energy-intensive sample preparation techniques like traditional Soxhlet extraction. They save time and energy because they are much faster, use less solvent, and run in softer environments [7]. On the instrumentation side, this idea is perfectly aligned with the development of portable, low-power analytical devices, which are occasionally combined with renewable power sources (such as solar-powered sensors).
7. **Use of Renewable Feedstocks**
   Renewable resources are used in GAC as solvents, sorbents, and even instrument parts. Examples include biobased solvents like NADES made from natural metabolites (such as sugars, organic acids, and amino acids) and biopolymer-based solid-phase extraction sorbents (chitosan, cellulose, and starch derivatives). These renewable resources provide biodegradability and lessen dependency on petroleum-based inputs [6].
8. **Reduce Derivatives**
   Derivatization has historically been used in many analytical techniques, such as silylation in gas chromatography. Derivatization, however, results in more waste and chemical use. GAC encourages the use of greener derivatization agents when necessary and direct analytical methods (such as DART-MS, or direct analysis in real-time mass spectrometry). For example, in peptide analysis, water-soluble carbodiimides can be used in place of hazardous reagents [4].
9. **Catalysis**
   In detection or sensor steps, catalysts—most notably enzymes, nanocatalysts, or heterogeneous catalysts—improve selectivity, lower reagent consumption, and increase reaction efficiency. Classical examples are biosensors based on enzymes. Electrochemical/optical sensors use nanomaterials (such as graphene and gold nanoparticles) to amplify signals with little reagent use.
10. **Design for Degradation**
    After use, solvents and analytical reagents ought to break down into safe byproducts. Biodegradable solvents like NADES or aqueous-based systems are preferred over persistent, halogenated solvents like chlorinated organics. Long-term environmental accumulation is decreased as a result.
11. **Real-Time Analysis for Pollution Prevention**
    The focus on on-site, real-time monitoring is arguably the most robust application of green principles in GAC. Pollutants can be directly detected in the field using portable and small devices like lab-on-a-chip technologies, field-deployable Raman and infrared spectrometers, and electrochemical sensors. This minimizes chemical handling and waste by avoiding delays related to laboratory preparation, transportation, and sampling. Clinical diagnostics, food

safety, and environmental analysis all benefit greatly from real-time monitoring [8].

12. **Inherently Safer Chemistry for Accident Prevention**
    The risk of exposure to dangerous chemicals is decreased by employing automated, sealed, and miniature systems (such as enclosed chip-based devices and microfluidics). Additionally, this improves laboratory safety by lowering the possibility of spills, inhalation, or operator error.

The analytical sciences are transformed by GAC's reinterpretation of the 12 Principles of Green Chemistry, which emphasizes sustainability, operator safety, ecological responsibility, and accuracy, precision, and reproducibility. This dual emphasis guarantees that analytical chemistry actively lessens its own environmental impact in addition to acting as a tool for monitoring environmental contaminants. The future of analytical practice will continue to be redefined as green technologies develop through the incorporation of safer solvents, miniaturized platforms, real-time monitoring tools, and renewable materials.

2 Metrics for Green Analytical Techniques

A key component of Green Analytical Chemistry (GAC) is quantifying sustainability because it allows researchers to assess, contrast, and enhance analytical techniques over time. While accuracy, precision, sensitivity, and robustness are the main criteria used in traditional evaluations of analytical quality, environmental performance must also be taken into consideration in contemporary methods. A number of metrics, from those initially created for industrial chemistry to more recent, analytical chemistry-specific indicators, have been put forth to accomplish this.

The Environmental Factor (E-factor), which represents the ratio of the mass of waste produced to the mass of product obtained, is among the oldest and most significant metrics. Although it was first created for large-scale chemical processes, analytical workflows—where solvents, reagents, and disposable materials significantly increase waste—are using it more and more.

Atom economy, a molecular-level idea that gauges how well atoms from reactants are integrated into the finished product, is closely related. In analytical chemistry, this is especially important during derivatization processes, which frequently lower atom efficiency. Reactions or techniques that reduce needless chemical transformations are reflected in high atom economy.

Process Mass Intensity (PMI), which builds on the E-factor, takes into account all materials used in a process, not just those that directly contribute to waste. When evaluating complete analytical workflows, like chromatography, which involve solvents, carrier gases, and calibration standards, this metric is extremely useful. Metrics specifically designed for the analytical sciences have been introduced more recently. By allocating penalty points for elements like waste, excessive energy use, and hazardous reagents, the Analytical Eco-Scale offers a semi-quantitative evaluation.

By producing a numerical score, this system makes it possible to compare methods directly. In addition, the Analytical GREEnness Evaluator tool (AGREE) provides a comprehensive digital platform for assessing methods in relation to the 12 GAC principles, producing a visual radar chart in addition to a numerical score.

These metrics work together to create a useful toolkit. Some emphasize waste generation (E-factor, PMI), while others focus on method-specific greenness (Eco-Scale, AGREE) or chemical efficiency (atom economy). Their combined use facilitates a more thorough, empirically supported assessment of analytical procedures, directing the shift toward analytical sciences that are safer, more economical, and ecologically sustainable.

2.1 *Environmental Factor (E-Factor)*

One of the most well-known sustainability metrics in chemistry is the E-factor, which was first proposed by Roger Sheldon in the early 1990s and refined in 2007 [9–11]. Its definition, which is the mass ratio of waste produced to product obtained, offers a straightforward yet effective method of calculating a process's environmental impact.

$$\text{E-factor} = \text{Total mass of waste(kg)}/\text{Mass of product(kg)}$$

When the E-factor value is near zero, it means that there is less waste produced in comparison to the useful product. Because so many solvents, reagents, and purification processes are used in industrial chemistry, fine chemicals and pharmaceuticals usually have much higher E-factors (25–100) than bulk chemical production (10–14).

The E-factor concept has been modified for use in analytical chemistry to assess complete workflows, including post-analysis waste, instrumental analysis, sample preparation, and sampling. Organic solvents, mobile phases, carrier gases, reagents, standards, and disposable consumables like syringes or filters are all considered "waste" in this context. For example, large solvent volumes in traditional liquid–liquid extraction or Soxhlet extraction methods usually result in high E-factors, but miniaturized techniques like solid-phase microextraction (SPME) or dispersive liquid–liquid microextraction (DLLME) produce very little waste.

The E-factor's advantages include its ease of calculation, ability to compare different approaches, and emphasis on the significance of waste reduction. It does not, however, take into account the waste's toxicity or environmental impact; instead, 1 kg of harmless salt solution and 1 kg of hazardous chlorinated solvent are treated equally. Because of this, the E-factor is frequently used in conjunction with complementary metrics like Eco-Scale or PMI.

A low E-factor is a useful guide when designing environmentally friendly analytical protocols because it not only lessens environmental harm but also lowers costs and increases method efficiency in the context of Green Analytical Chemistry.

2.2 Atom Economy

Trost [11] first proposed the idea of atom economy (AE) as a way to quantify the effectiveness of chemical reactions by counting the number of atoms of the starting materials that are included in the finished product [12]. Atom economy emphasizes the quantity of waste produced at the molecular level, in contrast to yield, which only shows how well reactants are converted into products.

Atom economy becomes especially crucial in the sample preparation and derivatization processes of analytical chemistry. Analytes must be chemically modified for many traditional analytical procedures (e.g., derivatization for gas chromatography or colorimetric detection). These reactions frequently result in byproducts, lower atom economy, and the consumption of additional reagents.

The general equation is:

$$\text{Atom Economy}(\%) = (\text{Molecular weight of desired product} / \text{Sum of molecular weights of reactants}) \times 100$$

When a reaction has 100% atom economy, no atoms are wasted and every atom from the reactants is present in the finished product. For instance, the highest atom economy is naturally attained by techniques that permit direct analysis of analytes without derivatization. On the other hand, multi-step derivatization reactions usually have low atom economy scores, despite occasionally increasing selectivity or sensitivity.

In order to increase atom economy, analytical chemists strive to create workflows that use few or no reagents. Direct injection methods, biosensors, microextraction, and ambient ionization mass spectrometry are a few examples that reduce the chemical transformation of analytes.

In addition to lowering chemical waste, atom economy in GAC is important for coordinating analytical techniques with sustainability objectives. Inherently, a high atom economy approach is safer, more straightforward, and more environmentally friendly.

2.3 Process Mass Intensity (PMI)

As a comprehensive indicator of material efficiency, Process Mass Intensity (PMI) was first presented by the American Chemical Society Green Chemistry Institute (ACS GCI) and then refined by Curzons et al. [12], and associates in the early 2000s [13]. PMI takes into consideration all materials used in a process, whether they become waste or are a component of the finished product, in contrast to the E-factor, which concentrates on the waste to product ratio. It is described as:

$$\text{PMI} = \text{Total mass of all input materials(kg)} / \text{Mass of product(kg)}$$

Every unit of material input becomes product with no excess when the PMI value is 1, which is the ideal value. Increased resource consumption and inefficient procedures are indicated by higher PMI values. PMI is now used as a standard in large-scale industrial applications to compare synthetic pathways in bulk chemicals, fine chemicals, and pharmaceuticals [10–12].

Because PMI accounts for the hidden resource costs that other metrics frequently miss, it is becoming more and more significant in the context of analytical chemistry. Chromatography-based techniques, for instance, need consumables (such as vials, columns, and syringes), mobile phases, calibration standards, and carrier gases in addition to solvent consumption. By combining all of these, PMI provides a more thorough understanding of resource utilization than the E-factor by itself [12–14].

Take the high-performance liquid chromatography (HPLC) method, where a large PMI is produced by the constant flow of solvents (acetonitrile, methanol, and water) even though the sample itself may be small. On the other hand, PMI values are significantly decreased by solvent-free or miniature methods (e.g., direct MS analysis, electrochemical sensors, or microfluidics) [13, 14].

The fact that PMI promotes efficiency throughout the workflow rather than just concentrating on waste at the end is one of its main advantages. The inability of PMI to distinguish between toxic and benign inputs is a drawback, though; using large amounts of ethanol or water is penalized equally with hazardous solvents. As a result, PMI is frequently combined with metrics that take into account qualitative aspects like safety and hazard potential, like the Eco-Scale or AGREE [14–17].

2.4 Analytical Eco-scale

In 2012, Gałuszka, Migaszewski, Konieczka, and Namieśnik presented the Analytical Eco-Scale, one of the first semi-quantitative instruments created specially to evaluate the environmental friendliness of analytical techniques. Although general green chemistry metrics like the PMI, atom economy, and E-factor offer insightful information, they were initially created for synthetic chemistry and aren't always appropriate for analytical workflows. By providing a scoring system specifically designed to meet the demands of analytical sciences, the Eco-Scale fills this gap.

The ideal "perfectly green" approach is represented by a baseline score of 100 points at the start of the concept. Penalty points are subtracted from this total according to a number of criteria:

- Using toxic or dangerous chemicals;
- Using solvents excessively;
- Using energy-intensive equipment; producing a lot of waste;
- Workplace dangers or safety issues.

Three categories are then created from the final score:

- >75 points → Outstanding green approach
- 50–75 points = Acceptable greenness
- <50 points = Inadequate (method needs to be improved)

For instance, a UV–Vis spectrophotometric technique that uses very little energy and only small amounts of water as a solvent would lose very few points, scoring above 80 and being deemed "excellent." On the other hand, a conventional HPLC method that uses a lot of acetonitrile or methanol, needs constant energy input, and produces solvent waste might only be 50–60% green.

The main advantages of the Eco-Scale are its adaptability, transparency, and simplicity. It can identify areas for improvement, like changing solvents or reducing sample preparation size, and enables researchers to compare analytical techniques directly using a single score [16, 18]. Its drawbacks include the scoring system's degree of subjectivity, as different users may assign different penalty points. Furthermore, it does not completely take into consideration the effects of reagents or instruments over their life cycles [14, 16].

The Analytical Eco-Scale is still a frequently used and useful tool for incorporating green chemistry concepts into routine laboratory work in spite of these drawbacks. It serves as a foundation for Green Analytical Chemistry (GAC) by bridging the gap between theoretical sustainability concepts and their practical application in the development of analytical methods [14, 16, 18].

2.5 *Analytical GREEnness Evaluator (AGREE)*

One of the newest and most sophisticated instruments created to evaluate the environmental sustainability of analytical techniques is the Analytical GREEnness Evaluator (AGREE). AGREE, which was first presented by Pena-Pereira and colleagues in 2020 [15, 17], is a more thorough, principle-driven, and automated evaluation system that expands on earlier ideas like the Analytical Eco-Scale.

The 12 Principles of Green Analytical Chemistry (GAC) are directly incorporated into the framework of AGREE, in contrast to single-metric approaches like the E-factor, atom economy, or PMI [2]. Every principle is assessed separately, given a score, and then combined into a circular radar chart that graphically illustrates a method's advantages and disadvantages. Additionally, the tool produces a numerical greenness score ranging from 0 to 1, where 1 denotes the highest level of sustainability [15, 17, 19].

With the help of the user-friendly and open-source AGREE software, researchers can enter specifics about their methodology and receive an evaluation right away. Its use in academic and industrial laboratories is encouraged by its accessibility. For instance, because it uses less energy, less solvent, and produces less waste, a portable electrochemical sensor for heavy metal detection usually receives a higher AGREE score than an ICP-MS method [3].

AGREE's holistic viewpoint is one of its key advantages. It avoids overemphasizing waste by analyzing each of the 12 principles separately and instead takes into account more general issues like energy consumption, safety, renewability, and miniaturization [1, 2, 15, 17]. Additionally, scientists can easily spot areas for improvement with the help of the visual radar chart.

However, AGREE also has some limitations. The tool does not fully address life cycle assessment (LCA) aspects, such as instrument manufacturing or end-of-life disposal, and the evaluation process may involve subjectivity in the classification of some inputs [14, 16]. In spite of this, it is a major step toward putting the GAC principles into practice.

All things considered, AGREE offers a comprehensive and open platform for evaluating how environmentally friendly analytical techniques are. It provides a more principle-centered, methodical framework for sustainable analytical chemistry and is a great addition to older metrics like the Eco-Scale due to its harmony of quantitative scoring and visual presentation (Fig. 2).

The AGREE tool provides the broadest sustainability coverage by integrating all 12 GAC principles, whereas E-factor and PMI mainly emphasize waste and material efficiency.

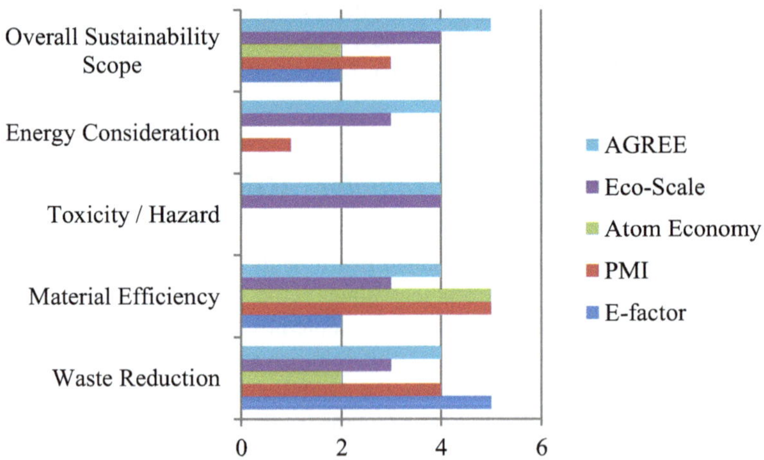

Fig. 2 Comparison of major green analytical metrics according to their focus areas. Scale 0 = not considered → 5 = strongly considered

3 Life Cycle Analysis in Analytical Chemistry

3.1 Introduction

A methodological framework for assessing the environmental effects of goods, procedures, or services over the course of their whole life cycle is called life cycle analysis (LCA), sometimes referred to as life cycle assessment. LCA offers an organized method for evaluating the upstream and downstream effects related to reagents, equipment, energy consumption, and waste management in analytical chemistry, in addition to the laboratory phases of an analytical procedure. Instead of just shifting the burden from one environmental domain to another, this holistic approach guarantees that "green" innovations in analytical chemistry actually contribute to sustainability [1, 2].

The International Organization for Standardization (ISO) established standardized LCA methodology through ISO 14040 and ISO 14044. These guidelines emphasize that the evaluation should cover all phases of a process, including resource extraction, production, use, and end-of-life treatment. For analytical chemistry, this means examining everything from solvent production and instrument energy demands to waste treatment and disposal.

3.2 Stages of LCA in Analytical Chemistry

i. **Goal and Scope Definition**

- The LCA's goal is made clear at this point. Comparing the environmental effects of two different sample preparation methods or assessing the total footprint of a laboratory workflow (e.g., HPLC vs. capillary electrophoresis) could be the objective for analytical chemistry.
- The boundaries of the system are also established, including whether the study solely looks at the laboratory phase or if it also takes into account upstream operations like solvent production and downstream operations like waste treatment.

ii. **Life Cycle Inventory (LCI)**

- Quantitative information on all inputs and outputs must be gathered for this. Solvents, reagents, consumables (columns, filters, vials), carrier gases, calibration standards, and electricity for instruments are examples of inputs. Emissions, waste solvents, byproducts, and CO_2 from electricity use are examples of outputs.
- Since many analytical chemistry techniques work at the micro- or nanogram level, but solvent and energy consumption can be disproportionately high, it can be difficult to obtain accurate inventory data.

iii. **Life Cycle Impact Assessment (LCIA)**

- Data from the inventory is now converted into impact categories, including acidification, eutrophication, photochemical smog, global warming potential, human toxicity, and ecotoxicity.
- For example, the extensive use of acetonitrile in HPLC adds to upstream environmental burdens associated with its petrochemical production in addition to local laboratory hazards.

iv. **Interpretation**

- "Hotspots" in the life cycle are found by analyzing the results. For example, a study may show that although solvent-free techniques reduce chemical waste, mass spectrometers' high electricity demand has a major impact on the climate.
- Improvement strategies, like switching to renewable solvents, employing energy-efficient equipment, or implementing smaller methods, are also guided by interpretation.

3.3 Applications of LCA in Analytical Chemistry

1. **Sample Preparation Methods**
   - Because of solvent production and waste management, traditional solvent-intensive techniques (such as liquid–liquid extraction and Soxhlet extraction) typically have greater environmental effects.
   - On the other hand, smaller methods such as microextraction in packed syringes (MEPS), dispersive liquid–liquid microextraction (DLLME), and solid-phase microextraction (SPME) use less solvent and therefore perform better in LCAs.
2. **Chromatographic Techniques**
   - Alternatives like supercritical fluid chromatography (SFC), which uses CO_2 as the mobile phase, or capillary electrophoresis, which uses minimal aqueous solutions, typically show better life cycle assessment (LCA) results.
   - High-performance liquid chromatography (HPLC) mainly relies on organic solvents like acetonitrile and methanol, which generate significant life cycle burdens.
3. **Analytical Solvents**
   - Deep eutectic solvents (DESs) and ionic liquids (ILs) are frequently marketed as environmentally friendly substitutes. LCA can, however, show that the energy-intensive synthesis of ILs may outweigh any local advantages.

- Since DESs are made of natural, biodegradable materials, they usually perform better in life cycle assessments (LCAs); however, the effects of large-scale production still need to be assessed.

4. **Instrumentation and Energy Use**
   - For pumps, detectors, and data systems, instruments like GC–MS, LC–MS, and ICP-MS need a lot of electricity. One of the main environmental hotspots identified by LCA is energy demand.
   - Because they require less energy and are easier to use, portable and point-of-care devices (such as paper-based devices and electrochemical sensors) exhibit lower life cycle burdens.

3.4 *Advantages of Using LCA in Analytical Chemistry*

- Holistic Evaluation: By taking into account every step of the process, LCA offers a comprehensive picture in contrast to single metrics (such as the E-factor).
- Prevents Burden Shifting: Prevents situations in which a reduction in solvents results in an increase in the demand for electricity or consumables.
- Assists in Decision-Making: Assists in selecting analytical techniques based on environmental sustainability in addition to sensitivity and accuracy.
- Supplementary to Green Metrics: Creates a comprehensive sustainability profile by collaborating with atom economy, PMI, Eco-Scale, and E-factor.

3.5 *Limitations and Challenges*

- **Data Availability**: It can be challenging to gather precise life cycle inventory data for analytical techniques.
- **Scale Issues**: Although analytical chemistry works on a small scale, the environmental effects can be overstated due to solvent and energy consumption.
- **Subjectivity in Boundaries**: The way system boundaries are established (e.g., whether instrument manufacture is included) affects the results.
- **Standardization is required** because LCA applications in analytical chemistry are still developing and do not yet have universal standards, in contrast to industrial chemistry.

3.6 *Conclusion*

An essential tool for integrating sustainability into analytical chemistry is life cycle analysis. It captures the actual worldwide impact of laboratory operations, going beyond regional evaluations of waste or hazard reduction. LCA makes sure that analytical chemistry not only yields accurate results but also fits with the larger objectives of resource conservation and environmental protection by locating environmental hotspots, evaluating alternative approaches, and directing methodological innovation. A more thorough framework for eco-efficient analytical science will be developed as the field develops by combining LCA with the current green metrics (Eco-Scale, AGREE, and PMI).

4 Green Solvents: Ionic Liquids, Deep Eutectic Solvents, and Water-Based Systems

Because solvents are necessary for sample preparation, extraction, separation, and detection, the choice of solvent is crucial in analytical chemistry. Solvents actually make up over 80% of the mass of chemicals used in industries and laboratories, according to estimates. For many years, traditional organic solvents like acetonitrile, methanol, hexane, toluene, chloroform, and dichloromethane have been used extensively because of their ability to dissolve analytes, compatibility with instruments, and predictable physicochemical characteristics. However, there are serious health, safety, and environmental issues associated with their widespread use. Since many of these solvents are flammable, toxic, volatile, and made from non-renewable petrochemical resources, their widespread use is not long-term viable.

- Reduction of toxicity and health risks to people.
- Lower emissions of volatile organic compounds (VOCs), enhancing lab safety and lowering air pollution.
- Preference for biodegradable and renewable feedstocks.
- By permitting milder extraction and analytical conditions, energy efficiency is increased.
- Adherence to environmentally friendly metrics like life cycle assessment, atom economy, Process Mass Intensity (PMI), and Eco-Scale.

Ionic liquids (ILs), deep eutectic solvents (DESs), and water-based systems are three solvent classes that have garnered a lot of interest as sustainable alternatives in analytical chemistry during the last 20 years. Every one of these groups contributes special qualities that fit the guidelines of green chemistry.

1. Ionic Liquids (ILs): They are referred to as "designer solvents" because they are made up of organic cations coupled with different anions, which allows for precise control over their polarity, hydrophobicity, and solvation capacity. They are desirable substitutes for volatile organic solvents because of their low vapor pressure, which reduces air pollution.
2. Deep Eutectic Solvents (DESs): These are created by combining a hydrogen bond acceptor (HBA) and a hydrogen bond donor (HBD), resulting in a eutectic mixture with a melting point significantly lower than the melting points of the constituent parts. DESs, particularly natural DESs (NADESs), are economical

and environmentally friendly because they are frequently made from renewable, biodegradable substances like sugars, amino acids, or organic acids.

3. Water-Based Solvent Systems: Because of its abundance, affordability, and non-toxicity, water is the best "green solvent." Its use for dissolving both hydrophilic and hydrophobic analytes has increased due to recent advancements such as micellar solutions, subcritical and supercritical water, and water-based DESs. These systems show how water's performance as a universal solvent in analytical chemistry can be improved by altering its physicochemical environment.

These environmentally friendly solvents have revolutionized analytical techniques. Their increasing use in spectroscopic techniques, solid-phase extraction, liquid–liquid microextraction, and chromatography demonstrates that analytical precision and sustainability can coexist. However, there are drawbacks to their adoption as well, such as problems with viscosity in DESs, biodegradability issues for some ILs, and solubility limitations in pure water systems, which call for careful consideration using life cycle analysis (LCA) and complementary green chemistry metrics.

In summary, green solvents mark a paradigm shift in analytical chemistry, where sustainability, safety, and environmental responsibility are now taken into consideration when choosing a solvent rather than just compatibility and efficiency. This change emphasizes how GAC principles are incorporated into routine laboratory operations, guaranteeing that analytical chemistry develops into a discipline that is both exacting and environmentally conscious (Fig. 3).

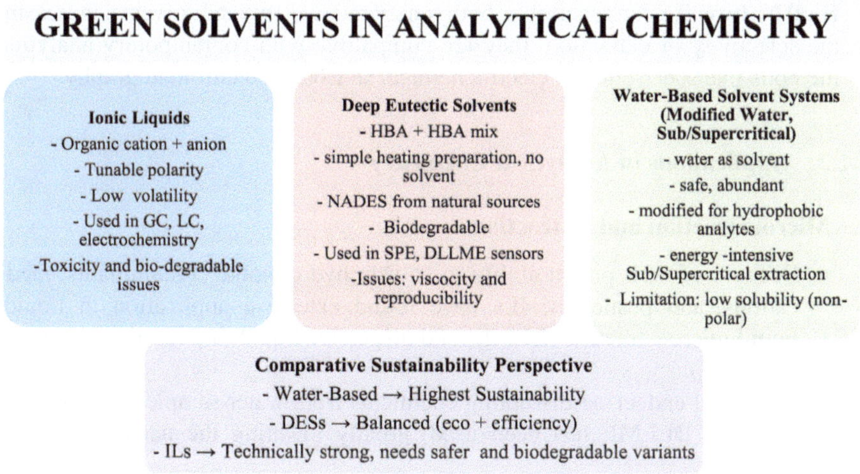

Fig. 3 Conceptual framework of green solvents used in analytical chemistry

4.1 *Ionic Liquids (ILs)*

Large, asymmetric organic cations (such as imidazo-lium, pyridinium, ammonium, and phosphonium) combined with different inorganic or organic anions (such as halides, tetrafluoroborate, hexafluorophosphate, and bis(trifluoromethylsulfonyl)imide) form ionic liquids (ILs). Because of the inadequate lattice packing of their large ions, ILs, in contrast to conventional salts, maintain their liquid state at or below 100 °C. They are known as "designer solvents" because of their distinctive structures, which allow them to have highly adjustable physicochemical characteristics.

In Green Analytical Chemistry (GAC), ILs are popular because of their low vapor pressure, which reduces emissions of volatile organic compounds (VOCs), and their capacity to dissolve a variety of polar and nonpolar substances. Additionally, because of their chemical adaptability, ILs can be specially made for particular analytical applications, providing benefits over traditional organic solvents.

4.1.1 Advantages of Ionic Liquids in Analytical Chemistry

- **Minimal Volatility**: lowers inhalation risks and air pollution.
- **Chemical and Thermal Stability**: appropriate for reactive or high-temperature systems.
- **Tunability**: By adjusting cation/anion combinations, one can modify solvation ability, polarity, viscosity, and miscibility.
- **Dual Nature**: ILs can function as both functional reagents and solvents, increasing the selectivity of extraction; they are compatible with contemporary analytical methods like spectrometry, electrochemical sensors, and chromatography.

4.1.2 Applications in Analytical Chemistry

1. **Microextraction and Extraction**

    - Because of their potent ability to solvate hydrophobic contaminants, medications, and pesticides, ILs have found extensive application in liquid–liquid microextraction (LLME) and dispersive liquid–liquid microextraction (DLLME).
    - To extract endocrine-disrupting chemicals from water samples, for instance, IL-based DLLME has been used, greatly lowering the need for organic solvents.

2. **Chromatography**

    - To improve separation effectiveness and peak resolution, these substances are added to liquid chromatography (LC) mobile phases.

- In gas chromatography (GC), where IL-based columns exhibit exceptional selectivity for polar analytes, they serve as stationary phases.

3. **Electrochemical Analysis**
   - ILs are used in modified electrodes for the detection of heavy metals, drugs, and biomolecules because of their high ionic conductivity and broad electrochemical windows, which make them excellent electrolytes.
4. **Aqueous two-phase systems (ATPS)**
   - ILs can form ATPS when combined with salts or polymers, enabling selective extraction of biomolecules such as proteins and enzymes with high efficiency.

4.1.3 Environmental Considerations

ILs are not always "green," despite their potential. Numerous ILs are toxic to aquatic life, according to studies, and because of their poor biodegradability, they may linger in the environment. Additionally, their synthesis may require hazardous precursors or energy-intensive steps. Thus, life cycle assessment (LCA) and green metrics (AGREE, Eco-Scale) must be used to assess the sustainability of ILs. The development of biodegradable ionic liquids (Bio-ILs), which combine the tunability of ILs with enhanced environmental compatibility, is the focus of recent research. Bio-ILs are made from natural feedstocks, such as amino acids, choline derivatives, and carbohydrates.

4.1.4 Conclusion

In analytical chemistry, ionic liquids have proven to be effective substitutes for conventional solvents, especially in the fields of extraction, chromatography, and electrochemistry. Although their environmental profile is still up for debate, their tunability, stability, and low volatility make them versatile. To guarantee that these "designer solvents" achieve their potential as genuinely green solvents in analytical chemistry, more research on bio-based and biodegradable ILs is essential.

4.2 Deep Eutectic Solvents (DESs)

A new class of solvents known as deep eutectic solvents (DESs) is created by combining a hydrogen bond donor (HBD) and a hydrogen bond acceptor (HBA) in a particular molar ratio. The mixture has a melting point that is significantly lower than either of the constituent parts due to extensive hydrogen bonding, producing a liquid that is at or close to room temperature. A typical example is the 1:2 mixture of urea (HBD) and choline chloride (HBA), which, although both substances are solids at room temperature, forms a liquid.

Although DESs and ionic liquids (ILs) are similar, DESs are usually less costly, simpler to make, and more biodegradable. DESs are especially appealing for green analytical chemistry (GAC) since they can be made by heating and stirring their constituents without the need for solvents, in contrast to ILs, which frequently require multistep synthesis.

4.2.1 Types of DESs

DESs are typically divided into four major groups:

1. Type I: Quaternary ammonium salts combined with metal salts (such as zinc chloride).
2. Type II: Hydrated salts plus metal halides.
3. Type III: HBDs (like urea, glycerol, and organic acids) combined with quaternary ammonium salts (like choline chloride).
4. Type IV: HBDs plus metal salts.

Type III DESs are the most extensively researched for analytical chemistry because of their affordability, simplicity in preparation, and utilization of renewable and biodegradable components.

Because they are made from natural metabolites (sugars, amino acids, organic acids, and cho-line) and offer excellent biocompatibility and biodegradability, a subcategory known as natural deep eutectic solvents (NADESs) has drawn interest.

4.2.2 Advantages of DESs in Analytical Chemistry

- Easy preparation without the need for solvents or purification procedures; Low toxicity and high biodegradability in comparison to most ILs.
- When NADESs are utilized, renewable feedstocks are utilized.
- Very low volatility, which lowers emissions into the atmosphere.
- Tunability: By adjusting the HBD/HBA combinations, solvent polarity, viscosity, and solvation ability can be customized.
- Biocompatibility and mild conditions make it compatible with biological samples.

4.2.3 Applications in Analytical Chemistry

1. Sample Extraction and Preparation

    - DESs have been extensively employed in microextraction methods like solid-phase extraction (SPE) and dispersive liquid–liquid microextraction (DLLME).

- Heavy metals, pesticides, medications, phenolic compounds, and food contaminants can all be effectively extracted by them [6].
- NADESs have demonstrated potential in the extraction of bioactive substances from foods and plants, providing sustainable substitutes for ethanol and methanol.

2. Chromatographic and Spectroscopic Analysis

   - DESs can improve spectroscopic sensitivity by acting as solubilizing agents for analytes that have low water solubility.
   - They are being researched as solvents for sample dissolution before GC–MS and LC–MS as well as mobile phase modifiers in liquid chromatography.

3. Applications of Electrochemistry

   - Because of their high ionic conductivity and electrochemical stability, DESs are utilized as electrolytes in sensors. They enhance the performance of the electrodes used to detect environmental pollutants, medications, and biomolecules.

4. Combining with Green Techniques

   - To increase efficiency while reducing solvent consumption, DESs are commonly used in conjunction with microwave-assisted extraction (MAE) and ultrasound-assisted extraction (UAE).

4.2.4 Challenges and Limitations

- One common disadvantage is high viscosity, which can reduce the effectiveness of extraction and mass transfer. Water is frequently added to reduce viscosity without sacrificing solvation ability, a process known as "aquo-DESs" [8].
- Absence of Toxicity Data: Although NADESs are thought to be safe, not all DESs have had their ecotoxicity and biodegradability thoroughly examined.
- Standardization Problems: Preparation conditions can affect the reproducibility of DES composition and performance.
- Scalability Issues: DESs' industrial-scale uses are still restricted when compared to ILs and traditional solvents.

4.2.5 Conclusion

One of the most promising types of green solvents for analytical chemistry is deep eutectic solvents (DESs). They are especially appealing for applications ranging from microextraction to electrochemical sensing because of their affordability, simplicity of preparation, biodegradability, and versatility. DESs are further aligned with sustainability goals by natural DESs (NADESs), which are made from renewable feedstocks. In green analytical workflows, DESs are quickly becoming recognized as viable substitutes for both organic solvents and ionic liquids, despite obstacles like high viscosity and the requirement for additional ecotoxicity research.

4.3 Water-Based Solvent Systems

Water's abundance, non-toxicity, non-flammability, biodegradability, and affordability make it the most sustainable solvent. Water is the "ideal solvent," according to Green Analytical Chemistry (GAC), for substituting dangerous organic solvents in sample preparation, extraction, and separation procedures. However, historically, its poor miscibility with many organic compounds, relatively high boiling point, and low solubility for hydrophobic analytes have limited its widespread use in analytical chemistry. In order to get around these restrictions, creative solutions have been created, making water a flexible medium for a variety of analytical uses.

4.3.1 Modified Water Systems

1. **Micellar Aqueous Solutions**
   - Micelles can encapsulate hydrophobic analytes when surfactants (such as sodium dodecyl sulfate or Triton X-100) are added to water.
   - Micelles function as pseudo-stationary phases in micellar electrokinetic chromatography (MEKC), increasing the efficiency of separation.
   - Additionally, micellar media make it possible to extract medications, pesticides, and food contaminants in an environmentally friendly manner.
2. **Subcritical and Supercritical Water**
   - Water's polarity dramatically reduces at high temperatures (100–374 °C) and pressures (to maintain liquid state), enabling it to dissolve hydrophobic analytes like hydrocarbons and PAHs.
   - Supercritical water has also been studied as a medium for the oxidation and degradation of hazardous compounds.
   - Subcritical water extraction (SWE) has been successfully used to isolate bioactive compounds, phenolics, and environmental pollutants without the use of organic solvents.
3. **Water-Based Deep Eutectic Solvents (Aquo-DESs)**
   - By reducing viscosity and improving mass transfer, water can be added to DESs to create hybrid water-based solvent systems.

- These systems combine the biodegradability of DESs with the adaptability of water, and they have been used in liquid–liquid microextraction of compounds like parabens and bisphenol A in edible oils.

4.3.2 Applications in Analytical Chemistry

Chromatography

- In order to lessen dependency on acetonitrile and methanol, aqueous mobile phases are becoming more and more important in high-performance liquid chromatography (HPLC).
- Green reversed-phase HPLC is one technique that uses buffered water with few organic modifiers.

Sample Preparation and Extraction

- For the isolation of natural products, food contaminants, and medications, subcritical water extraction is a promising substitute for Soxhlet extraction and organic solvent-based techniques.
- Micellar water systems enhance analyte solubilization, especially in food and environmental analysis.

Bio-analytical Applications

- Water-based NADESs offer biocompatible solvents for protein, nucleic acid, and metabolite extraction. When non-toxic solvents are needed for clinical and pharmaceutical analyses, these systems are extremely pertinent.

Electrochemical and Sensor Technologies

- Water-based electrolytes are integral to the development of portable electrochemical sensors for environmental monitoring, reducing the need for hazardous solvents.

4.3.3 Advantages and Limitations

Advantages

- Environmentally benign (safe, biodegradable, and non-toxic).
- Affordable and easily accessible.
- When modified (e.g., micelles, subcritical water), it can be used with both polar and hydrophobic analytes.
- Adaptable to a variety of analytical methods.

Limitations

- Needs specific equipment for subcritical or supercritical water applications (high-pressure vessels);
- Has poor solubility for highly hydrophobic analytes in ambient conditions.
- The possibility of analyte deterioration during SWE at high temperatures.
- Not as widely used in standard labs as organic solvents.

4.3.4 Conclusion

One of the most promising avenues for sustainable analytical chemistry is the use of water-based solvent systems. Water's inherent limitations are addressed by techniques like micellar solutions, subcritical/supercritical water, and aquo-DESs, which enable it to replace hazardous organic solvents in a wide range of applications. Analytical chemistry can attain high performance while upholding the values of environmental responsibility, safety, and renewability by incorporating water-based systems into chromatographic, spectroscopic, and extraction workflows. Water's position as the "green solvent of the twenty-first century" will be further strengthened by upcoming developments in equipment design and solvent modification techniques.

4.4 Comparative Evaluation and Sustainability

One of the most important developments in Green Analytical Chemistry (GAC) is the creation of green solvents, such as water-based systems, ionic liquids (ILs), and deep eutectic solvents (DESs). Although each class of solvent has distinct advantages, their sustainability profiles differ, so performance standards and environmental factors must be taken into account when choosing one. When developing environmentally friendly techniques, analytical chemists can make well-informed decisions with the aid of a comparative analysis.

4.4.1 Comparative Properties

i. **Ionic Liquids (ILs)**

Advantages: Great tunability, low vapor pressure, and broad use in electrochemistry, chromatography, and extraction.

Weaknesses: Energy-intensive synthesis, poor biodegradability, and toxicity concerns. Long term sustainability concerns are raised by the persistence of certain ILs in aquatic environments.

ii. **Deep Eutectic Solvents (DESs)**

Advantages: Inexpensive, simple to make, frequently biodegradable, and able to be made from renewable resources (NADESs). They are particularly well-suited for food and bio-analytical applications due to their low volatility and biocompatibility.

Drawbacks: Toxicity and biodegradability data for many synthetic DESs are still lacking, and high viscosity can impede mass transfer.

iii. **Water-Based Systems**

Advantages: Most readily available, non-toxic, biodegradable, safest, and most plentiful solvent. Its applicability to both hydrophilic and hydrophobic analytes is expanded by modified water systems (micellar, subcritical, and aquo-DESs).

Drawbacks include the need for specialized equipment, the possibility of analyte degradation at high temperatures during subcritical/supercritical extraction, and the limited solubility of nonpolar analytes under ambient conditions.

4.4.2 Sustainability Assessment Using Green Metrics

E-Factor and PMI

When compared to conventional organic solvents, DESs and ILs usually result in lower waste volumes. However, because DESs are easier to synthesize without the need for dangerous precursors, they frequently perform better than ILs when PMI (process mass intensity) is taken into account.

Atom Economy

Relevant during solvent synthesis but not directly related to solvent use. Because of their multi-step synthesis, ILs typically exhibit lower atom economy, whereas NADESs made from natural metabolites frequently have higher scores.

Eco-scale and Agree Tools

Due to their low toxicity and renewable sources, DESs and water-based systems usually receive higher Eco-Scale scores. AGREE, which incorporates all 12 GAC principles, frequently emphasizes that while DESs and water achieve greater alignment with green principles, ILs need to improve their biodegradability.

Life Cycle Assessment (LCA)

- Research indicates that ILs might have unreported upstream effects (such as energy-intensive synthesis or the use of halogenated precursors) that outweigh their laboratory benefits.
- LCA offers the most comprehensive sustainability assessment, accounting for effects from the sourcing of raw materials to disposal.

- Because they are renewable and biodegradable, DESs—especially NADESs—generally perform better in LCA.
- Water is still the most sustainable solvent throughout its life cycle because it is non-toxic and renewable, but modified water systems (like supercritical water) need energy-intensive conditions, which lessens their overall benefit.

4.4.3 Comparative Applications in Analytical Chemistry

- ILs are excellent as task-specific solvents in microextraction, electrochemistry, and advanced separations (GC stationary phases, LC additives).
- As DESs are compatible with biological samples, they are becoming more and more popular for sample preparation, environmental contaminant microextraction, and bioanalytical applications.
- Micellar electrokinetic chromatography (MEKC), subcritical water extraction for pollutants and natural products, and green HPLC mobile phases are all dominated by water-based systems.

4.4.4 Future Outlook

Hybrid strategies that integrate the benefits of multiple systems are the way of the future for green solvents. For instance, bio-based ILs use renewable feedstocks to address toxicity concerns, while aquo-DESs combine the tunability of DESs with the safety of water. Furthermore, task-specific solvents with optimal performance and sustainability profiles will be designed with the aid of computational modeling and artificial intelligence integration.

From the standpoint of sustainability:

- Before being widely used, ILs require considerable advancements in toxicity and biodegradability testing.
- When it comes to balancing cost, safety, and performance, DESs are currently the most promising.
- Although technological advancements are required to increase their applicability, water-based systems continue to be the gold standard for sustainability.

4.4.5 Conclusion

In conclusion, there isn't a single class of green solvent that is always better. Water-based systems, deep eutectic solvents, and ionic liquids all have unique advantages and disadvantages. ILs struggle with toxicity and life cycle issues despite their unmatched versatility. Water-based systems continue to be the gold standard of green chemistry because of their inherent safety and renewability, but DESs—especially natural DESs—offer a great balance between performance and sustainability. To make sure that solvent selection is in line with both analytical performance and environmental responsibility, a method-specific assessment utilizing green metrics (Eco-Scale, AGREE, LCA, and PMI) is necessary.

5 Designing Eco-friendly Analytical Methods

One of the most significant uses of Green Analytical Chemistry (GAC) is the development of environmentally friendly analytical techniques. Large volumes of samples and reagents, energy-intensive processes, and toxic solvents have long been the mainstays of traditional analytical techniques, which generate a substantial amount of hazardous waste. The goal of eco-friendly method design is to reduce negative effects on the environment and human health without sacrificing analytical precision, accuracy, or dependability. To achieve this, it is necessary to balance three equally significant dimensions rather than concentrating only on analytical performance:

1. Efficiency: accuracy, precision, sensitivity, and selectivity.
2. Sustainability, which includes energy efficiency, the use of renewable solvents and reagents, and less waste.
3. Applicability: usefulness, affordability, and ease of use for regular use in actual labs.

The idea of White Analytical Chemistry (WAC), which combines the "red" (analytical efficiency), "green" (environmental sustainability), and "blue" (applicability) dimensions into a comprehensive framework, is increasingly capturing this holistic viewpoint.

5.1 Principles Guiding Eco-friendly Method Design

The design of eco-friendly methods adapts the 12 principles of green chemistry and the 12 principles of green analytical chemistry to real-world workflows. One of the most important components is the reduction or elimination of toxic reagents; this involves:

i. Using green solvents (such as ethanol, ionic liquids, deep eutectic solvents, and water-based systems) instead of dangerous solvents like acetonitrile, benzene, and chloroform.
ii. Reduction of sample size and reagent volumes: to attain high sensitivity with low consumption, microfluidics, lab-on-a-chip platforms, and microextraction techniques are used.
iii. Waste prevention and treatment: creating strategies that produce little to no waste while including options for biodegradation or recycling in cases where waste cannot be prevented.

iv. Energy efficiency: using quicker and cleaner technologies like microwave-assisted extraction (MAE), ultrasound-assisted extraction (UAE), or supercritical fluids in place of time-consuming heating or reflux steps.
v. Safety and operator health: selecting techniques that reduce exposure to high-energy processes, fumes, and hazardous chemicals.
vi. Utilizing renewable resources: giving solvents, reagents, and materials made from natural or bio-based feedstocks priority.
vii. Real-time and direct analysis: to save time and chemicals, prefer techniques that require little to no sample preparation, such as ambient ionization MS and direct spectroscopy.

5.2 *Strategies for Designing Eco-friendly Methods*

The foundation of Green Analytical Chemistry (GAC) is the development of environmentally friendly techniques. Without sacrificing analytical precision and accuracy, the strategies seek to reduce environmental impact, maximize resource utilization, and enhance operator safety. Key tactics for achieving greener analytical practices are listed below.

5.2.1 Greener Solvent Selection

One of the main causes of analytical chemistry's environmental impact is the use of solvents. Acetonitrile, hexane, and dichloromethane are examples of conventional petroleum-based organic solvents that are frequently hazardous, combustible, and challenging to properly dispose of. Therefore, choosing greener solvents is essential.

1. Low-toxicity and renewable solvents: In extractions and separations, petroleum-based solvents can frequently be replaced by ethanol, water, and bio-derived solvents (such as glycerol and ethyl lactate). In high-performance liquid chromatography (HPLC), ethanol–water mixtures are being used more and more as environmentally friendly substitutes for acetonitrile.
2. Alternative solvent systems: Due to their tunable characteristics, potential for recycling, and low vapor pressure, ionic liquids (ILs) and deep eutectic solvents (DESs) show promise. The biodegradability of natural DES (NADES), which are made from sugars, amino acids, or organic acids, makes them especially appealing.
3. Micellar systems: To lessen dependency on dangerous organic solvents, chromatography also uses aqueous solutions based on surfactants.
4. Evaluation of a holistic solvent: In addition to solubility and extraction efficiency, the choice of solvent should take into account factors such as flammability, volatility, biodegradability, toxicity, and life cycle impacts (production → use → disposal). Making educated decisions is aided by resources like the GHS hazard assessments and the CHEM21 solvent selection guide.

5.2.2 Miniaturization and Automation

Miniaturization is the process of scaling down analytical processes in order to reduce the consumption of solvents and reagents.

- **Methods of Microextraction**: Analyte isolation using microliter-scale solvent volumes is made possible by techniques like solid-phase microextraction (SPME), dispersive liquid–liquid microextraction (DLLME), and microextraction in packed syringe (MEPS) that preserve sensitivity. When compared to conventional extraction, these methods can cut solvent consumption by over 90%.
- **Lab-on-a-Chip Systems and Microfluidics**: Microfluidic devices significantly reduce sample size and reagent use by combining several processes (sample preparation, separation, and detection) into a single microsystem. These systems are particularly useful for point-of-care testing and clinical diagnostics.
- **Automation**: Automated sample preparation and analysis reduces needless waste production, improves reproducibility, and reduces human error. For example, microextraction and robotic autosamplers increase throughput and decrease variability while conserving resources.

5.2.3 Energy-Efficient Technologies

In addition to using more resources, energy-intensive processes produce waste and excess heat. Greener technologies are made to speed up analysis while using less energy.

- By using dielectric heating to quickly transfer energy into the matrix and solvent, microwave-assisted extraction (MAE) cuts down extraction times from hours to minutes. Food, environmental, and pharmaceutical analysis all make extensive use of MAE.
- Cavitation effects are used in ultrasound-assisted extraction (UAE) to break up matrices and enhance analyte mass transfer. UAE is a cheap and widely available green technology that cuts down on extraction time and solvent usage.
- Carbon dioxide above its critical temperature and pressure is used as a solvent in supercritical fluid extraction (SFE). Supercritical CO_2 is easily accessible, recyclable, non-flammable, and non-toxic. It frequently offers greater selectivity than traditional solvents and is particularly useful for the analysis of pharmaceutical compounds, food ingredients, and natural products.

When combined, these technologies allow for analytical workflows that are quicker, cleaner, and more energy-efficient.

5.2.4 Direct and In-Field Analysis

Conventional workflows frequently entail time-consuming sample preparation, transportation, and sampling, all of which contribute to increased chemical use and waste.

Real-time monitoring is made possible by direct and in-field analysis, which resolves these problems.

- **Ambient Ionization Mass Spectrometry (AIMS)**: Analytes can be directly analyzed from surfaces or bulk samples with little to no preparation thanks to methods like desorption electrospray ionization mass spectrometry (DESI-MS) and direct analysis in real time mass spectrometry (DART-MS).
- **Portable Devices**: In situ monitoring of pollutants, food contaminants, or clinical biomarkers is made possible by portable chromatographs, handheld Raman and infrared spectrometers, and electrochemical sensors.
- **Benefits**: By avoiding pre-treatment and transportation, these methods save resources and facilitate quick decisions in clinical diagnostics, food safety, and environmental monitoring.

5.2.5 Use of Computational and Quality-by-Design (QbD) Approaches

Digital tools are essential for creating sustainable and successful methods.

- **Chemometrics**: Design of experiments (DoE) and multivariate analysis optimize analytical parameters (such as solvent ratio, extraction time, and temperature) while reducing the number of experiments. This increases method robustness while using less energy and material.
- **Quality-by-Design (QbD)**: QbD frameworks use methodical planning to guarantee that techniques fulfill predetermined quality standards. QbD improves reproducibility and reduces trial-and-error development by identifying critical process parameters and their relationships.
- **Predictive Modeling**: Before a method is actually put into practice, computational tools can model its greenness. For instance, solvent–analyte interactions can be predicted using molecular modeling, and the environmental impact can be virtually assessed using greenness metrics software (such as AGREE, GAPI, and Eco-Scale).
- **Sustainability Forecasting**: By incorporating life cycle assessment (LCA) into the design of the method, environmental effects are taken into account right away.

5.3 Evaluation Tools for Eco-friendly Methods

While creating more environmentally friendly analytical techniques is crucial, it's just as critical to assess their sustainability impartially. "Green" claims could continue to be arbitrary or inconsistent in the absence of standardized evaluation. In order to address this, a number of metrics and tools have been put forth to measure greenness and offer frameworks for making decisions. These consist of life cycle assessment (LCA), AGREE, GAPI, and Analytical Eco-Scale.

- Analytical Eco-Scale: gives out penalty points according to waste production and hazard; higher scores denote more environmentally friendly practices.
- A color-coded pictogram called the Green Analytical Procedure Index (GAPI) allows users to visually evaluate how environmentally friendly each method step—sample collection, preparation, analysis, and waste—is.
- The Analytical GREEnness Metric, or AGRE, is a software program that combines the 12 GAC principles into a single numerical score (0–1).
- Life Cycle Assessment (LCA): assesses the entire environmental impact, including hidden effects from waste treatment, instrument energy use, and solvent production.

5.4 Advantages and Challenges

Advantages

- Benefits include decreased waste and toxicity as well as lower operating expenses due to less energy and solvent use.
- More secure working conditions for lab employees.
- Adherence to global sustainability objectives (UN SDGs, circular economy principles).

Challenges

- Biodegradability is still an issue with some green solvents, such as ionic liquids.
- The initial cost of specialized equipment, such as supercritical fluid systems and microwave extractors, is high.
- Reproducibility may be impacted by a lack of laboratory standardization.
- There are still situations where sustainability and analytical performance must be traded off.

5.5 Conclusion

Rethinking conventional laboratory workflows from a sustainability perspective is necessary to design environmentally friendly analytical techniques. Choosing safer solvents, downsizing procedures, utilizing energy-efficient technologies, and incorporating sustainability assessment tools are all part of a comprehensive approach to method design that goes beyond merely substituting hazardous reagents. Analytical science can produce accurate results and be in line with environmental stewardship and societal responsibility if modern laboratories adopt a balance of performance, safety, and applicability, as embodied in ideas like White Analytical Chemistry.

6 Tools for Predicting Greenness

The incorporation of sustainability principles into analytical procedures is the focus of Green Analytical Chemistry (GAC). Although the 12 Green Chemistry principles offer general conceptual direction, they don't provide numerical standards for determining whether a particular approach is more environmentally friendly than another. Because of this, chemists have created standardized instruments and measurements that can be used consistently throughout analytical processes. By predicting, assessing, and comparing the greenness of methods, these tools enable researchers to make sure that sustainability objectives are taken into account in addition to analytical performance. The Analytical Eco-Scale, Analytical GREEN-ness metric (AGREE), and Green Analytical Procedure Index (GAPI) are some of the most popular frameworks. Because each tool offers a distinct viewpoint—whether it be through digital integration based on principles, visual representation, or numerical scoring—they are complementary rather than antagonistic.

6.1 Analytical Eco-scale

One of the earliest widely used instruments for evaluating greenness was the Analytical Eco-Scale. It was created by Gałuszka et al. [17] and is based on a system of penalty points, with 100 points going to the best green method. Penalty points are deducted for things like:

- Using dangerous or toxic chemicals (like chloroform or acetonitrile).
- Overuse of energy (e.g., high-pressure systems, lengthy heating times).
- The production of a lot of waste.
- The working conditions are unsafe.

The degree of sustainability is reflected in the final score:

- >75 points: Excellent green approach.
- 50–75 points: Good, but needs work.
- <50 points indicate poor sustainability.

For instance, a liquid chromatography method with little waste and ethanol–water as the mobile phase would perform better than one that uses a lot of acetonitrile. The Eco-Scale's primary benefits are its simplicity, transparency, and suitability for standard laboratory procedures; however, it has drawbacks in that it is unable to visually represent the full analytical workflow.

6.2 Green Analytical Procedure Index (GAPI)

The Eco-Scale's shortcomings are addressed by the Green Analytical Procedure Index (GAPI), which was introduced in 2018 and assesses an analytical method's whole life cycle, from sampling to waste management.

A multi-sectioned, color-coded pentagram (pictogram) serves as a representation of the assessment. Every segment aligns with a phase of the analytical process:

1. The gathering and transportation of samples.
2. Sample preparation (equipment, energy consumption, and solvents).
3. Instrumentation and chemicals.
4. Execution of the method (energy, safety, throughput).
5. The handling and disposal of waste.

Every factor has a rating of red (high impact), yellow (medium impact), or green (low impact). An immediate visual depiction of the areas with the greatest environmental burden is provided by the overall pentagram.

For example, if cryogenic cooling is required, a microextraction technique that uses water as the solvent may appear green in most sections but red for energy. Because it simplifies the greenness assessment, facilitates comparisons between different approaches, and effectively conveys results to policymakers and non-specialists, GAPI is especially useful.

6.3 Analytical GREEnness Metric (AGREE)

Launched in 2020, the AGREE tool is a software-based digital framework that introduces greenness assessment into the computational chemistry era. AGREE is completely in line with the 12 tenets of Green Analytical Chemistry, in contrast to Eco-Scale or GAPI, which are semi-quantitative or visual.

The user enters method details into an open-source software program that powers the tool. It produces:

- One numerical score, ranging from 0 to 1, with 1 being the most sustainable.
- A circular, color-coded diagram with 12 segments, each of which represents a different GAC principle.

Whereas deviations (such as excessive waste or hazardous reagents) show up as yellow or red sections, a greener method yields a consistent green circle.

AGREE's advantages include:

- Automation and digital accessibility.
- Modifiable weighting to highlight particular ideas.
- Simple integration for high-throughput screening with sizable datasets.
- Standardization, which makes it possible to compare the greenness of methods globally.

For instance, AGREE can quickly evaluate ten distinct chromatographic techniques and highlight the best one in terms of both overall performance and particular drawbacks (such as solvent usage or energy consumption). For laboratories creating new environmentally friendly processes, this makes AGREE a potent decision-support tool.

6.4 Comparative Evaluation

A critical comparison highlights their complementary roles:

Tool	Format	Strengths	Limitations
Eco-scale	Numerical (penalty point)	Simple, easy to apply, quantitative	Limited scope, does not visualize workflow
GAPI	Visual (pictogram)	Covers entire workflow, intuitive	Qualitative, less precise for fine comparisons
AGREE	Digital (score + diagram)	Principle-based, automated, holistic	Requires software, can be complex for non-experts

Researchers are increasingly advising combining these tools. For instance, AGREE incorporates principles for a comprehensive evaluation, GAPI identifies workflow hotspots, and Eco-Scale offers a rapid numerical benchmark.

6.5 Greenness Predictions for Future Trends

The field is quickly moving toward hybrid and specialized tools:

- AGREEprep: This stage, which is frequently the most resource-intensive, focuses exclusively on sample preparation.
- Based on toxicity, biodegradability, and renewability, the Green Solvent Selection Tools (GSST) assist in forecasting the sustainability of solvent selections.
- Machine learning models: These are being created to automatically forecast greenness ratings based on operational and chemical data.
- Whiteness and Blueness metrics: These are extensions of greenness that combine analytical performance (red) and method applicability (blue) to provide a comprehensive White Analytical Chemistry (WAC) score.

6.6 Conclusion

Integrating sustainability into laboratory science requires the ability to forecast how environmentally friendly analytical techniques will be. The way chemists quantify and convey sustainability has changed as a result of tools like the Eco-Scale, GAPI, and AGREE. Although each has advantages and disadvantages, when combined, they offer a multifaceted viewpoint that includes workflow visualization, numerical scoring, and principle-based holistic evaluation. Greenness prediction is expected to become even more precise, automated, and incorporated into routine analytical procedures with the continued development of digital and AI-driven methodologies.

References

1. Anastas, P. T., & Warner, J. C. (1998). *Green Chemistry: Theory and Practice.* Oxford University Press.
2. Koel, M., & Kaljurand, M. (2006). Application of the principles of green chemistry in analytical chemistry. *Pure and Applied Chemistry, 78(11), 1993–2002.* https://doi.org/10.1351/pac200678111993
3. Branković, S. (2023). Miniaturized analytical techniques: advancing green analytical chemistry. *Journal of Analytical Science and Technology, 14(1), 65.* https://doi.org/10.1186/s40543-023-00465-3
4. Pena-Pereira, F., Armenta, S., & de la Guardia, M. (2022). Green analytical chemistry: principles, applications, and metrics. *TrAC Trends in Analytical Chemistry, 148, 116530.* https://doi.org/10.1016/j.trac.2021.116530
5. Ražić, S., Onjia, A., & Đurđević, P. (2023). Sustainable solvents in green analytical chemistry. *Sustainability, 15(9), 7654.* https://doi.org/10.3390/su15097654
6. Martínez-Pérez-Cejuela, H., & Gionfriddo, E. (2024). Natural deep eutectic solvents and biopolymer-based sorbents: sustainable solutions for sample preparation. *Analytica Chimica Acta, 1299, 341-356.* https://doi.org/10.1016/j.aca.2023.341356
7. Meher, S., & Zarouri, R. (2025). Green and energy-efficient sample preparation techniques in analytical chemistry. *TrAC Trends in Analytical Chemistry, 168, 117174.* https://doi.org/10.1016/j.trac.2023.117174
8. Sheldon, R. A., & Woodley, J. M. (2018). Role of biocatalysis in sustainable chemistry. *Chemical Reviews, 118(2), 801–838.* https://doi.org/10.1021/acs.chemrev.7b00203
9. Armenta, S., Garrigues, S., & de la Guardia, M. (2008). Green analytical chemistry. *TrAC Trends in Analytical Chemistry, 27(6), 497–511.* https://doi.org/10.1016/j.trac.2008.05.003
10. Sheldon, R. A. (1992). Organic synthesis—past, present and future. *Chemistry & Industry, 23(1), 903–906*. [Introduced the concept of the E-factor for waste evaluation.]
11. Sheldon, R. A. (2007). The E factor: fifteen years on. *Green Chemistry, 9(12), 1273–1283.* https://doi.org/10.1039/B713736M [Expanded analysis of E-factor applications across industries.]
12. Trost, B. M. (1991). The atom economy—a search for synthetic efficiency. *Science, 254(5037), 1471–1477.* https://doi.org/10.1126/science.1962206
13. Curzons, A. D., Constable, D. J. C., & Cunningham, V. L. (2001). Metrics to 'green' chemistry—which are the best? Green Chemistry, *3(4), 156–164.* https://doi.org/10.1039/B1027010K
14. Sheldon, R. A. (2017). The E factor 25 years on: the rise of green chemistry and sustaina-bility. *Green Chemistry, 19(1), 18–43.* https://doi.org/10.1039/C6GC02157C
15. ACS Green Chemistry Institute. (2018). Process Mass Intensity (PMI) Calculator. American Chemical Society. https://www.acs.org/content/acs/en/greenchemistry/research-innovation/tools-for-green-chemistry/process-mass-intensity.html
16. Tobiszewski, M., Marć, M., Gałuszka, A., & Namieśnik, J. (2015). Green chemistry met-rics with special reference to green analytical chemistry. *Molecules, 20(6), 10928–10946*. https://doi.org/10.3390/molecules200610928

17. Pena-Pereira, F., Wojnowski, W., & Tobiszewski, M. (2020). AGREE—Analytical GREEnness Metric Approach and Software. *Analytical Chemistry, 92(14), 10076–10082.* https://doi.org/10.1021/acs.analchem.0c01887
18. Gałuszka, A., Migaszewski, Z., Konieczka, P., & Namieśnik, J. (2012). Analytical Eco-Scale for assessing the greenness of analytical procedures. *TrAC Trends in Analytical Chemistry, 37, 61–72.* https://doi.org/10.1016/j.trac.2012.03.013
19. Gałuszka, A., Migaszewski, Z., & Namieśnik, J. (2013). The 12 principles of green analytical chemistry and the significance mnemonic of green analytical practices. *TrAC Trends in Analytical Chemistry, 50, 78–84.* https://doi.org/10.1016/j.trac.2013.04.010

Applications of Green Analytical Chemistry in Key Sectors

Minal R. Ghante, Komal D. Patil, and Rupali M. Barhate

Abstract Green Analytical Chemistry (GAC) has emerged as a new set for sustainable development, aligning chemical processes with the 12 principles of green chemistry to minimize environmental impact while preserving analytical performance. This review explores the transformative applications of GAC across five critical sectors: pharmaceuticals, food safety, environmental monitoring, industrial processes, and forensic science. In the pharmaceutical sector, GAC significantly reduces waste generation during API synthesis, sample preparation, and quality control, with notable examples including Pfizer's greener Pregabalin route (E-factor reduced from 82 to 17) and Merck's Letermovir synthesis achieving 90% less water usage. Eco-friendly assays employing water-based or solvent-free techniques, combined with direct spectroscopic methods, demonstrate superior sustainability without compromising accuracy. The food sector also has become beneficial from green nutritional and contaminant analysis that replaces toxic solvents with biodegradable alternatives and minimizes sample pretreatment. Environmental applications emphasize real-time, low-impact pollutant monitoring and reduced reagent consumption. In industry, GAC drives process optimization, waste reduction, and energy efficiency, while forensic science adopts sustainable crime scene investigation protocols that limit exposure to hazardous solvents and reagents. Various greenness assessment tools—lifecycle assessment (LCA), Analytical Eco-Scale, GAPI, AGREE, RGB-12 algorithm, and others—are enlisted with their strengths and limitations. The review briefs that widespread adoption of GAC principles mitigates ecological harm and enhances cost-effectiveness, safety, and regulatory compliance, paving the way for a truly sustainable analytical future across disciplines.

Keywords Green analytical chemistry (GAC) · Sustainability · Greenness assessment tools · Process optimization · Eco-friendly analytical techniques

M. R. Ghante (✉) · K. D. Patil
Department of Pharmaceutical Chemistry, Sinhgad Technical Education Society's, Smt. Kashibai Navale College of Pharmacy, Kondhwa, Pune, India
e-mail: mrghante@gmail.com

R. M. Barhate
Department of Pharmaceutical Chemistry, Sinhgad Technical Education Society's, Sinhgad College of Pharmacy, Vadgaon, Pune, India

V. Bhusari and M. Ghante (eds.), *Green Analytical Chemistry in India*,
https://doi.org/10.1007/978-3-032-24564-9_2

Abbreviations

VOC Volatile organic compounds
HAP Hazardous air pollutants
CFC Chlorofluorocarbons
LCA Life cycle assessment
RGB Red green blue 12 algorithm
API Active pharmaceutical ingredient

1 Introduction

The term "Green Analytical Chemistry" has brought by J. Namieśnik and is wide topic to explain [1]. But before we penetrate into it, let's have quick mind dwelling question to understand what we truly expect. The synthesis of Sitagliptin which is useful for the treatment of type II diabetes by Merck is significantly expresses the dependency on green method of present and future. The first conventional route involved the aqueous waste of 65 kg and organic waste was of 205 kg with yield of 45%, another partially applied green chemistry route showed the aqueous waste of 2 kg, and organic waste was of 65 kg with yield of 84% [2]. The question arises about which route to be followed. As human tendency, we may choose second category. Binding ourselves with Green Analytical Chemistry is the need of today's world and is essential for sustainable ecosystem development.

Green Analytical Chemistry is truly advancing in different fields where chemistry, science, and technology is involved which covers all sectors with negligible exceptions. Though P.T. Anastas showed vibrant nature of Green Analytical Chemistry in 1998, but true roots lie in 1975 or before that [3]. Twelve principles of green chemistry act as guidance for designing of any chemical products and any processes to decease or eradicate the usage and production of hazardous substances. There are many sectors which are primarily or indirectly impacted by Green Analytical Chemistry. As illustrated in Fig. 1, authors describe the applicability of green principles in key sectors which increases its importance for the society and well-being of every individual present on the planet [4–10]. The idea of Green Analytical Chemistry lies in resolving the problems covered under principles and catering sectors affected directly or indirectly into it.

The tools for analyzing efficiency of Green Analytical Chemistry are better of consideration now which directly correlates with efficiency of methods [11]. The tools used are lifecycle assessment (LCA), Red Green Blue Algorithm (RGB) to give more information on method efficiency and advancing Green Analytical Chemistry field acceptance globally [12, 13]. Table 1 represents brief information about various tools employed for measurement of Greenness of the method [11–26].

These tools have pros and cons; but choosing best one after understanding the constraints is optimistic way. Methods are qualitative way of selection of database.

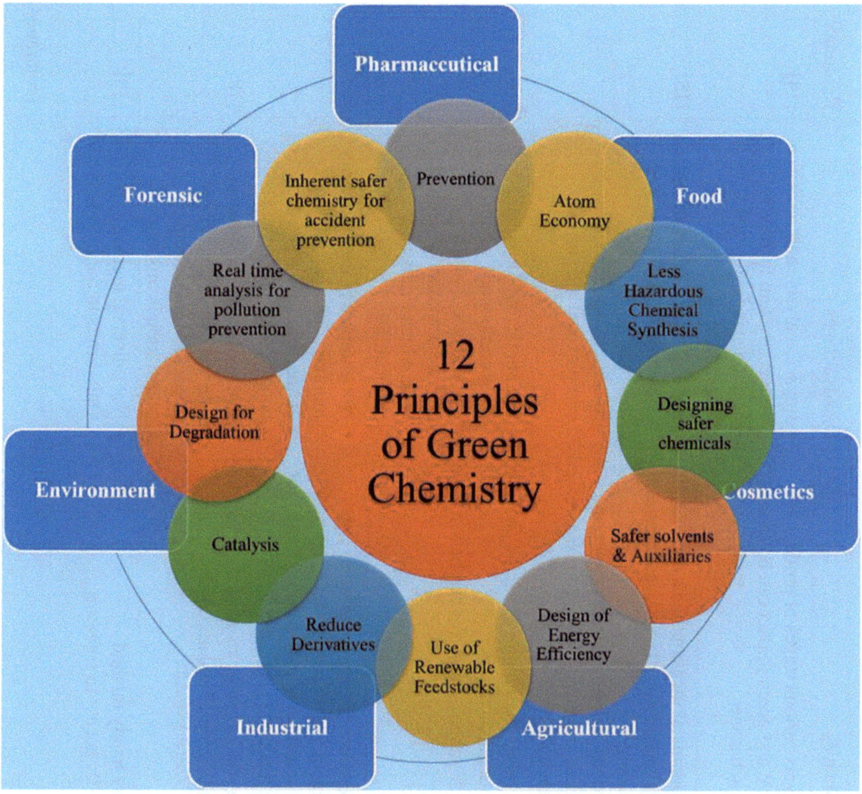

Fig. 1 The applicability of green chemistry in various sectors

The primary challenge is selecting metrics for analytical methods to ensure about sustainable approach utilized in the reaction is quite tough. The impact of green chemistry is vast and developing method which can measure all aspects with one method is really challenging but it's evolving with the time. Even *newer techniques can be developed* to evaluate method on green parameters. However, metrics have limitations but it becomes urge to find greener methods for simple chemical reactions for the global benefits. Finding better sources rather than depending upon conventional methods or sources available so far which are not going to make life sustainable. Green chemistry is vast even for operations, for assessment of factors affected by it and also for converting simple and complex chemical reactions into simple, cost-conservative, time-conservative green methods is really intrudable. As green chemistry impacts various sectors among all, its broad term for discussion. However, five sectors primarily impacted by Green Analytical Chemistry are enlisted here and they are:

a. Pharmaceutical;
b. Food safety;

Table 1 Greenness assessment tools

Name of the technique	Purpose	Advantages	Disadvantages	References
Life cycle assessment (LCA)	To assess the environment aspect in product life cycle	• Covers various stages of product life cycle and green impact can be measured at each stage	• Impact analysis shall be driven with ecosystem factors • Method standardization is incomplete	[14]
RGB 12 algorithm	To assess the red, green and blue impact of every simple reaction with algorithms	• Economical measurements such as cost efficiency, time efficiency can be easily measured • Toxicity level and direct impact on human and environment can be assessed at each stage	• Method integration, automation of techniques is not assessed completely • Specificity and criticality are complex and intervened	[15]
Analytical method volume intensity (AMVI)	To assess liquid chromatography techniques	• Mixture of components can be analyzed simultaneously • Wide range of analytes can be screened	• Not applicable for volatile and other compounds which are by-products • Sample preparation is liquid form which limits its applicability	[16]
Chemical hazard evaluation for management strategies (CHEMS)	To assess the chemicals based on the assessment value	• The value gives numbering to every reagent, solvent used which is helpful for selection of particular chemical for the reaction	• Not completely useful as only number cannot measure the entire impact of green or nongreen techniques	[17]
Green Analytical Procedure Index (GAPI)	To assess the green aspect in each stage of analytical methodologies	• Greenness can be measured at each step • Quite popular and acceptable	• Not applicable for synthesis • Can not measure complete impact as initial sample preparation is missing	[18]
High performance liquid chromatography—environmental assessment tools (HPLC-EAT)	To assess liquid chromatography mobile phases	• Very handy method • Easily access to software	• Not universal method • Only limits to the mobile phase	[19]

(continued)

Table 1 (continued)

Name of the technique	Purpose	Advantages	Disadvantages	References
Analytical eco-scale	To quantify greenness of method parameters- *Scale 0–100* determines the greenness	• Easy to use and comparable with other methodologies • Semi-quantitatively measures the chemicals and wastes generated • Evaluation of different environmental impact and their assessments can be quite possible	• No proper information on waste generation • Not able to assess the workability in microscale and macroscale methods • No measurement of negative impact on environment	[20]
Analytical greenness (AGREE) calculator metric	To assess analytical procedures for green impact	• Scale of 1–12 assess the procedures based on green principles • Quite better way of predicting the information	• Does not consider Synthesis part which actually involves impurities, toxic agents and hazardous substances	[21]
National Environmental Methods Index (NEMI)	To assess the methods and its parameters	• Pictogram can easily conclude about toxicity and safety information	• Output is non-quantifiable • No information about waste and Toxic chemical generated in the methodology	[22]
Preference ranking organization method for enrichment evaluations (PROMETHEE)	To compare methods and applicability	• Multiple criterions are involved for assessment which gives better result • Decision maker can choose method based on their observation	• Complex methodology compared to other techniques • Time consuming	[23]
Green certificate-modified eco-scale	To quantify Green Parameters of the Method and updated version of Eco-scale	• Measures toxicity, waster generation and Hazardous substances involved in the methodology • Used reagents, solvents are quantifiable	• Only number cannot assess the overall impact	[24]

(continued)

Table 1 (continued)

Name of the technique	Purpose	Advantages	Disadvantages	References
HEXAGON	To assess the Greenness in piles or hexagonal way	• Quality attributes measurable • Energy-efficient measurements	• Need of standardization • Labor-intensive as well as focused on special integration	[25]
Technique for order of preference by similarity to ideal solution (TOPSIS)	To assess analytical procedures	• To choose between ideal and non-ideal solutions • Scalable and versatile method	• Based on weights which limits usage, any small change in weight can misinterpret the results • Correlation between criterions are not considered which impact the results	[26]

c. Environmental science;
d. Industrial;
e. Forensic sciences.

2 Applications of Green Chemistry to Reform Pharmaceutical Field

2.1 Need in the Pharmaceutical Field

Pharmaceutical sector is most advanced combination of Chemistry, Science, and Technology, which is growing day-by-day. In addition to the growth, it is having impacts on Green Analytical Chemistry. The amount of waste generated in any pharmaceutical operation is making question on the efficiency of methods and urges the need of adoption of Green Analytical Chemistry in the field. The Green Analytical Chemistry is usable in different stages of Pharmaceutical Sector. Figure 2 provides different subsections of pharmaceutical field encompassing the need of Green Analytical Chemistry.

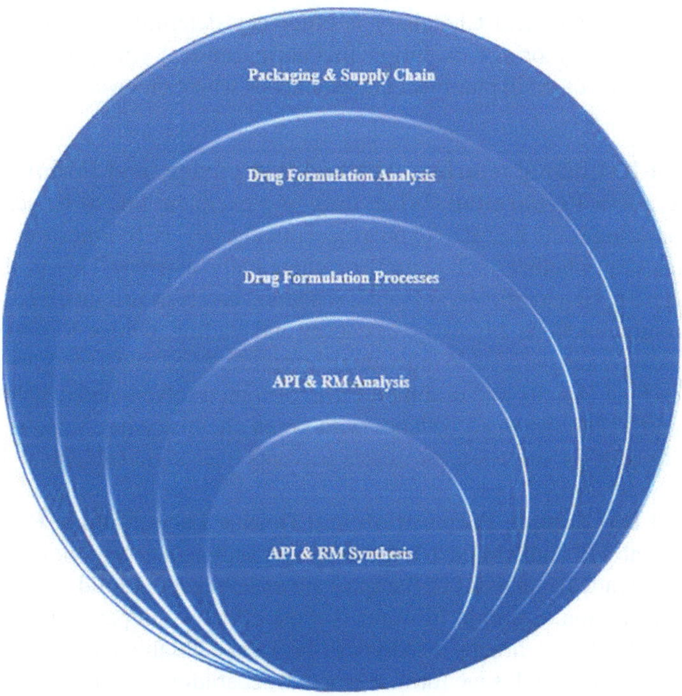

Fig. 2 Representation of the different subsections of pharmaceutical field encompassing the need of Green Analytical Chemistry

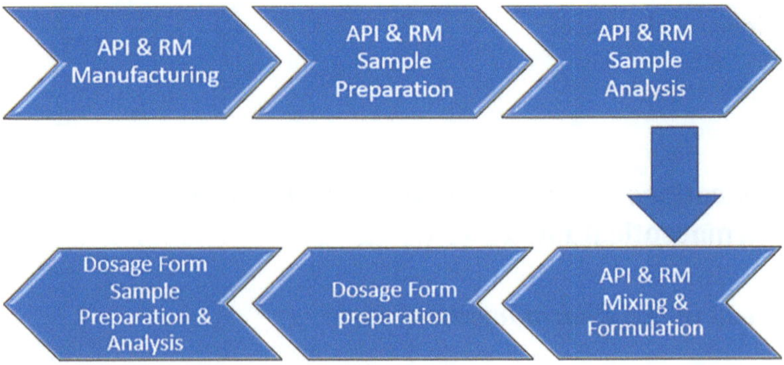

Fig. 3 Representation of steps involved in pharmaceutical analysis where green chemistry is widely applicable

Synthesis of active pharmaceutical ingredient (API) or any excipient involves different manufacturing steps where Green Analytical Chemistry is helpful. The synthesis of Pregabalin API with Green Analytical Chemistry is reported and followed by Pfizer for commercials since September 2006 and also further advances to decrease *Efactor* from 17 to 8 [2]. Similar kind of work was carried out for Antiviral drug Letermovir synthesis by Merck led to 90% less usage of water making it lucrative [27]. Green Analytical Chemistry is efficiency determination method. Green Analytical Chemistry brings evolution in Pharmaceutical Analysis [1]. Figure 3 represents steps involved in pharmaceutical analysis where green chemistry is widely applicable.

1. **API and RM manufacturing, API and RM mixing, and Formulation and Dosage Form Preparation**: When drug and excipient has to be screened, its manufacturing is the first step in pharmaceutical analysis. Synthesis or extraction is initial step whether it is natural or chemically derived. This is the crucial step where toxic chemicals, solvents, reagents, catalysts are involved with high temperature reactions, big usage of energy and leading to huge waste generation which is toxic and harmful to individuals in nature. Most of solvents are classified based on their toxicity and usability. Pfizer designed solvent and reagent tools for usage and scalability in accordance with green chemistry [9]. Various zones are shown to help for selection for usable solvents over toxic solvents, which were in use so far.
2. **API and RM Sample Preparation and Dosage Form Sample Preparation**: When Final API and RM have to be assessed, sample preparation is second step in pharmaceutical analysis [1]. Samples are usually kept in solid or solution form before separation. Solution in water is most effective preparation in making it greener or loading it directly also bypass the usage of highly reactive solvents and toxic chemicals. For dosage form, it is actual separation of active pharmaceutical ingredient which is quite complex and it's carried out based on the solubility, partitioning effect where various solvents are utilized. Modifying methods in such

way that direct sample introduction into spectroscopic techniques such as Raman Spectroscopy and solventless or water-based solution preparation methods make process greener.

3. **API and RM Sample Analysis and Dosage Form Analysis**: This phase has involvement of spectroscopic techniques and other detection techniques which we can say is greener already. Newer spectroscopic techniques and currently existing techniques all are less harmful to environmental. Identification or quantification is last step where computer-based programming or instrument-based outputs which is advanced on the verge of green methods.

It's very clear from above information that step 1 and step 2 methods needs to be made green so as to make field more sustainable. Considering the cost, time, efficiency, regulatory acceptability, scalability of methods, green methods turn really sustainable. When selecting analytical method, scientist must ensure that sample preparation needs to be direct or with water, separation should be with minimal toxic solvents and reagents also with waste generation and prevention of the same [1]. This will transit simple analytical method into Green Analytical Method which will be more sustainable and adaptable to the industry.

After manufacturing, packaging and distribution play important role. Packaging materials often seen as non-biodegradable and prominent player as waste generator. Glass is used in many preparations which is fragile and have limited applicability, preparation involved high temperature which adverse on its selectivity. Polyvinyl polymers are quite easy to use, readily available but most dreadful to environmental health. The third and widely acceptable packaging material is aluminum based on its acceptance. Use and development of biodegradable packaging material is wide research area open for all in Green Analytical Chemistry.

The distribution of pharmaceutical products is also crucial step as well as load to environmental health. Long distance shipments are reliable on factors which causes impact on environmental health. Developing low-cost, high-speed transition with minimal waste the area is quite far away from research and development.

However, adopting Green Analytical Chemistry is *not completely successful yet* but it is under progression and also positively accepted by many organizations.

2.2 *Eco-friendly Assay*

Assay is quantitative as well as qualitative determination of drug present in the formulation or any dosage form using different instrument or non-instrument techniques. When the term is derived as eco-friendly assay which means quantitative determination of drug in dosage form considering ecological benefits from the same.

The method should be

- Reproducible;
- No toxic reagents and chemicals utilized in any of the process;

- Non harmful and acceptable for upcoming years;
- Not dependent on conventional energy sources such as petrochemicals;
- Time and cost should be minimal compared to traditional methods of analysis;
- Sustainable.

This is wide research area which is constantly triggering researchers over the years. Every molecule which is used in pharmaceutical preparations should be assessed using Green Analytical Chemistry and there should be minimal waste generation to make an environment eco-friendly. Many pharmaceutical eco-friendly assay methods reported and used in the industrial scale and Table 2 enlists few of them.

2.3 *Quality Control in Greener Ways*

Synonym for quality is standard, which further extends meaning of quality control as maintaining the standards of products without affecting its inherent properties or strength. Table 3 gives an information about few QC examples in greener ways.

2.4 *Recent Advances and Future Remarks*

Overall Green chemistry represents a paradigm shift in pharmaceutical manufacturing, offering an innovative and environmentally conscious approach to drug development and production. By concentrating on the product design and processes design, it helps to reduce the formation and use of hazardous substances, also reduces environmental harm enabling in the increase in efficiency and sustainability of pharmaceutical operations.

As the global demand for safe, effective, and sustainable healthcare solutions continues to rise, integrating green chemistry into the pharmaceutical sector has become both an ethical imperative and a strategic necessity. The fundamental principles of green chemistry—such as atom economy, the use of renewable feedstocks, safer solvents, and energy-efficient synthesis—serve as powerful tools for transforming traditional manufacturing practices. Incorporating these approaches can reduce waste creation, toxic emissions, and resource consumption, while simultaneously improving process yields and product quality.

Replacing conventional organic solvents with greener alternatives (water, supercritical carbon dioxide, or ionic liquids) considerably reduces environmental hazards. Similarly, the use of biocatalysts and enzyme-mediated reactions allows for milder reaction conditions, higher selectivity, and reduced energy requirements. These advancements can truly attribute in a sustainable and economically viable pharmaceutical production model [63].

Despite these benefits, the widespread adoption of green chemistry in the pharmaceutical industry faces several challenges. High initial costs associated with

Table 2 Ecofriendly assay methods

Sr. No.	Assay	Eco-friendly approach	Instrumental method conditions	Green metrics	References
1.	Apremilast, its enantiomer and its seven impurities using Chiral RP-HPLC	Use of ammonium buffer and acetonitrile in mobile phase	Technique: HPLC Stationary phase: aAmylose Chiralpack IA-3 (250 mm × 4.6 mm, 3 μm) Mobile phase: buffer (0.01 M NH_4HCO_3, pH 8.0) and acetonitrile 50:50 (v/v) Flow rate: 0.4 ml/min Temperature: 25 °C *Impurity recovery = 96.1–102.1%*	AGREE score = 0.66	[28]
2.	Atorvastatin calcium by analytical quality by design (AQbD) and HPLC	Use of ethanol and acetic acid in mobile phase	Technique: HPLC Stationary phase: ZORBAX Eclipse plus C18 (4.6 × 150 mm, 5 μm) Mobile phase: 0.5% v/v aqueous acetic acid & ethanol (42:58 v/v) Flow rate: 1 mL/min Temperature: 40 °C	Eco Scale score = 90	[29]
3.	Atorvastatin (ATO), ezetimibe (EZB), and fenofibrate (FF)	Green solvent and reagent such as ethanol and triethylamine in buffer	Technique: HPLC Stationary phase: Inertsil ODS 3 (250 mm × 4.6 mm), 5 μm Mobile phase: buffer (0.1% triethanolamine in water) & ethanol (10:90 v/v)	Eco Scale score = 91	[30]

(continued)

Table 2 (continued)

Sr. No.	Assay	Eco-friendly approach	Instrumental method conditions	Green metrics	References
4.	Atenolol, amiloride and chlorthalidone by UV	Less solvent usage	Technique: UV UV–VIS double-beam spectrophotometer (JASCO V-630) Cuvette: quartz cells (1.00 cm) Wavelength: 230 nm Solvent: methanol	GCC score = 89	[31]
5.	Benzodiazepines	Green reagents	Technique: HPLC Stationary phase: monolithic C_{18} column Mobile phase: sodium dodecyl sulfate (SDS): polyoxyethylene (23) lauryl ether (Brij-35) 50:50% v/v Flow Rate: 1.0 ml/min	Eco Scale score = 95	[32]
6.	Brivaracetam detection together with piracetam and carbamazepine	Use of green mobile phase	Technique: HPLC Stationary phase: RP C_{18} column Mobile phase: (70:30v/v) mixture of 1% of triethylamine in water and acetonitrile pH 6.5 Flow rate: 0.6 ml/min	Eco Scale score = 85	[33]

(continued)

Table 2 (continued)

Sr. No.	Assay	Eco-friendly approach	Instrumental method conditions	Green metrics	References
7.	Benidipine hydrochloride (BEN) and chlorthalidone (CTD) by HPLC	Green solvent and reagent such as ethanol and phosphate buffer	Technique: HPLC Stationary Phase: Agilent Eclipse Plus (C18, 250 mm × 4.6 mm i.d, 5 μm) Mobile phase: ethanol and potassium dihydrogen orthophosphate (orthophosphoric acid to 3.5) in a ratio of 40:60 v/v Flow Rate: 1.0 ml/min	Eco Scale score = 93	[34]
8.	Betamethasone dipropionate (BD) and calcipotriene (Cal) using a UHPLC	Central composite design Green solvent and reagent such as ethanol and phosphate buffer	Technique: UHPLC Stationary Phase: Dikma Endeversil C18 ODS (2.1 × 50 mm, 1.8 μm) column Mobile phase: ethanol and potassium dihydrogen phosphate (3.0 pH) buffer 51:49 (v/v) Flow Rate: 0.31 ml/min	AGREE Metric score = 0.89	[35]
9.	Chlorthalidone and clinidipine	Use of propylene carbonate	Technique: UV Instrument: Lab India UV VIS spectrophotometer (3092 double beam) India with UV Win Lab Version 5.1.1 software	AGREE Metric score = 0.91	[36]

(continued)

Table 2 (continued)

Sr. No.	Assay	Eco-friendly approach	Instrumental method conditions	Green metrics	References
10.	Chrysin	Use of green mobile phase	Technique: UHPLC Stationary phase: C-8 column, 50 × 2 (mm), particle size 2 μm (Capcell Pak) Mobile phase: acetonitrile and water in 80:20 v/v Flow Rate: 0.5 ml/min	Eco Scale score = 81	[37]
11.	Clorsulon, albendazole, triclabendazole and ivermectin	Use of green components	Technique: HPLC Stationary Phase: Anonyx monolithic C18 column (100 × 4.6 mm) Mobile phase: micellar mobile phase consisted of 120 mM sodium dodecyl sulfate, 15% n-propanol and 15 mM phosphate buffer, adjusted to pH 5.5 with 0.1 M sodium hydroxide Flow Rate: 1.4 ml/min Temperature: 40 °C	Eco Scale score = 75	[38]
12.	Crotamiton and hydrocortisone	Use of green mobile phase	Technique: RP-UPLC Stationary phase: waters X-bridge (C18 50 × 2.5 mm i.d, 1.7 μm) Mobile Phase: ethanol: 0.01 M ammonium acetate buffer (pH 3.0) 30:70 v/v Flow rate: 0.3 ml/min	AGREE score = 0.83	[39]

(continued)

Table 2 (continued)

Sr. No.	Assay	Eco-friendly approach	Instrumental method conditions	Green metrics	References
13.	Delafloxacin	Green technique and mobile phase	Technique: UPLC-MS Stationary Phase: BEH C_{18} column (50 × 2.1 mm; 1.7 μm) Mobile Phase: 0.1% formic acid in acetonitrile and 0.1% formic acid in water	AGREE score = 0.78	[40]
14.	Escitalopram and etizolam in the formulation by quality by design-based HPLC	Green solvent and reagent such as ethanol and phosphate buffer	Technique: RP-HPLC Stationary phase: phenomenex column C18 Mobile phase: ethanol: ethanol and phosphate buffer (60: 40% v/v) Flow rate: 0.3 ml/min	AGREE score = 0.78	[41]
15.	Favipiravir	Green solvents	Technique: RP-HPLC & ANOVA Stationary PHASE: C18 column (4.6 mm × 150 mm, 3 μm) Mobile Phase: methanol, ethanol, and water (25: 35: 40 v/v/v Flow Rate: 0.8 ml/min	Eco Scale score = 92	[42]
16.	Fluticasone propionate and azelastine	Use of green reagents	Technique: UV Instrument: UV VIS spectrophotometer	Eco Scale score = 92	[43]

(continued)

Table 2 (continued)

Sr. No.	Assay	Eco-friendly approach	Instrumental method conditions	Green metrics	References
17.	Fluorescein sodium and benoxinate hydrochloride	Use of green solvent and reagent in mobile phase	Technique: RP-UPLC Stationary phase: waters X-bridge (C18 50 × 2.5 mm i.d, 1.7 μm) mobile phase: ethanol: 0.01 M ammonium acetate buffer (pH 3.0) 30:70 v/v Flow Rate: 0.3 ml/min	Eco Scale score = 90	[44]
18.	Fenoverine	Use of green solvent and reagent in mobile phase	Technique: RP-UPLC Stationary phase: Spherisorb C18 column (150 × 4.6 mm, 3 μm) Mobile phase: methanol: ammonium acetate buffer 20 mM, 81:19% v/v Flow rate: 0.3 ml/min	EAT Score = 41.82	[45]
19.	Isosorbide dinitrate and hydralazine hydrochloride	Use of green solvent and reagent in mobile phase	Technique: UPLC Stationary Phase: Spherisorb C18 column (150 × 4.6 mm, 3 μm) Mobile phase: methanol: ammonium acetate buffer 20 mM, 81:19% v/v Flow Rate: 0.3 ml/min	Eco Scale score = 96	[46]
20.	Ketoconazole	Use of green solvent and reagent	Technique: UV Instrument: UV VIS spectrophotometer	Eco Scale score = 96	[47]

(continued)

Table 2 (continued)

Sr. No.	Assay	Eco-friendly approach	Instrumental method conditions	Green metrics	References
21.	Lamivudine, zidovudine and nevirapine	Use of green solvent and reagent in mobile phase	Technique: HPLC Stationary phase: C18 column (ARV4 5 μm 250 × 3.0 mm, Interchim Mobile PHASE: 0.1 M ammonium acetate buffer (pH 4.5) and ethanol—gradient flow rate: 0.45 ml/min	Eco Scale score = 75	[48]
22.	Lesinurad and febuxostat	Use of green solvent and reagent in mobile phase	Technique: HPLC Stationary phase: hypersil BDS C18 column Mobile Phase: Acetonitrile: potassium phosphate buffer (30.0 mM; pH 5.5, 32.2:67.8% v/v) Flow Rate: 1.0 ml/min	Eco Scale score = 75	[49]
23.	Levetiracetam	Use of green solvent and reagent in mobile phase	Technique: HPLC Stationary Phase: Lichrosorb RP-18 (250 × 4.6 mm i.d., 5 μm) Mobile phase: phosphate buffer (pH 3.1) and acetonitrile (87:13 v/v) Flow Rate: 1.0 ml/min	AGREE Score = 0.82	[50]

(continued)

Table 2 (continued)

Sr. No.	Assay	Eco-friendly approach	Instrumental method conditions	Green metrics	References
24.	Lenvatinib	Use of green solvent and reagent	Technique: HPTLC Mobile phase: ethanol and water (60:40, v/v)	AGREE Score = 0.82	[51]
25.	Zonisamide by HPLC	Green and safe mobile phase of ethanol and water	Stationary phase: kromasil C_{18} column (150 mm × 4.6 mm, 5 μm) Mobile phase: ethanol: Water (30:70 v/v) Flow rate: 1 ml/min Temperature: 35 °C	Eco Scale score (91) > Conventional method score (81)	[52]

process redesign, regulatory complexities, and scalability limitations often hinder the transition from traditional to sustainable manufacturing systems. Moreover, the lack of standardized green metrics and limited awareness among industry professionals further complicate implementation. However, recent progress in computational modeling, process intensification, and green solvent design has begun to address these barriers, demonstrating that sustainability and profitability can coexist within pharmaceutical manufacturing. Government incentives, academic–industry collaborations, and the growing emphasis on environmental, social, and governance criteria are also accelerating the shift toward greener practices. The future of green pharmaceutical chemistry depends on sustained innovation, interdisciplinary collaboration, and comprehensive education. Researchers, business frontrunners, and administrative authorities must work together to establish framework encouraging sustainable innovation and ensure regulatory alignment with emerging technologies. Academic institutions play a crucial role in fostering this transition by embedding green chemistry concepts into pharmaceutical science curricula, equipping future scientists with the knowledge and motivation to design environmentally responsible solutions [64]. Furthermore, transparent sustainability reporting and lifecycle assessments can help industries evaluate and communicate the environmental impact of their products, reinforcing accountability and consumer trust.

In conclusion, green chemistry offers a transformative pathway toward a more sustainable and responsible pharmaceutical industry. Its principles provide practical solutions to minimize waste and environmental hazards and align with global sustainability goals such as the United Nations Sustainable Development Goals (SDGs) [65]. As scientific innovation continues to advance, the convergence of chemistry, biotechnology, and environmental science will drive the next era of pharmaceutical

Table 3 QC tests with green approach

Sr. No.	Details	Green approach	References
1	Lead determination in water samples	Stripping voltammetry—low detection limits, sensitive, simple, rapid, reliable, inexpensive, possible to use in-situ	[53]
2	Mercury determination in soil samples	Graphite furnace atomic absorption spectroscopy—direct, cost and time-effective method	[54]
3	Glucose determination in blood	Biosensor—low amount, direct and less-time consuming method, accurate and precise determinations	[55]
4	Determination of ascorbic acid and folic acid in the pharmaceutical formulation	Derivative spectrophotometric method for simultaneous determination—Analysis using UV spectrophotometer with the usage of phosphate buffer as diluent	[56]
5	Separation and analysis of antihyperlipidemic using UPLC	UPLC analysis	[57]
6	Rifaximin quantification in oral solid dosage form	UV spectrophotometric determination—less solvent consumption	[58]
7	Potency determination of ampicillin sodium	Turbidimetric assay	[59]
8	Caffeine in emulsions	HPLC assay—use of green mobile phase	[60]
9	Quantification of Doxorubicin in human plasma	HPLC assay	[61]
10	Assessment of API and excipients in QC	NIR method	[62]

development. Embracing green chemistry is no longer a choice—it is an essential commitment to safeguarding both human health and the planet for generations to come.

3 Utilization of Green Chemistry in Food Sector

3.1 Necessity to Use in Food Sector

Food is primitive need of human being or any organism living on the earth. For human, food safety becomes utmost concerned area for various regulatory authorities worldwide. Raw food items such as milk, grains, etc. have concerns about storage, manufacturing processes and handling whereas processed food items such as cheese, grain flour are having concerns related to additives and their safety, quality of the additives, storage, manufacturing processes, and shelf life of the food product. Various food borne pathogens, aerobic and anaerobic microorganisms, fungal contamination affects quality of the food product and consequently human health [66]. Degradation of food over time is also concern for food industry where traditionally antioxidants and different preservatives added which has toxicological profile-raising safety concerns, use of natural antioxidants and preservatives have given relief with greener approach [67].

Food products are evaluated for [68]

- Physico-chemical composition.
- Determination of macronutrients.
- Determination of micronutrients.
- Determination of nutritional components.
- Determination of bioactive or marker compounds.
- Determination of microbial contaminants and pathogens.
- Determination of origin of food (animal/plant/others).
- Determination of adulterants.
- Determination of shelf life of the product.

Various traditional methods are employed for the evaluation of above-mentioned parameters and now there is paradigm shift observed for employment of Green Techniques for the same. Food sector needs to be green for better sustainable growth of the field which is never ending industry among all.

3.2 *Green Nutritional Analysis*

Nutritives are the components of food product which cause nutritive value in it. For example, one glass of milk gives few calories of carbohydrate and proteins which means this is nutritive involved in milk and beneficial for health. Various food ingredients are from natural sources such as animal, plant, mineral, soil, etc. Various components are extracted from the natural sources. Traditional method of extraction involves Soxhlet extraction, maceration, and hydro distillation which has limitations and opposite to green chemistry principles. However green analytical techniques of extraction involving:

- Microwave-assisted extraction.
- Enzyme-assisted extraction.
- Ultrasound-assisted extraction.
- Pulse electric field extraction.
- Supercritical fluid extraction.

Either solvent free or use of green solvents such as water, CO_2, ethanol, glycerol, n-hexane and supramolecular solvents, etc. [69–71].

Table 4 Green methods employed for analysis of nutritives

Sr. No.	Details	Green approach	References
1	Detection of fucocoumarins and other components	Photobiological assay method	[72]
2	Gallic acid determination in food	Gallic acid—food additive determination using electrochemical sensors	[73]
3	Spectinomycin and lincomycin determination in poultry eggs	Accelerated solvent extraction coupled with GC—tandem MS	[74]
4	Amino acid determination in triticum wheat species	HILIC UHPLC-MS/MS	[75]
5	Green tea and grapes seeds extraction	Green methods of Extraction	[76]

Traditional method of analysis involves usage of hazardous solvents, tedious and multiple steps in processing, and hazardous waste generation. However, there are green methods, which overcome all the problems associated with traditional method of analysis. Table 4 shows green methods for analysis of nutritives.

3.3 *Green Contaminant Analysis*

In Table 5 green methods for analysis of contaminant in food is depicted.

Table 5 Food analysis by green approaches

Sr. No.	Details	Green approach	References
1	Degradation of aflatoxin M_1 in milk by ultraviolet energy	Use of UV irradiation to identify and analysis milk degradation	[77]
2	Nanoscale purifications of food	Use of nanomaterials for purification	[78]
3	Copper nanoparticles for aquatic pathogens	Use of anti-biofilm effect	[79]
4	Determination of heavy metals in water, food, and feed samples	ELISA and other chemiluminescent enzyme immunoassay	[80]
5	Detection of ochratoxin and neomycin in food	Advanced fluorescence assay	[81]

3.4 Combating with Traditional Methods

The transition from conventional to green methodologies in the food sector is not merely a technological advancement but an essential shift toward sustainability, safety, and environmental preservation. Traditional analytical and processing techniques, while effective in determining food quality, have long relied on hazardous solvents, high energy inputs, and multi-step procedures that generate significant chemical waste. Such practices can negatively impact human health and the environment by generating hazardous byproducts and elevating the overall carbon emissions of food production systems. The principles of green chemistry help address these issues by promoting safer, more sustainable, and resource-efficient alternatives that minimize ecological harm.

Similarly, analytical techniques have evolved toward greener paradigms, emphasizing miniaturization, automation, and solvent-free approaches. The implementation of Green Analytical Chemistry principles cut down waste generation through innovations. These innovations enhance analytical accuracy while ensuring environmental safety. Moreover, integrating renewable energy sources and adopting process intensification strategies in food analysis and preservation have significantly reduced energy demands and emissions focused on food safety, consumer health, and sustainable industrial practices.

By combating the limitations of traditional methods through the incorporation of green approaches, the food industry moves toward a future where quality, safety, and sustainability co-exist harmoniously. This shift ensures better environmental health and a resilient food system capable of meeting global nutritional demands responsibly.

4 Approaches of Green Chemistry for Betterment of Environmental Sciences

4.1 Uses of Green Chemistry in Environmental Causes

Uses of conventional methods had created huge issues such as climate change, energy production, water pollution, food pollution, and also increased toxicity [82]. Use of Fossil fuels, release of chemicals such Chlorofluorocarbon and other toxic reagents has made an impact on Ozone layer depletion, increment in sea water level which has created big changes on climate. Water pollution and food pollution are also brought by insufficient water treatment, release of toxic effluents into natural water sources and use of pesticides, insecticides into farming, use of preservatives and other additives lead to decrease in quality of food which a directly or indirectly affecting environmental health of the society. When environmental sciences are affected, it brings green chemistry into picture for betterment of environmental health. The use of green chemistry is associated with sustainable approaches guided by 12 principles of green chemistry. Since, 1980 efforts are undertaken to make environment greener. Environmental health is affected by use of reactive substances and could be better if there has been usage of non-reactive substances.

4.2 Reducing Pollutants and Pollution in Green Ways

Making changes at ground level will always help to build sustainable base for the future. When pollutants are controlled, pollution will be controlled in greener ways which will reduce environmental load. Industrial pollutants such as hazardous solvents, release of volatile organic compounds (VOC) or hazardous air pollutants (HAPs) or toxic chemicals. Toxic chemical is broad term which include flammable, carcinogenic, mutagenic chemicals as well. Table 6 depicts the applications of green chemistry in environmental health.

Before controlling the pollutants and reducing pollution, it's utmost important to monitor pollutants and pollution.

Catalysis is a prominent principle in green chemistry which is helpful for monitoring pollutants with safer products and byproducts [92]. Formetanates are water contaminants affecting lives which are determined using catalysis reduces toxic generation. In-line TiO2-catalyzed photodegradation for p-aminophenol (PAP) catalysis is reported [97].

Use of simple hydroxide solutions for treatment of heavy metals. Heavy metal waste determination and removal of waste is tactic as it comes non-degradable category. However, with the use of NaOH and Fe (III), Hg monitoring is possible with co-precipitation of Hg [98].

Methods which can be used without any reagent are also useful for online monitoring of pollutants. One of the toxic substances Iprodione is commonly present in solid pesticide formulations [99]. FT-Raman spectroscopy directly utilizes sample for determination of Iprodione which reduces the reagent uses and no waste is generated.

The newer techniques such as plasma chemistry and sensors utilization are also widely useful in monitoring environmental pollutants. The best and ecofriendly application reported and widely used in many countries is Ozone and light utilized methods for removal of water-soluble pollutants [100]. Sensors are also useful in controlling pollutants and pollution using green methods [101].

4.3 Overall Impact on Environment

Moreover, green chemistry drives innovation in catalysis, renewable energy, biodegradable materials, and pollution monitoring, ensuring that environmental health and industrial progress move hand in hand. It bridges the gap between scientific innovation and environmental responsibility, providing a framework for sustainable

Table 6 Applications of green chemistry in environmental health

Environmental pollutants	Green alternatives	Achievements	Remarks	References
Carbon monoxide, CFCs, alkyl derivatives, acyl derivatives	Supercritical carbon dioxide	Textile dry Cleaning	Replacement of perchloroethylene is remarkable green change	[83]
		Supercritical CO_2 reaction medium	Various organic reactions are with SCO_2	[3]
Organic solvents	Water	Many complex organic reactions are possible in water with catalytic activities	Green catalysis	[84]
	Ionic liquids	Development of newer cationic and anionic liquids	Control on environmental depletion	[85]
Petroleum chemicals	Biobased feedstocks	Newer sources of sustainable energy	Easily available and renewable sources over petroleum chemicals	[86, 87]
Synthetic waste in many chemical processes	Green catalysts and green reactions	Eliminated waste generations, improved user and employer safety	Replacement of toxic chemicals into safer reactants and products	[88–90]
	Genetically engineered reactions	Use of non-virulent strains of microorganisms	Highly active reactants to easily degradable byproducts	[91]
		Use of H_2O_2 for delignification process	No more use of chlorine hypochlorite solution in paper industry	[92]
	Recyclable catalysis	Use of FBC ligands for easier separation of product and catalyst	Simple method of separation	[93]
Analytical process monitoring	Green reactants	Elimination of generation of pollutants and release of safer biodegradable waste	Redesigning of old generational reactions into safer, green reactions with no waste	[94–96]

industrial development. The overall impact extends beyond laboratories and factories—it supports healthier ecosystems, reduces the toxic load on human populations, and promotes a greener, more resilient future. Thus, green chemistry stands as a cornerstone of sustainable development, aligning environmental health with economic growth and societal well-being.

5 Applicability of Green Chemistry Principles in Industrial Sector

5.1 General Overview on Process Optimization and Waste Reduction

Process Optimization and waste reduction are linked with each other. Optimization of process leads to waste reduction and waste reduction is one of the parts of process optimization. This statement gives direct impact of process optimization on waste reduction which is primary principle of Green Analytical Chemistry.

5.2 Comparison of Traditional and Green Chemistry for Process Optimization and Waste Reduction [102]

The Green way or eco-efficient way of process optimization with multiple processes simulation using multi-objective optimization is recorded.

The approach was used in the case study of Acrylonitrile manufacturing process. This process is of four steps such as process modeling and analysis, identification of process parameters, development of process alternatives and multi-objective optimization—applying sequential quadratic programming (SQP) within ASPEN PLUS to optimize for maximum profit and minimum waste simultaneously. Environmental performance was further evaluated using the waste reduction algorithm (WAR).

The study found that replacing the conventional continuous stirred tank reactor (CSTR) with a plug flow reactor (PFR) significantly improved process efficiency. The optimized process demonstrated significant improvements, including a 69% rise in overall cost savings (from $\$9.0 \times 10^7$ to $\$3.59 \times 10^8$ over a seven-year period), a 38% decrease in waste output (from 3053 kg/h to 1884 kg/h), and substantial reductions in several environmental impact indicators such as global warming potential, acidification, and human toxicity.

The research demonstrated that multi-objective optimization effectively identifies trade-offs between profitability and environmental performance, providing a systematic tool for sustainable process design. The resulting profit–waste trade-off curve allows decision-makers to select optimal process conditions according to economic and ecological priorities.

5.3 *Recent Improvements in Industrial Sector [103]*

The field is naïve to green chemistry however use of newer techniques for process optimization is recent advancement in the field. Artificial intelligence and machine learning are growing nowadays in the society. Use of machine learning for process optimization has been employed. The process optimization is dependent on two things:

(a) Change in production parameters;
(b) No change in production parameters.

Change in production parameter involves alteration in process steps, usage of instruments, variables in techniques such as change in mixing time, use of algorithms for steps modification. The use of machine learning is exceptionally improving the quality of production as well as reducing waste generation or reducing the processes, which lead to waste generation using different algorithms.

Root cause analysis, predetermination of production outcomes, and diagnostic systems are parts of process optimization without change in production parameters. Root cause analysis involves inspection of primary steps of processes, which affect yield or quality of the product. This method helps to identify gaps in processes; modification of small variables also leads to big changes in production.

Early prediction of manufacturing outcomes through machine learning enables the identification of potential deviations even before the process is executed, thus allowing preventive corrections. Predictive models analyze vast historical datasets to recognize patterns and relationships between variables such as temperature, mixing speed, or pH, and the final product quality. These insights facilitate real-time decision-making and adaptive control, ensuring consistent production with minimal variability.

Diagnostic systems form another crucial component in process optimization without altering the existing parameters. These systems continuously monitor production lines, detect abnormalities, and suggest corrective measures through automated alerts or intelligent dashboards. By integrating such systems, industries can achieve a higher level of process transparency and reliability.

Overall, integrating artificial intelligence (AI) and machine learning (ML) into process optimization represents a major advancement toward more sustainable and efficient manufacturing practices. When aligned with green chemistry principles, these technologies enhance efficiency and product quality and contribute to environmental preservation by minimizing energy consumption, solvent use, and chemical waste. This synergy between digital innovation and green chemistry promotes the transition from traditional, resource-intensive production methods to smart, eco-efficient manufacturing systems.

6 Applications of Green Chemistry in Forensic Science

6.1 Forensic Experts and Multidimensional Risks

Forensic analyses involve physical, biological, chemical, and nuclear hazards. Risks include exposure to infectious bodily fluids, unsanitary materials, contaminated weapons, and sharp objects, beginning at the crime scene and continuing through laboratory analysis. Developing risk management procedures, training, and contingency plans in advance is essential for protecting experts [104]. Several essential measures must be taken, including protective clothing, decontamination zones, and proper tools for collecting evidence. The integrity of the evidential record must be maintained. Analyses should progress from non-destructive to destructive techniques to preserve evidence for future re-examination, ensuring repeatability and admissibility. Preparatory steps involve sample selection, quantity, and homogeneity. Visual, physical, and sensory examinations are performed without damage, followed by non-destructive methods such as microscopy and light-based techniques [105]. All procedures, including destructive analyses, are documented in written and digital form. In addition to public scrutiny, public expectations increase in forensic science, increasing the complexity of the role of the expert.

6.2 Crime Scenes: The Starting Point of Forensic Analysis

Forensic investigations are based on crime scenes, and accurate evidence analysis requires an understanding of crime scene investigation (CSI). Legal, scientific, and physical factors are all taken into account when processing a crime scene. Experts in crime scene processing and reconstruction, forensic specialists are frequently trained in law enforcement or CSI [104]. To properly handle evidence, one must be aware of the protocols for crime scene protection, evidence detection, collection, preservation, packaging, and documentation.

Strict adherence to these processes, commonly summed up as the chain of custody, is necessary for the admissibility and validity of evidence in court. Given that these elements can impact analysis, forensic specialists must take into account the circumstances surrounding the discovery of the evidence, its duration of presence, and specifics like packaging and transportation. Skilled professionals can detect fiber or gunshot residues, use insect activity or blood color to estimate the time of a crime, and analyze shoe prints to determine weight or movement [105]. Forensic experts are capable of deciphering evidence and reaching significant conclusions even when they are not physically present at the scene.

Green Analytical Chemistry in forensics

Green Analytical Chemistry develops and applies analytical methods that reduce environmental and occupational hazards while maintaining accuracy and precision

[106]. It is guided by the 12 principles of green chemistry [107]. In forensic science, these principles guide the design of methods that are efficient, cost-effective, and sustainable, promoting minimal reagent use and reduced exposure to toxic substances [108].

6.3 Green Approaches in Forensic Science

Portable spectroscopic instruments such as Raman, FTIR, and X-ray fluorescence (XRF) allow on-site, non-destructive analysis, reducing the need for excessive sample transport and laboratory processing. Energy-efficient detectors and hybrid analytical systems further enhance sustainability by shortening analysis time and lowering energy use [106]. Traditional forensic analyses—such as fingerprint development, drug identification, and DNA extraction—often involve hazardous chemicals like silver nitrate, nitric acid, and volatile solvents. Green approaches aim to replace these with safer alternatives, including solvent-free and microextraction techniques, ionic liquids, and bio-based solvents [105]· These innovations reduce both chemical waste and analyst exposure without compromising analytical sensitivity.

6.4 Sustainable Crime Scene Analysis

The sequential use of non-destructive to destructive techniques ensures that evidence remains suitable for re-examination and legal proceedings. Portable and digital forensic tools enable real-time data transfer and remote consultation, minimizing repeated sample handling and transportation—thus lowering the overall environmental footprint [105].

Crime scene analysis forms the foundation of forensic investigation. Applying green principles here involves reducing contamination risks, minimizing waste, and preserving evidence integrity [108].

Science for Justice

The forensic laboratory plays a pivotal role as the central point where evidence is examined, procedures are standardized, and experts collaborate to ensure that materials presented in court meet admissibility requirements [106]. Forensic science brings together diverse scientific fields to uphold justice, addressing not only criminal cases such as homicide, arson, sexual assault, burglary, and drug trafficking, but also broader issues like medical negligence, food fraud, environmental contamination, and even war crimes [104].To effectively contribute to justice, forensic practitioners must have a solid understanding of both scientific techniques and the legal framework within which evidence is evaluated.

Fieldwork often presents additional risks, including chemical exposure, biological hazards, or active dangers like fire and structural collapse [105].

While some preliminary analyses can be carried out at the scene of an incident—such as roadside drug checks or rapid biological screening—complex, instrument-based investigations typically require the controlled setting of a laboratory [108].

Green Instrumental Methods and Detection

The application of green instrumental methods in forensic science aims to balance analytical reliability with environmental responsibility. Conventional instruments often consume large volumes of solvents, generate hazardous waste, and require high energy input. In contrast, green approaches emphasize techniques that are miniaturized, solvent-free, or use safer alternatives, thereby reducing both ecological and occupational risks [1]. Portable spectroscopic tools such as Raman, FTIR, or X-ray fluorescence allow rapid, non-destructive screening at the crime scene, minimizing the need for sample transport and excessive laboratory processing [104]· Similarly, advances in microextraction, solid-phase extraction with biodegradable sorbents, and chromatographic systems operating with water or bio-based solvents have significantly lowered the chemical burden of forensic analyses [108].

Green Materials in Forensic Applications

The adoption of biodegradable and renewable materials has revolutionized forensic practices. Biopolymers such as chitosan, polylactic acid (PLA), polyhydroxybutyrate (PHB), and cellulose derivatives are used in fingerprint visualization and explosive detection. Green carbon dots derived from biowaste exhibit excellent fluorescence, low toxicity, and high versatility in detecting drugs, explosives, and counterfeit materials [105]. These advancements support sustainability while maintaining analytical efficiency.

6.5 Overview of Green Chemistry in Forensic Sciences

Green Analytical Chemistry represents a transformative step in modern forensic science, combining sustainability with systematic approach. Forensic labs may drastically lower environmental and occupational dangers by implementing green materials, energy-efficient equipment, and solvent-free analytical techniques. In addition to maintaining the quality of the evidence and analytical accuracy, this sustainable method guarantees that forensic science will continue to serve justice in an ethical manner, protecting the environment and society.

7 Conclusion

The integration of Green Analytical Chemistry into pharmaceutical, food, environmental, industrial, and forensic sectors represents a sustainable alternative along with an ethical and scientific imperative for the twenty-first century. Traditional analytical protocols, often reliant on volatile organic solvents, excessive energy, and hazardous reagents, generate substantial waste and pose long-term risks to human health and ecosystems. By contrast, GAC delivers equivalent or superior analytical performance while dramatically reducing environmental footprints, as evidenced by reduced E-factors, minimal solvent consumption, and biodegradable methodologies across all examined fields.

The consistent applicability of the 12 principles of green chemistry to these diverse sectors—prevention, atom economy, safer solvents, energy efficiency, renewable feedstocks, and real-time monitoring—confirms that sustainability is not a compromise but an enhancement of scientific rigor. Emerging assessment tools, despite individual limitations, collectively provide robust frameworks for researchers and industry to select and validate greener methods objectively.

Continued advancement demands interdisciplinary collaboration, regulatory support for green metrics, and educational emphasis on sustainability from the earliest stages of chemical training. The transition to Green Analytical Chemistry is no longer optional; it is the only viable pathway toward analytical practices that protect both human health and planetary integrity. Embracing GAC today ensures that future generations inherit not only accurate data but also a habitable world.

References

1. Koel, M., & Kaljurand, M. (2006). Application of the principles of green chemistry in analytical chemistry. *Pure and Applied Chemistry*, *78*(11), 1993–2002
2. Dunn, P. J. (2012). The importance of green chemistry in process research and development. *Chemical Society Reviews*, *41*(4), 1452–1461.
3. Armenta, Sergio, Salvador Garrigues, and Miguel de la Guardia. "Green analytical chemistry." *TrAC Trends in Analytical Chemistry* 27.6 (2008): 497–511.
4. Zade, A. (2024). Green Chemistry in Pharmaceutical Development: Approaches, Challenges, and Future Directions. *International Journal of Phytochemistry and ChemAnalysis (IJPCA)*, *1*(01), 34–39.
5. Islam, F., Imran, A., Khalid, M. A., Afzaal, M., Fatima, M., & Chauhan, A. (2023). Green energy process "Ozonation" and food safety: a comprehensive review. *Current Research in Nutrition and Food Science Journal*, *11*(2), 488–503.
6. Alviri, H., Lynes, J., & Habib, K. (2025). Beyond green chemistry: a comprehensive review of how sustainability has been integrated into cosmetic research. *Global Sustainability*, 1–52.
7. Chojnacka, K. (2024). Sustainable chemistry in adaptive agriculture: A review. *Current Opinion in Green and Sustainable Chemistry*, *46*, 100898.
8. Sheldon, R. A., & Brady, D. (2022). Green chemistry, biocatalysis, and the chemical industry of the future. *ChemSusChem*, *15*(9), e202102628.
9. Herrmann, J. M., Duchamp, C., Karkmaz, M., Hoai, B. T., Lachheb, H., Puzenat, E., & Guillard, C. (2007). Environmental green chemistry as defined by photocatalysis. *Journal of hazardous materials*, *146*(3), 624–629.
10. Ünal, B. K., Uysal, S., & Uzun, L. (2025). Greener approaches/materials for forensic sciences. *Essential Chem*, *2*(1), 1–21.
11. Lankey, Rebecca L., and Paul T. Anastas. "Life-cycle approaches for assessing green chemistry technologies." *Industrial & engineering chemistry research* 41.18 (2002): 4498–4502.
12. Sajid, M., & Płotka-Wasylka, J. (2022). Green analytical chemistry metrics: A review. *Talanta*, *238*, 123046.

13. Martínez, J., Cortés, J. F., & Miranda, R. (2022). Green chemistry metrics, a review. *Processes, 10*(7), 1274.
14. Kralisch, D., Ott, D., & Gericke, D. (2015). Rules and benefits of life cycle assessment in green chemical process and synthesis design: a tutorial review. *Green Chemistry, 17*(1), 123–145.Rgb
15. Nowak, P. M., Wietecha-Posłuszny, R., & Pawliszyn, J. (2021). White analytical chemistry: an approach to reconcile the principles of green analytical chemistry and functionality. *TrAC Trends in Analytical Chemistry, 138*, 116223.
16. Hartman, R., Helmy, R., Al-Sayah, M., & Welch, C. J. (2011). Analytical Method Volume Intensity (AMVI): A green chemistry metric for HPLC methodology in the pharmaceutical industry. *Green Chemistry, 13*(4), 934–939.
17. Swanson, M. B., Davis, G. A., Kincaid, L. E., Schultz, T. W., Bartmess, J. E., Jones, S. L., & George, E. L. (1997). A screening method for ranking and scoring chemicals by potential human health and environmental impacts. *Environmental toxicology and chemistry, 16*(2), 372–383.
18. Płotka-Wasylka, J. (2018). A new tool for the evaluation of the analytical procedure: Green Analytical Procedure Index. *Talanta, 181*, 204–209.
19. Gaber, Y., Törnvall, U., Kumar, M. A., Amin, M. A., & Hatti-Kaul, R. (2011). HPLC-EAT (Environmental Assessment Tool): a tool for profiling safety, health and environmental impacts of liquid chromatography methods. *Green chemistry, 13*(8), 2021–2025.
20. Gałuszka, A., Migaszewski, Z. M., Konieczka, P., & Namieśnik, J. (2012). Analytical Eco-Scale for assessing the greenness of analytical procedures. *TrAC Trends in Analytical Chemistry, 37*, 61–72.
21. Pena-Pereira, F., Wojnowski, W., & Tobiszewski, M. (2020). AGREE—Analytical GREEnness metric approach and software. *Analytical chemistry, 92*(14), 10076-10082.
22. Moema, D., Makwakwa, T. A., Gebreyohannes, B. E., Dube, S., & Nindi, M. M. (2023). Hollow fiber liquid phase microextraction of fluoroquinolones in chicken livers followed by high pressure liquid chromatography: Greenness assessment using National Environmental Methods Index Label (NEMI), green analytical procedure index (GAPI), Analytical GREEnness metric (AGREE), and Eco Scale. *Journal of Food Composition and Analysis, 117*, 105131.
23. Tobiszewski, M., & Orłowski, A. (2015). Multicriteria decision analysis in ranking of analytical procedures for aldrin determination in water. *Journal of Chromatography A, 1387*, 116–122.
24. Gallart-Mateu, D., Cervera, M. L., Armenta, S., & De La Guardia, M. (2015). The importance of incorporating a waste detoxification step in analytical methodologies. *Analytical Methods, 7*(13), 5702–5706.
25. Ballester-Caudet, A., Campíns-Falcó, P., Pérez, B., Sancho, R., Lorente, M., Sastre, G., & González, C. (2019). A new tool for evaluating and/or selecting analytical methods: summarizing the information in a hexagon. *TrAC Trends in Analytical Chemistry, 118*, 538–547.
26. Al-Hazmi, H., Namiesnik, J., & Tobiszewski, M. (2016). Application of TOPSIS for selection and assessment of analytical procedures for ibuprofen determination in wastewater. *Current Analytical Chemistry, 12*(4), 261–267.
27. Humphrey, Guy R., et al. "Asymmetric synthesis of letermovir using a novel phase-transfer-catalyzed aza-Michael reaction." *Organic Process Research & Development* 20.6 (2016): 1097–1103.
28. Vijaykumar, C., Kumar, Y. R., Aparna, P., & Marisetti, V. M. (2022). Enantioselective green HPLC method for simultaneous determination of enantiomer, and potential impurities in Apremilast drug substance. *Analytical Chemistry Letters, 12*(6), 691-714.
29. Kokilambigai, K. S., & Lakshmi, K. S. (2022). Analytical quality by design assisted RP-HPLC method for quantifying atorvastatin with green analytical chemistry perspective. *Journal of Chromatography Open, 2*, 100052.

30. Lotfy, H. M., Obaydo, R. H., & Mohamed, E. H. (2023). Environmentally sustainable computationally spectrophotometric resolution strategy for analysis single-tablet regimens of antihypertension with overlapped spectra. *Talanta Open*, *7*, 100226.
31. Sukumar, V., Chinnusamy, S., Chanduluru, H. K., & Rathinam, S. (2023). Method development and validation of Atorvastatin, Ezetimibe and Fenofibrate using RP-HPLC along with their forced degradation studies and greenness profiling. *Green Chemistry Letters and Reviews*, *16*(1), 2198651.
32. Elmansi, H., & Belal, F. (2019). Development of an Eco-friendly HPLC method for the simultaneous determination of three benzodiazepines using green mobile phase. *Microchemical Journal*, *145*, 330–336.
33. Mansour, N. M., El-Sherbiny, D. T., Ibrahim, F. A., & El Subbagh, H. I. (2021). Development of an Inexpensive, sensitive and green HPLC method for the simultaneous determination of brivaracetam, piracetam and carbamazepine; application to pharmaceuticals and human plasma. *Microchemical Journal*, *163*, 105863.
34. Dharuman, N., Lakshmi, K. S., & Krishnan, M. (2023). Environmental benign RP-HPLC method for the simultaneous estimation of anti-hypertensive drugs using analytical quality by design. *Green Chemistry Letters and Reviews*, *16*(1), 2214176.
35. Kannaiah, K. P., & Sugumaran, A. (2021). Environmental benign analytical quality by design and UPLC method development for Betamethasone and Calcipotriene in ointment. *Journal of Taibah University for Science*, *15*(1), 1156–1167.
36. Kumar Chanduluru, H., Sugumaran, A., & Kannaiah, K. P. (2022). Multiple spectrophotometric determinations of Chlorthalidone and Cilnidipine using propylene carbonate–As a step towards greenness. *Analytical Biochemistry*, *657*, 114890.
37. Sharma, T., Jain, A., Saini, S., Katare, O. P., & Singh, B. (2020). Implementation of analytical quality-by-design and green analytical chemistry approaches for the development of robust and ecofriendly UHPLC analytical method for quantification of chrysin. *Separation Science Plus*, *3*(9), 384–398.
38. Rashed, N. S., Zayed, S., Abdelazeem, A., & Fouad, F. (2020). Development and validation of a green HPLC method for the analysis of clorsulon, albendazole, triclabendazole and ivermectin using monolithic column: Assessment of the greenness of the proposed method. *Microchemical Journal*, *157*, 105069.
39. Kannaiah, K. P., Sugumaran, A., & Chanduluru, H. K. (2023). Simultaneous estimation of crotamiton and hydrocortisone by RP-UPLC using green analytical principles embedded analytical quality by design (AQbD) method. *Microchemical Journal*, *184*, 108166.
40. Iqbal, M., Ezzeldin, E., Anwer, M. K., & Imam, F. (2021). Eco-friendly UPLC-MS/MS quantitation of delafloxacin in plasma and its application in a pharmacokinetic study in rats. *Separations*, *8*(9), 146.
41. Perumal, D. D., Krishnan, M., & Lakshmi, K. S. (2022). Eco-friendly based stability-indicating RP-HPLC technique for the determination of escitalopram and etizolam by employing QbD approach. *Green Chemistry Letters and Reviews*, *15*(3), 671–682.
42. Chakraborty, S., & Mondal, S. (2023). A Green Eco-friendly Analytical Method Development, Validation, and Stress Degradation Studies of Favipiravir in Bulk and Different Tablet Dosages Form by UV-spectro-photometric and RP-HPLC Methods with their Comparison by Using ANOVA and in-vitro Dissolution Studies. *International Journal of Pharmaceutical Investigation*, *13*(2).
43. Fawzy, M. G., Saleh, H., Reda, A., & Bahgat, E. A. (2022). A green spectrophotometric method for the simultaneous determination of nasal binary mixture used in respiratory diseases: applying isosbestic point and chemometric approaches as a resolving tool, greenness evaluation. *Spectrochimica Acta Part A: Molecular and Biomolecular Spectroscopy*, *283*, 121585.
44. Muchakayala, S. K., Katari, N. K., Saripella, K. K., Schaaf, H., Marisetti, V. M., Kowtharapu, L. P., & Jonnalagadda, S. B. (2022). AQbD based green UPLC method to determine mycophenolate mofetil impurities and Identification of degradation products by QToF LCMS. *Scientific Reports*, *12*(1), 19138.

45. Saroj, S., Jairaj, V., & Rathod, R. (2017). Applying green analytical chemistry for development and validation of RP-HPLC stability indicating assay method for estimation of fenoverine in bulk and dosage form using quality by design approach. *Journal of Liquid Chromatography & Related Technologies*, *40*(7), 340–352.
46. Chanduluru, H. K., & Sugumaran, A. (2021). Eco-friendly estimation of isosorbide dinitrate and hydralazine hydrochloride using Green Analytical Quality by Design-based UPLC Method. *RSC advances*, *11*(45), 27820–27831.
47. Kannaiah, K. P., & Sugumaran, A. (2022). Eco-friendly multivariant green analytical technique for the estimation of ketoconazole by UV spectroscopy in bulk and cream formulation. *Química Nova*, *45*(01), 23–30.
48. Vieira-Sellaï, L., Quintana, M., Diop, O., Mercier, O., Tarrit, S., Raimi, N.,... & Galmier, M. J. (2022). Green HPLC quantification method of lamivudine, zidovudine and nevirapine with identification of related substances in tablets. *Green chemistry letters and reviews*, *15*(3), 695–704.
49. Magdy, G., Abdel Hakiem, A. F., Belal, F., & Abdel-Megied, A. M. (2021). A novel quality by design approach for development and validation of a green reversed-phase HPLC method with fluorescence detection for the simultaneous determination of lesinurad, febuxostat, and diflunisal: application to human plasma. *Journal of separation science*, *44*(11), 2177–2188.
50. Michael, A. M., Lotfy, H. M., & Nessim, C. K. (2023). Greenness profile and whiteness assessment of the stability-indicating HPLC method for the assay of levetiracetam. *Microchemical Journal*, *190*, 108669.
51. Bang, P. P., & Bhatt, H. G. (2023). Development of green RP-and green NP-HPTLC methods for estimation of lenvatinib and comparative evaluation by AGREE. *ACS Sustainable Chemistry & Engineering*, *11*(6), 2249–2263.
52. Elsheikh, S. G., Hassan, A. M., Fayez, Y. M., & El-Mosallamy, S. S. (2023). Green analytical chemistry and experimental design: a combined approach for the analysis of zonisamide. *BMC chemistry*, *17*(1), 38.
53. Chen, J., Xiao, S., Wu, X., Fang, K., & Liu, W. (2005). Determination of lead in water samples by graphite furnace atomic absorption spectrometry after cloud point extraction. *Talanta*, *67*(5), 992–996.
54. Resano, M., Garcia-Ruiz, E., Aramendía, M., & Belarra, M. A. (2005). Solid sampling-graphite furnace atomic absorption spectrometry for Hg monitoring in soils. Performance as a quantitative and as a screening method. *Journal of Analytical Atomic Spectrometry*, *20*(12), 1374–1380.
55. Yoo, E. H., & Lee, S. Y. (2010). Glucose biosensors: an overview of use in clinical practice. *Sensors*, *10*(5), 4558–4576.
56. Aziz, K. S., & Qader, H. A. (2024). Derivative spectrophotometric method for simultaneous determination of ascorbic acid and folic acid in the pharmaceutical formulation; Comparative greenness assessment. *Green Analytical Chemistry*, *11*, 100157.
57. Al-Shatti, B. J., Alsairafi, Z., & Al-Tannak, N. F. (2024). Green Chemistry and its Implementation in Pharmaceutical Analysis: Green Pharmaceutical Analysis.
58. Kogawa, A. C., & Salgado, H. R. N. (2016). Quantification of rifaximin in tablets by spectrophotometric method ecofriendly in ultraviolet region. *Scientifica*, *2016*(1), 3463405.
59. Tótoli, E. G., & Salgado, H. R. N. (2013). Development and validation of a rapid turbidimetric assay to determine the potency of ampicillin sodium in powder for injectable solution. *Analytical Methods*, *5*(21), 5923–5928.
60. Spagnol, C. M., Isaac, V. L. B., Corrêa, M. A., & Salgado, H. R. N. (2016). Validation of HPLC–UV assay of caffeic acid in emulsions. *Journal of chromatographic science*, *54*(3), 305–311.
61. Sandhya, J., & Gopukumar, S. T. (2025). Application Of Green Analytical Chemistry In The Quantification Of Anticancer Agents: Eco-Friendly Method Development And Validation. *Journal of Applied Bioanalysis*, *11*(4), 624632.
62. Sarraguça, M. C., & Lopes, J. A. (2009). Quality control of pharmaceuticals with NIR: From lab to process line. *Vibrational Spectroscopy*, *49*(2), 204–210.

63. Venkatesan, K., Sundarababu, J., & Anandan, S. S. (2024). The recent developments of green and sustainable chemistry in multidimensional way: current trends and challenges. *Green Chemistry Letters and Reviews*, *17*(1), 2312848.
64. Bryan, M. C., Dunn, P. J., Entwistle, D., Gallou, F., Koenig, S. G., Hayler, J. D.,... & Weiberth, F. J. (2018). Key Green Chemistry research areas from a pharmaceutical manufacturers' perspective revisited. *Green Chemistry*, *20*(22), 5082–5103.
65. Kurul, F., Doruk, B., & Topkaya, S. N. (2025). Principles of green chemistry: building a sustainable future. *Discover Chemistry*, *2*(1), 68.
66. Iammarino, M., Panseri, S., Unlu, G., Marchesani, G., & Bevilacqua, A. (2022). Novel chemical, microbiological and physical approaches in food safety control. *Frontiers in Nutrition*, *9*, 1060480.
67. Lourenço, S. C., Moldão-Martins, M., & Alves, V. D. (2019). Antioxidants of natural plant origins: From sources to food industry applications. *Molecules*, *24*(22), 4132.
68. Pallone, J. A. L., dos Santos Caramês, E. T., & Alamar, P. D. (2018). Green analytical chemistry applied in food analysis: alternative techniques. *Current Opinion in Food Science*, *22*, 115–121.
69. Choton, S., Bandral, J. D., Sood, M., Gupta, N., Singh, J., Langeh, A., & Choudhary, A. N. (2023). Green extraction technology and its application in food industry. *Chemical Science Review and Letters*, *12*(46), 79–88.
70. Tiwari, B. K. (2015). Ultrasound: A clean, green extraction technology. *TrAC Trends in Analytical Chemistry*, *71*, 100–109.
71. Carpentieri, S., Soltanipour, F., Ferrari, G., Pataro, G., & Donsì, F. (2021). Emerging green techniques for the extraction of antioxidants from agri-food by-products as promising ingredients for the food industry. *Antioxidants*, *10*(9), 1417.
72. Ashwood-Smith, M. J., Poulton, G. A., Ceska, O., Liu, M., & Furniss, E. (1983). An ultrasensitive bioassay for the detection of furocoumarins and other photosensitizing molecules. *Photochemistry and Photobiology*, *38*(1), 113–118.
73. Ghaani, M., Nasirizadeh, N., Ardakani, S. A. Y., Mehrjardi, F. Z., Scampicchio, M., & Farris, S. (2016). Development of an electrochemical nanosensor for the determination of gallic acid in food. *Analytical Methods*, *8*(5), 1103–1110.
74. Wang, B., Wang, Y., Xie, X., Diao, Z., Xie, K., Zhang, G.,... & Dai, G. (2020). Quantitative analysis of spectinomycin and lincomycin in poultry eggs by accelerated solvent extraction coupled with gas chromatography tandem mass spectrometry. *Foods*, *9*(5), 651.
75. Tsochatzis, E., Papageorgiou, M., & Kalogiannis, S. (2019). Validation of a HILIC UHPLC-MS/MS method for amino acid profiling in triticum species wheat flours. *Foods*, *8*(10), 514.
76. Perumalla, A. V. S., & Hettiarachchy, N. S. (2011). Green tea and grape seed extracts—Potential applications in food safety and quality. *Food Research International*, *44*(4), 827–839.
77. Yousef, A. E., & Marth, E. H. (1985). Degradation of aflatoxin M1 in milk by ultraviolet energy. *Journal of Food Protection*, *48*(8), 697–698.
78. Jeon, C. S. (2016). *Surface Functionalization of Bioanalytical Applications: Virus-decorated Gold Microshells and Modified Synaptic Cell Adhesion Molecules* (Doctoral dissertation, 서울대학교 대학원).
79. Chari, N., Felix, L., Davoodbasha, M., Ali, A. S., & Nooruddin, T. (2017). In vitro and in vivo antibiofilm effect of copper nanoparticles against aquaculture pathogens. *Biocatalysis and agricultural biotechnology*, *10*, 336–341.
80. Xu, L., Suo, X. Y., Zhang, Q., Li, X. P., Chen, C., & Zhang, X. Y. (2020). ELISA and chemiluminescent enzyme immunoassay for sensitive and specific determination of lead (II) in water, food and feed samples. *Foods*, *9*(3), 305.
81. Varriale, A., Staiano, M., Iozzino, L., Severino, L., Anastasio, A., Cortesi, M. L., & D'Auria, S. (2009). FCS-based sensing for the detection of ochratoxin and neomycin in food. *Protein and Peptide Letters*, *16*(12), 1425–1428.
82. Ogbuagu, OLUCHUKWU OBINNA, et al. "Sustainable pharmaceutical supply chains: Green chemistry approaches to drug production and distribution." *IRE Journals* 8.4 (2024): 761–767.

83. Sutanto, Stevia. "Textile dry cleaning using carbon dioxide: process, apparatus and mechanical action." *Master Sci Chem Eng, TU Delft, Delft, April* 7 (2014).
84. Hitzler, Martin G., and Martyn Poliakoff. "Continuous hydrogenation of organic compounds in supercritical fluids." *Chemical Communications* 17 (1997): 1667–1668.
85. Kitanosono, Taku, et al. "Catalytic organic reactions in water toward sustainable society." *Chemical Reviews* 118.2 (2018): 679–746.
86. Forsyth, Stewart A., Jennifer M. Pringle, and Douglas R. MacFarlane. "Ionic liquids—an overview." *Australian Journal of Chemistry* 57.2 (2004): 113–119.
87. Szmant, H. H. Organic Building Blocks of the Chemical Industry; Wiley: New York, 1989; p 4. 13
88. Chemicals, and Fuels. In The Presidential Green Chemistry Challenge Awards Program: Summary of 1996 Award Entries and Recipients; EPA744-K-96-001, U.S. Environmental Protection Agency, Office of Pollution Prevention and Toxics: Washington, DC, 1996; p 7.14
89. Monsanto Company. The Catalytic Dehydrogenation of Diethanolamine.In the Presidential Green Chemistry Challenge Awards Program: Summary of 1996 Award Entries and Recipients; EPA744-K-96–001; U.S. Environmental Protection Agency, Office of Pollution Prevention and Toxics: Washington, DC, 1996; p 2. 15
90. Trost, B. M.; Pinkerton, A. B. A Three-Component Coupling Approach to Cyclopentanoids. J. Org. Chem. **2001**, 66, 7714–7722. 16
91. Habermann, J.; Ley, S. V.; Scott, J. S. Synthesis of the potent analgesic compound (()-epibatidine using an orchestrated multistep sequence of polymer supported reagents. J. Chem. Soc., Perkin Trans. 1 **1999**, 1253–1256.
92. Anastas, P. T., Kirchhoff, M. M., & Williamson, T. C. (2001). Catalysis as a foundational pillar of green chemistry. *Applied Catalysis A: General*, *221*(1–2), 3–13.
93. J.M. Thomas, Chemical and Engineering News 78 (27) (2000) 7.
94. J.-M. Vincent, A. Rabion, V.K. Yachandra, R.H. Fish, Angew. Chem. Int. Ed. Engl. 36 (21) (1997) 2346.
95. Fouda-Mbanga, B. G., Onotu, O. P., & Tywabi-Ngeva, Z. (2024). Advantages of the reuse of spent adsorbents and potential applications in environmental remediation: A review. *Green Analytical Chemistry*, *11*, 100156.
96. Mollah, M. Y. A., Schennach, R., Patscheider, J., Promreuk, S., & Cocke, D. L. (2000). Plasma chemistry as a tool for green chemistry, environmental analysis and waste management. *Journal of hazardous materials*, *79*(3), 301–320.
97. Escuriola, M. J., Morales-Rubio, A., & De la Guardia, M. (1999). A clean analytical method for the spectrophotometric determination of formetanate incorporating an on-line microwave assisted hydrolysis step. *Analytica chimica acta*, *390*(1–3), 147–154.
98. Inoue, Y., & Munemori, M. (1979). Coprecipitation of mercury (II) with iron (III) hydroxide. *Environmental Science & Technology*, *13*(4), 443–445.
99. Khairmode, P. R., & Darade, R. B. (2025). Recent Advances in Green Analytical Chemistry. *Asian Journal of Pharmaceutical Research and Development*, *13*(1), 231–237.
100. Guzel-Seydim, Z. B., Greene, A. K., & Seydim, A. C. (2004). Use of ozone in the food industry. *LWT-Food Science and technology*, *37*(4), 453–460.
101. Brett, C. M. (2007). Novel sensor devices and monitoring strategies for green and sustainable chemistry processes. *Pure and applied Chemistry*, *79*(11), 1969–1980.
102. Shadiya, O. O., Satish, V., & High, K. A. (2012). Process enhancement through waste minimization and multiobjective optimization. Journal of Cleaner Production, 31, 137–149.
103. Mulholland, K. L., & Dyer, J. A. (2001). Process analysis via waste minimization: Using DuPont's methodology to identify process improvement opportunities. *Environmental progress*, *20*(2), 75–79.
104. Wahid, M., Ahmad, F., & Ahmad, N. (2017, July). Green chemistry: Principle and its application. In 2nd International Conference on Advancement in Engineering, Applied Science and Management.
105. Ünal, B. K., Uysal, S., & Uzun, L. (2025). Greener approaches/materials for forensic sciences. Essential Chemistry, 2(1), 1–21.

106. Koel, M. (2024). Developments in analytical chemistry initiated from green chemistry. Sustainable Chemistry for the Environment, 5, 100078.
107. Ivanković, A., Dronjić, A., Bevanda, A. M., & Talić, S. (2017). Review of 12 principles of green chemistry in practice. International Journal of Sustainable and Green Energy, 6(3), 39–48.
108. Mehta, M., Mehta, D., & Mashru, R. (2024). Recent application of green analytical chemistry: Eco-friendly approaches for pharmaceutical analysis. Future Journal of Pharmaceutical Sciences, 10(1), 83.

Challenges and Future Trends in Green Analytical Chemistry

Lakshmi Madhavi Chitra, Dinakaran SathisKumar, and Avasarala Harani

Abstract Green Analytical Chemistry (GAC) has emerged as a pivotal framework for aligning analytical procedures with principles of environmental sustainability. This chapter examines the challenges and future trends within the GAC, with special emphasis on the Indian context. The ongoing efforts to balance cost-effectiveness, analytical efficiency, and environmental sustainability are investigated critically, with relevance to pharmaceutical industries, food safety, and environmental monitoring. The regulatory approaches that govern the green analytical methods, both globally and nationally, are analyzed that shape their adoption. Through the real-world case studies and practical implementation models, the operational and financial challenges encountered when scaling laboratory-developed green techniques to industrial applications are addressed. Integrating artificial intelligence (AI) and machine learning (ML) in optimizing green analytical workflows leads to a transformative approach. The future trends include bio-based materials, solvent-free techniques, and zero-waste processes. This chapter offers a comprehensive guide for researchers, practitioners, and policymakers to navigate the transition to a sustainable future in analytical science.

Keywords Green analytical chemistry · Sustainable analytical practices · Regulatory frameworks · Artificial intelligence (AI) · Machine learning (ML) · Bio-based materials · Zero-waste processes

L. M. Chitra · D. SathisKumar (✉)
Department of Pharmaceutical Analysis, School of Pharmacy, Aditya University, Surampalem 533437, Andhra Pradesh, India
e-mail: satmpdina@yahoo.co.in; satmpdina@gmail.com

A. Harani
Department of Pharmaceutics, School of Pharmacy, Aditya University, Surampalem 533437, Andhra Pradesh, India

V. Bhusari and M. Ghante (eds.), *Green Analytical Chemistry in India*,
https://doi.org/10.1007/978-3-032-24564-9_3

1 Introduction

Analytical chemistry has always been the key to scientific discovery, product development, and quality control. It ensures evidence-based decision-making across various industries, from monitoring air and water pollutants to guaranteeing the safety of medicines and food [1, 2]. But for many years, the field has neglected the environmental and human health hazards caused by the toxic solvents, hazardous wastes, and energy-intensive equipment while focusing on its performance for maximizing accuracy, sensitivity, and efficiency [3–5].

Green Analytical Chemistry (GAC) evolved as a solution to this imbalance. The principles of Green Analytical Chemistry are applied in analytical science to enhance laboratory safety and lessen the environmental effects by reducing the waste at its source, choosing safer solvent alternatives, using energy-efficient methods, and using miniaturized equipment [6–8].

Even with these advancements, GAC still faces challenges. In laboratories, the balancing of cost, efficiency, and sustainability is essential [9]. Regulatory frameworks have not been fully adopted [10], and it is still a challenge to scale up green methods from research settings to industrial practice [11].

Meanwhile, the interventions of artificial intelligence (AI) and machine learning (ML), and with innovations like bio-based materials and zero-waste analytical processes, are developing new opportunities for the future [8–12].

This chapter explores these interconnected dimensions. It addresses the need to balance sustainability with the economic and technical performance, reviews the regulatory perspectives, examines the scaling-up issues of GAC, promotes the use of AI and ML in optimization, and inspects the new trends like bio-based sorbents, zero-waste processes, and integrated real-time monitoring systems. Together, they offer a path for transforming analytical chemistry into a scientifically sound and environmentally sustainable discipline [19].

2 Balancing Cost, Efficiency, and Sustainability

2.1 High Energy Consumption and Costs

The conventional analytical techniques often demand high energy, generating waste that burdens the environment. These techniques depend on power-intensive equipment that increases the operational costs and carbon emissions [20].

- **Sample Preparation**: Acid digestion procedures demand high temperatures, a long time, and a high amount of electricity [21]. Another procedure is gravimetric analysis, where the drying of samples requires large ovens that operate with high energy.

- **Instrumentation**: Advanced analytical instruments, such as gas chromatography (GC) and high-performance liquid chromatography (HPLC), require a lot of energy to maintain the ovens at high temperatures and run pumps, respectively. The most energy-intensive instruments are mass spectrometers, as they operate with powerful vacuum pumps and heating ion sources [22].
- **Ancillary Equipment**: Energy-intensive HVAC systems are required to maintain stable lab conditions by reducing the heat generated by the advanced instruments. Energy loss also occurs by using the fume hoods continuously to protect personnel while handling hazardous solvents [23].

To meet the demands of such energy-intensive setups, fossil fuels are sourced, which directly contribute to increased costs and also the carbon footprint of the analytical process [24]. This cumulative energy demand globally across several industrial, academic, and government facilities urges the need for adopting cost-effective, green analytical alternatives [25].

2.2 *Trade-Offs in Analytical Performance*

Efficiency and reliability are important considerations while implementing GAC. The major challenge is a perception that using green methods might compromise analytical performance.

- **Sensitivity**: Traditional methods involve the use of non-polar solvents such as hexane, which aid in the efficient extraction of non-polar compounds like pesticides or polychlorinated biphenyls (PCBs), whereas the green methods employ green solvents like water or ethanol, which have less extraction efficiency for such non-polar compounds [26]. This might impact the sensitivity of the extraction method. The researchers solved this challenge by adding small amounts of co-solvents or by using advanced methods of extractions like stir-bar sorptive extraction (SBSE) [27]. These strategies increase the sensitivity of the extraction method.
- **Selectivity**: In GAC, hazardous chemicals are replaced with safe alternatives. But these alternatives may respond to multiple metal ions instead of the selective ions, and give false positive results, within complex mixtures. Researchers suggested that by adding a pre-separation step, like solid-phase extraction (SPE) in the method, or by using a green reagent together with a high-resolution detector like a mass spectrometer, helps to identify the individual compounds accurately [28].
- **Reproducibility**: As GAC methods use new materials under less regulated conditions, they may not be reproducible. When compared to silica-based sorbents, a bio-based sorbent may exhibit an uncontrollable variability. Moreover, the new methods might not be tested for inter-laboratory studies, which leads to lower robustness and reproducibility [29]. To overcome this, strict validation studies and standard protocols are to be developed for the production and use of green materials [30].

2.3 Generation of Hazardous Waste

Sustainability is another major factor. Conventional analytical methods generate large volumes of hazardous waste, particularly during sample preparation, which is often a solvent-intensive process.

- Liquid-liquid extraction (LLE) can consume hundreds of milliliters of organic solvents like chloroform and dichloromethane in a single run [31]. In a typical high-throughput lab, thousands of samples are run per day, which leads to the generation of thousands of liters of waste.
- High-performance liquid chromatography (HPLC) presents a similar challenge, in which the mobile phases contain significant proportions of organic solvents like acetonitrile and methanol [32]. Though these solvents are generally less toxic than the chlorinated compounds, their poor biodegradability and their potential to contaminate water sources on their disposal lead to environmental challenges [33].
- Multi-step, solvent-heavy clean-up and concentration procedures increase waste volume, and their disposal requires specialized and costly methods such as incineration, which may again lead to harmful atmospheric pollution [9].

These issues highlight the need to reduce solvent use and adopt green alternatives.

2.4 Case Studies of Compromises in the Pursuit of Sustainability

- **Compromise in Cost**: Small or medium-sized laboratories face difficulties in adopting new green technologies like supercritical fluid chromatography (SFC) since they require higher investment costs. But in the long term, the operational costs are reduced as the solvent purchasing and waste-disposal fees are decreased [34]. Thus, adopting green technologies should be based on cost-benefit analysis.
- **Compromise in Speed**: The sample preparation process for a traditional method of extraction is faster, but employing a green method like microwave-assisted extraction (MAE) with green solvents can be highly efficient and generate less waste [35]. For a high-throughput lab, for which the critical performance criterion is speed, a compromise can become a deciding factor between rapidity and sustainability.
- **Compromise in Sensitivity**: In the analysis of polychlorinated biphenyls (PCBs), the green methods replace the large volumes of hazardous consumption with small volumes of less hazardous solvent mixtures or by an alternative solid-phase microextraction (SPME) technique [36]. But, this trade-off may result in a slight increase in detection limit, yet in routine monitoring programs where the concentration of interest is well above the limits, these greener alternative methods prove to be effective and environmentally sustainable.

Obtaining the balance between cost, efficiency, and sustainability is essential. The objective of GAC is to provide high performance and environmental sustainability through innovative and less costly green solutions [37].

2.5 Drivers for Adopting GAC

Due to the environmental, economic, and health demands, a significant change in analytical procedures is required, which drives the shift toward Green Analytical Chemistry (GAC).

- **Environmental Drivers**: The traditional methods produce a lot of toxic waste, like dichloromethane and chloroform, and their treatment also adds to the environmental impact [7]. By using green solvents such as water or supercritical fluids or by solvent-free methods, GAC eliminates this problem while protecting the ecosystem and reducing waste at its source [38].
- **Economic Drivers**: GAC technologies, like supercritical fluid chromatography (SFC) and microfluidics, require an initial investment, but they significantly reduce solvent use and hazardous waste disposal, which reduces long-term costs. Therefore, increases sustainability and profitability [39].
- **Health Drivers**: GAC promotes closed systems, automation, and safer chemicals to improve workplace safety by reducing exposure of employees to toxic volatile organic compounds (VOCs) and flammable chemicals, which are released by conventional procedures [8].

Together, these drivers demand the need for cost-effective, efficient, and sustainable solutions that prove that GAC is essential for transforming analytical science.

3 Regulatory Perspectives on Green Analytical Methods

3.1 Standardization on Green Metrics

The lack of globally accepted standards to evaluate the "greenness" of the analytical methods is the main challenge for the adoption of GAC. The qualitative assessments that have been used in the past lack consistency.

Quantitative techniques like AGREE Index and GAPI are developed recently.

- AGREE Index provides a detailed evaluation of greenness of the method, and visually represents its performance by a color-coded radial diagram [25, 40].
- Green Analytical Procedure Index (GAPI) considers each and every step of the analytical process including sample collection to waste disposal for the evaluation [41].

The global harmonization is still lacking even with the development of these tools.

3.2 Regulatory and Standardization Challenges

The adoption and implementation of Green Analytical Chemistry (GAC) also involves regulatory and standardization challenges [42].

- **Pace of Regulatory Change**: Regulatory compliance is consistently delayed, and this represents a significant barrier to the implementation of GAC. Although many regulatory authorities acknowledge the importance of sustainability, the process of formally validating and approving new green methods remains gradual and systematic because the regulators prioritize consumer protection and product quality [43]. Any new green method must be rigorously proven to be accurate, precise, and robust compared to the traditional methods. This process is governed by widely accepted guidelines, such as ICH, Pharmacopeia like United States Pharmacopeia (USP) and European Pharmacopeia (Ph.Eur.), which demand comprehensive and expensive studies of a method's performance [44]. The methods published in the guidelines are official for any given drug or substance. Modifying these established protocols is a time-consuming process that requires a periodic review, inter-laboratory studies, and official review. The absence of formal approval from key authorities generates a "catch-22" situation, in which laboratories are hesitant to invest in or adopt green methods without regulatory approval [45].
- **Harmonization of Global and National Standards**: The absence of a unified global framework for GAC leads to a serious challenge for global enterprises and international laboratories. The different regulatory bodies of different countries function with their own set of guidelines and standards for analytical methods. This fragmentation makes it difficult for a company to adopt a single green method for an internationally marketed product [46]. For example, a drug manufacturer has to employ different analytical methods to test the same drug, as per the regulatory requirements of different regions, like Europe and the United States, to market in those regions. Such differences lead to increased operations and costs, and also result in an environmental footprint. Thus, there is a need for global harmonization of the standards and metrics of GAC. This includes:

- **Consensus on Greenness Metrics**: Establishing an internationally accepted system like the AGREE index for measuring and reporting the greenness of a method allows impartial comparisons and encourages advancements [25, 47].
- **Unified Validation Guidelines**: A uniform framework for validating green methods, which is approved by all regulatory bodies across the nations, accelerates the approval process and saves time and money.

For the global adoption of GAC, international cooperation between the scientific communities and regulatory agencies, such as ICH, FDA, and EMA, is essential [48].

3.3 The Role of GAC in India: A Regional Perspective

The environmental challenges and a rapidly growing industrial sector urge the adoption of Green Analytical Chemistry (GAC) in India.

- **Specific Environmental Issues**: The pressing environmental concerns, such as water pollution in major rivers like the Ganga, the Yamuna, and severe air pollution in the urban cities of India, can be addressed by the GAC by integrating sustainable analytical methods to achieve advancing environmental protection goals [49].
- **Challenges and Opportunities in Key Sectors**: While the Pharmaceutical sectors offer a hurdle to the adoption of GAC, particularly in the generic drug manufacturing, the Food industries offer prospects for GAC adoption to ensure quality and safety across India's large and diverse supply chain [50].
- **Government Regulations and Education**: The Central Pollution Control Board (CPCB), the Bureau of Indian Standards (BIS), and other regulatory bodies govern to shape the sustainable practices, and the academic institution in education and research contribute to laying a strong foundation for a sustainable analytical community in India [51].

4 Challenges in Scaling up Green Techniques

The technical, economic, and regulatory challenges outweigh the benefits of Green Analytical Chemistry, so its implementation needs to overcome these challenges [11].

4.1 Technical and Operational Hurdles

4.1.1 The Search for Greener Solvents

The main aim of GAC is to replace hazardous organic solvents with greener alternatives. But this search creates numerous technical challenges due to the high efficiency of organic solvents [52]. The organic solvents have better chromatographic characteristics, less viscosity, and high solubilizing power. Finding a "neoteric" solvent with such characteristics is a major challenge [53].

Green Solvent Alternatives and Their Limitations: In search for greener solvents, better alternatives were developed with both advantages and limitations.

- **Ionic Liquids (ILs)**: These organic solvents, often referred as "designer solvent" are liquid at room temperature and have turnability, which allows researchers to alter them into cations or anions to modify their polarity, viscosity, and other physicochemical properties these are also characterized by negligible vapor pressure, non-flammable, and can be used in various extraction and chromatographic applications [54]. But these solvents are also known to have high viscosity and back pressure, which leads to technical problems in chromatographic separations. A few ionic liquids were not widely used in high-volume analytical applications as they are poorly biodegradable, toxic, and expensive [55].
- **Supercritical Fluids (SCFs)**: Green solvents like supercritical carbon dioxide ($scCO_2$) offer diffusivity like a gas and solvating power like a liquid and are also readily available, non-toxic, and non-flammable. Supercritical fluid chromatography (SFC) provides rapid responses and efficient separations by using less organic solvents, but it requires specialized and expensive equipment to keep the fluids in a supercritical state. The low polarity of $scCO_2$ mandates the addition of polar co-solvents such as methanol to analyze a wide range of analytes, which results in a slight compromise in the method's "greenness" [56].
- **Deep Eutectic Solvents (DESs)**: DESs are a novel class of solvents formed by combining two or more substances in a solid state. These solvents are biodegradable, generally non-toxic, and synthesized from renewable sources [57]. Their viscosity is similar to Ionic Liquids (ILs), due to which they do not show good column performance in chromatographic techniques. Due to their aqueous miscibility, the sample preparation is challenging as it prevents phase separation. In addition, though they are non-toxic and biodegradable, they are not thermally stable, and their analytical performance is still an area of research [58].

4.1.2 Miniaturization and Sample Preparation

The miniaturization of analytical procedures serves as a key principle of GAC, as it reduces sample and reagent consumption, which leads to reduced wastage and expenses [59]. Customized solutions are required to deal with the challenges caused by the shift from macro-scale analysis to miniaturization analysis.

- **Scaling Down Analytical Methods**: Because of the technical challenges, highly precise advanced instruments are required to deal with micro- and nanoliters of components. For example, traditional pipettes lack accuracy, which demands the need to be replaced by automated pipettes. In the miniature devices, the risk of cross-contamination is persistent as the total mass of the analyte is small, and even a minute amount of impurity from the environment, fabrication components, or carryovers can influence the signal-to-noise ratio, which leads to inaccurate results [60].
- **Green Sample Preparation Techniques**: In analytical methods, the sample preparation method requires more labor and generates more waste. The sample collection, extraction, and separation steps are combined into a single solvent-free

step by the green sample preparation methods, like solid-phase microextraction (SPME). But its adoption on a large scale is challenging [61].

- **Fiber Robustness and Capacity**: Although the fibers main components of SPME, their efficiency in handling a wide range of samples is still limited. Developing a universal fiber that can efficiently extract analytes from foods, biological fluids, and soil samples remains a critical challenge. The commercially available fibers lack robustness and are not durable [62].
- **Matrix Effects:** The heterogeneous sample matrices may compete for the binding sites on fibers with the specific analytes, which may result in inaccurate quantification. This "matrix effect" may reduce the reproducibility of the method and demand precise optimization and validation for every new application [63].
- **Regulatory Approval**: Reproducibility is a fundamental criterion for regulatory approval. The inconsistencies in outcomes, arising due to matrix interferences and fiber variability, often lead to difficulties for SPME techniques to meet the validation standards imposed by regulatory authorities like the Food and Drug Administration (FDA) and the Environmental Protection Agency (EPA), which depend on traditionally established methods. In spite of these challenges, ongoing research is focused on developing innovative coatings for SPME fibers, like molecularly imprinted polymers and carbon nanotubes, to enhance their specificity, robustness, and precision [64].

4.1.3 Method Validation and Reproducibility

The analytical methods have a framework of validated and reproducible methodologies. This challenges the new GAC approaches, because they must prove their environmental sustainability and show their analytical compatibility or superiority over the well-established traditional methods [65].

- **Validating Novel Green Methods**: Method validation guidelines such as ICH Q2 (R1) are maintained by regulatory authorities like the United States Food and Drug Administration (USFDA), European Medicines Agency (EMA), and various pharmacopeias. These frameworks require extensive validation parameters like accuracy, precision, linearity, limit of detection (LOD), and robustness [66]. Validating a new analytical method that uses a green solvent, a miniaturized device, and a new sorbent under these stringent standards can be time-consuming, expensive, and technically challenging. For example, a method utilizing a bio-based sorbent instead of a traditional silica-based sorbent must be similar in performance and show the least batch-to-batch variability [67]. Moreover, the lack of historical data for such new methods demands generating large data sets and conducting extensive inter-laboratory validation studies, which is a burden to academic and small to medium-sized facilities.
- **Skepticism and Resistance from Industry Professionals**: Many industrial professionals stay cautious in regard of replacing well-established methods with new GAC methods, due to several factors:

- **Fear of Non-compliance**: In highly regulated industries, non-validated methods may have regulatory non-compliance, which results in product recalls, legal liabilities, or severe financial penalties [68]. The conservative nature of such industries makes them reluctant to adopt new and unproven methods.
- **Cost of Re-validation**: Green methods, though they demonstrate analytical validity, their revalidation process require personnel time, technical expertise, and laboratory resources, leading to temporary disruption to routine laboratory operations [69].
- **Uncertainty Regarding Long-Term Performance**: Traditional methods have the benefit of largely accumulated data over the decades, which proves their robustness under day-to-day, high-throughput laboratories. Whereas the new green methods lack such data, which raises concerns about their reliability, durability, and consistency [70]. Industry professionals evidence that such methods are scientifically reliable and environmentally sustainable.

To overcome these challenges, a collaborative approach is required to develop robust validation data through comprehensive inter-laboratory studies and engage regulatory authorities in order to establish efficient and transparent pathways to support the adoption of green analytical methods [71].

4.2 *Economic and Financial Barriers*

The economic and financial barriers prohibit the implementation of Green Analytical Chemistry. Though the long-term benefits like sustainability, reduced solvent use, and lower disposal costs are evident, the high initial investments and showcasing economic rationale often act as major deterrents [72].

4.2.1 High Initial Investment

The capital expenditure required for adopting the advanced green analytical methods is a major burden for small laboratories, academic institutions, and industries in developing countries. For example, the acquisition of a supercritical fluid chromatography (SFC) system, which runs with benign CO_2 as the primary mobile phase, requires a relatively high investment when compared to a traditional high-performance liquid chromatography (HPLC) system due to its reliance on specialized high-pressure pumps, back-pressure regulators, and precision components essential to maintain the supercritical state of the mobile phase [73]. So, for the institutions with limited financial resources, it is difficult to adopt the green methods even if they recognize the long-term benefits.

4.2.2 Cost-Benefit Analysis

Though Green Analytical Chemistry (GAC) has long-term financial advantages, they are often diminished by the short-term expenses. A clear and strong financial justification is essential to support the transition from traditional methods to greener practices [74]. This requires a comprehensive cost-benefit analysis that considers both financial and environmental returns on investment.

A comprehensive assessment should extend beyond the initial purchase price and instead evaluate the Total Cost of Ownership (TCO). The TCO of a traditional method includes:

- **Direct Operational Costs**: The substantial expense of acquiring large quantities of analytical-grade solvents, combined with the escalating fees of hazardous waste disposal. For example, the disposal cost of hazardous solvent waste may vary from $0.80 to $4 per pound (lb), or even higher for halogenated solvents [75]. In large laboratories, such expenses can account for millions every year.
- **Indirect Costs**: Costs incurred due to energy-intensive equipment, staff training, and regulatory fines for non-compliance lead to indirect costs.

GAC-compliant method offers strong financial incentives like:

- **Long-Term Saving**: 90% of solvent consumption can be reduced by using the supercritical fluid chromatography (SFC) method when compared to a traditional HPLC method. Therefore, solvent purchasing costs and hazardous waste disposal fees are reduced in the long-term [76].
- **Improved Efficiency**: In some cases, compared to HPLC, SFC systems can reduce the analysis time by several times. This improvement leads to low viscosity and high diffusivity of supercritical fluids, which results in better sample handling capacities and increases profitability [77].
- **Enhanced Corporate Image and Market Advantage**: Implementing GAC shows a commitment toward sustainability. It serves as a key differentiator and a source of strategic benefit in an environment where customers and stakeholders value sustainable practices abundantly [78].

Therefore, the successful implementation of GAC demands a comprehensive cost-benefit analysis that measures all these elements.

5 The Role of Artificial Intelligence (AI) and Machine Learning (ML) in GAC

Green Analytical Chemistry can undergo a potential transformation in the future by integrating modern technologies. Traditional trial-and-error methods are upgraded to predictive and automated analytical methods by the integration of artificial intelligence (AI) and machine learning (ML) with GAC principles [13].

5.1 Method Optimization and Predictive Modeling

The development of a high-performing and green analytical method is time-consuming and tedious. This process can be accelerated by AI and ML algorithms, which can analyze vast historical experimental datasets to predict the optimal conditions for the analytical method.

For example, in High Performance Liquid Chromatography (HPLC) method development, the mobile phase composition, column type, and flow rate parameters can be predicted using an ML model trained on extensive data of past experiments, to attain the desired separation with the least amount of solvent and run time. The number of experimental trials will be reduced by this in silico method. The AI models the separation in minutes by saving time and money, and also the wastage of chemicals and energy due to the failed experiments conducted by chemists over weeks [14].

5.2 In Silico Design of Green Solvents

The development of non-toxic and biodegradable green solvents from renewable resources that are novel and high-performing in nature is a major challenge in GAC. The computational design of these solvents is made possible by AI and ML to accelerate this process. Algorithms like Quantitative Structure-Property Relationship (QSAR) modeling can predict the physicochemical properties (such as polarity, viscosity, biodegradability, and toxicity) of the novel molecules before their laboratory synthesis [15].

For example, deep eutectic solvents (DESs) derived from natural components can be designed and screened by computational methods by the use of AI. The promising candidates that can meet the criteria of green solvents can be identified using AI by stimulating a wide range of potential structures and combinations. Therefore, in silico designing of the green solvents helps to reduce the time and resources required to discover and develop a sustainable solvent [15].

5.3 Automated and Smart GAC Workflows

AI-powered automation and robotics are the main features of GAC labs. Fully automated systems require less human assistance and can handle the entire analytical workflow, from sample preparation to data analysis. AI-powered robotic systems can perform tasks like operating miniaturized instruments, dispensing minute amounts of solvents, and real-time data interpretation.

While maintaining the analytical performance of the "smart labs," the greenness metrics (like lower energy consumption, reduced solvent usage) can also be maintained by continuously modifying the parameters to optimize them with AI algorithms. For example, to monitor real-time pressure and temperature of a chromatographic system, a robotic system can be used, which can be integrated with an AI algorithm to make minute modifications in flow rates, to ensure optimal separation with less energy consumption. To automatically identify the analytes and warn of any possible problems, AI can be used to interpret complex data, such as mass spectra, and reduce human error and increase efficiency. With this level of automation and intelligence, the GAC procedures are becoming more widely available and accessible [16].

GAC is greatly assisted by AI and ML, as they offer predictive optimization, automation, miniaturization, and computational solvent design. With the combined integration into GAC, the cost is reduced, efficiency is improved, and environmental impact is minimized, making GAC scalable and accessible.

6 Future Trends

6.1 Next-Generation Materials and Techniques

With the development and application of next-generation materials and techniques, the GAC is about to undergo a significant transformation. The key component to lay the pathway to a circular and eco-friendly analytical workflow is the use of sustainable materials [17].

6.1.1 Bio-based Materials

As an alternative to the traditional petroleum-derived sorbents and reagents, the bio-based materials are developed, which are biodegradable and renewable in nature. Our dependence on fossil fuels will be reduced by this approach, and the goal of the circular economy to valorize waste is also attained [18].

- **Sorbents and Reagents**: From natural resources like cellulose, chitin, and chitosan, the sorbents for solid phase extraction (SPE) are developed. The second most abundant biopolymer is chitin, after cellulose. It is a significant source of waste obtained from the seafood industry, as it is a major component of the exoskeletons of crustaceans like shrimp and crab. Due to the chemical properties of chitosan, it is effective among all the other chitin derivatives. Chitosan allows the heavy metal ions like lead (Pb^{2+}), copper (Cu^{2+}), and cadmium (Cd^{2+}) to bind and chelate with the amino and hydroxyl functional groups on its backbone, thereby making it an excellent sorbent for environmental analysis. The utilization of renewable resources and providing solutions for critical environmental problems can be attained by this strategy. To develop low-cost and high-performance sorbents for various pollutants, other bio-based materials derived from agricultural wastes like lignin from corn cobs or bagasse are being explored [18].

- **Their Use in Sample Preparation and Chromatography**: Bio-based materials are being used in various GAC procedures other than the traditional sorbents. They lead to efficient separations as stationary phases in chromatography when compared to the traditional silica or polymer-based phases. For example, in high-performance liquid chromatography (HPLC), the cellulose-based sorbents offer excellent separations when compared to traditional sorbents. These bio-based materials are best suited for miniature techniques like solid-phase microextraction (SPME) and lab-on-a-chip systems, as they require very little or no organic solvents. By replacing silica-based materials and synthetic polymers with bio-based materials, a step toward the circular economy is attained. These materials are efficiently used and eventually biodegraded by reducing waste and closing the loop on material usage [18].

6.1.2 Zero-Waste and Solvent-Free Processes

Eliminating waste and the use of hazardous solvents is the main aim of Green Analytical Chemistry (GAC). By altering the way of carrying out the analytical measurements with the help of several next-generation methods and platforms, this can be achieved.

- **Direct Analysis in Real Time Mass Spectroscopy (DART-MS) and Desorption Atmospheric Pressure Chemical Ionization (DAPCI)**: These methods represent the "zero-waste" analysis. With little or no sample preparation and without the need for solvents, these methods allow direct analysis of samples (solids, liquids, or gases) in open air and at ambient temperature, and are therefore called ambient ionization methods. In front of the inlet of the mass spectrometer, the sample is placed directly. A stream of heated inert gas desorbs and ionizes the analyte molecules from the surface in DART-MS and provides instantaneous results. The need for chromatographic separation or extraction is eliminated, which is a tedious and solvent-intensive process. DART-MS has a wide range of varied applications in quality control of the pharmaceutical industry to verify the tablet's authenticity, in forensics to detect narcotics on surfaces, and in food safety to screen for pesticides on fruits and vegetables [79].
- **The Potential of Miniaturized Lab-on-a-Chip Systems**: For GAC, Lab-on-a-chip (LOC) systems in which multiple analytical steps are integrated on a single microchip are ideal. They are a category of microfluidics, which is a science of fluid manipulation at the micro and nanoscale. LOC devices use small volumes of samples and reagents in microlitre or nanolitre ranges, thereby generating a very less amount of waste. In a single self-contained device, usually the size of a credit card, these systems perform multi-step analysis from sample preparation to separation and detection. With a high degree of miniaturization and integration for zero-waste analysis a wide range of applications, including point-of-care medical diagnostics, on-site environmental monitoring, and rapid food pathogens

detection, are possible. The LOC systems are the future of GAC, because of their potential for zero-waste, portable, and highly efficient analysis, while their development faces challenges such as the cost of fabrication and the need for standardized methods [80].

6.1.3 Process Analytical Technology (PAT) and In-Line Monitoring

By measuring the critical process parameters and quality attributes in real-time, the regulatory bodies like the US FDA framed the Process Analytical Technology (PAT) to design, analyze, and control manufacturing processes. A revolutionary shift in the functioning of industry is observed with the integration of GAC principles into this framework. The analytical science is integrated into production by the conjugation of PAT with GAC, leading to a sustainable and efficient industrial ecology [81].

- **Real-Time, In-Process Analysis**: The off-line sampling and lab-based traditional analytical methods are eliminated by PAT, through integrating GAC-compliant analytical instruments directly into the manufacturing practices. This philosophy also supports the "Quality by Design" (QbD) approach, which emphasizes building quality into products from the start rather than relying on final product testing. For example, to monitor the progress of a chemical reaction, a crystallization process, or blending of a tablet in real-time, non-destructive spectroscopic sensors such as Near-Infrared (NIR) or Raman spectroscopy can be placed in the reaction vessel or on a conveyor belt. As these technologies provide instant results without the need for solvents, reagents, or sample preparation, they are called "green" methods [82]. By enabling process modification to ensure that the final product meets quality standards and reduces the risk of expensive batch failures by monitoring real-time data.
- **Reducing Off-Line Sampling and Lab-Based Methods**: With the integration of PAT and GAC, frequent off-line sampling, which acts as a major source of waste and energy, can be significantly reduced. The lab-based traditional techniques for quality control are energy-intensive chromatographic methods that require large volumes of solvents for extraction and generate hazardous waste. These procedures can be eliminated by PAT, which leads to positive effects on the environment and the economy through regular monitoring [83]. Continuous monitoring and real-time product release can be achieved by this strategy, which is the main goal for pharmaceutical regulators. Batch-to-batch variation can be reduced, overall yield can be increased, and more efficient and sustainable production can be permitted by continuously monitoring the product quality in the product manufacturing industries.

6.2 Key Challenges and Opportunities

There are significant challenges and transformative opportunities in the journey to attain complete analytical sustainability. Technical, economic, and regulatory are the key challenges. Technical challenges include difficulties in replacing high-performing and established yet hazardous solvents with "neoteric" alternatives [1, 2, 5, 7], ensuring the reproducibility and robustness of the miniaturized methods [84–86], and validating new methods according to the existing stringent regulatory standards [88–90]. For example, highly polar compounds require the addition of significant volumes of organic co-solvents to supercritical carbon dioxide ($scCO_2$), which serves as the mobile phase in supercritical fluid chromatography (SFC), leading to a reduction in the greenness of the method [56, 73]. Economic challenges include the initial high investments for the green equipment like SFC systems, advanced mass spectrometers for DART, etc., and also difficulties in conducting a transparent cost-based analysis [9, 69, 78]. Though the long-term solvent purchasing and waste disposal costs are significantly less, the companies struggle to justify the upfront expenditure [9, 87]. Furthermore, the regulatory challenges, like the lack of a globally harmonized regulatory framework and the slow pace of validation of green methods, officially prevent their wide adoption, resulting in a "catch-22" scenario where companies hesitate to invest in officially unapproved new methods [44, 88].

However, the challenges are outweighed by the enormous opportunities. For method optimization, predictive modeling is enabled by the integration of artificial intelligence (AI) and machine learning (ML) to transform GAC [12–16]. AI can speed up research and reduce waste caused by the trial-and-error method by forecasting the efficient green solvent, column, and operating conditions for separation [14, 15]. The development of "smart labs" and the "in silico" design of new green solvents makes green methods economical and cost-efficient [15, 16]. Employing bio-based sorbents and reagents obtained from renewable resources leads to a circular analytical economy [17, 89]. The emergence of zero-waste and solvent-free methods like Direct Analysis in Real Time (DART-MS) and the miniaturization provided by lab-on-chip systems led to more effective and less resource-intensive analysis [18, 23, 79, 90]. The integration of GAC principles with Process Analytical Technology (PAT) has the potential to transform industrial process monitoring and control by shifting from reactive, end-of-line testing to proactive, real-time quality assurance [80, 81] (Table 1).

6.3 Recommendations for Policymakers, Researchers, and Industry Practitioners

For a sustainable analytical shift, all the stakeholders have to work together. Each group can contribute in a specific way to create a collaborative ecosystem that values sustainability, efficiency, and safety.

Table 1 Key challenges and opportunities in GAC adoption

Category	Specific aspect	Challenge	Opportunity
Technical	Method development	Difficult to obtain sensitivity and reproducibility like traditional methods	Artificial intelligence (AI) and machine learning (ML) can be used to optimize
	Solvents and reagents	Bio-based and alternative green solvents are costly and scarcely available	Development of new, non-toxic bio-based materials for extraction and separation of analytes
Economical	Investment cost	High initial investment costs for new technology and equipments	Long-term cost savings by reduced solvent use, increased efficiency, and less waste disposal fees
Regulatory	Standardization	Lack of globally accepted standardized methods and green metrics	Collaboration among regulatory, industry and academic bodies to frame standardized GAC guidelines and green metrics like AGREE, GAPI

- **For Policymakers**: The future of analytical science can be influenced by the establishment of laws and policies by policymakers. To provide financial incentives, such as tax credits or grants, the policies have to be framed to aid the industries and laboratories to invest in green analytical equipment and methods [88]. These incentives lessen the burden of the high initial investment associated with technologies like supercritical fluid chromatography (SFC) or other automated systems [9, 87]. To develop and enhance the validation of new methods, policymakers have to collaborate with key regulatory bodies. Creating a transparent, risk-based framework for assessing the greenness of a method without compromising on safety and quality will encourage innovation in the sector [44, 45]. Initiatives to harmonize the national and international GAC standards will reduce the fragmentation and promote their widespread adoption, which is a major barrier for multinational companies [91, 92].
- **For Researchers**: Research has to be comprehensive to create innovative green techniques throughout all stages of the analytical process, from preparation of the sample to disposal of waste [1, 2, 7, 31]. This requires a multidisciplinary approach by encouraging collaboration among analytical chemists, computer scientists (for AI/ML-driven predictive modeling), material scientists (for new bio-based materials), and engineers (for miniaturized devices and robotics) [12–16]. Researchers should also keep improving and standardizing the quantitative measure of "greenness," such as the AGREE Index [25, 40]. A transparent and objective means to compare and evaluate green methods should be provided to prevent "greenwashing" and give a common language for regulators and industries [19, 93].

Table 2 Recommendation for stakeholders to adopt GAC

Stakeholder group	Key recommendations	Actionable steps
Policymakers	Promote incentives and regulations to support GAC	Implement grant programs for green laboratory initiatives and develop regulatory frameworks to recognize and adopt GAC guidelines
Researchers	Develop GAC techniques that are robust and practicable	Prioritize research on green solvents, green techniques, miniature and real-time monitoring devices
Industry practitioners	Incorporate GAC principles into routine lab procedures	Conduct the assessments of greenness of existing methods; make investments in green equipment, and train lab personnel

- **For Industry Practitioners**: A long-term strategy needs to be adopted by the industry to understand that despite the initial investment in GAC being potentially high, the economic and environmental advantages, such as low operational costs, brand recognition, and a healthy workplace attained in the long run [9, 78, 87]. Before scaling up, GAC is to be implemented in a targeted manner using a "start small and scale small" approach [87]. For example, an industry can start by replacing an extremely risky preparative HPLC process with a greener SFC method for a single process and use it as evidence to show the cost savings and improved efficiency to support further investment. Establishing a sustainable culture and providing resources and training for the employees to adopt green methods enables them to take an active role in the transition [69, 78] (Table 2).

6.4 The Future of Analytical Science: A Sustainable Paradigm

Analytical science is not only about performing right measurements but also to do it in an integrated, intelligent, and environmentally responsible manner [2, 6]. The analytical chemist's traditional role as an isolated practitioner is evolving. To design new materials to enhance sustainability, the analytical chemists will collaborate with the engineers to develop in-line monitoring systems that are seamlessly integrated into production and use AI-driven systems to forecast results [12–16].

The sustainable global economy is one in which analytical science is a potent tool for environmental stewardship, public health, and economic efficiency, but no longer a source of waste and pollution [2, 3]. "Green" is not just an idiom but the standard where every analytical measurement is a step toward a healthier planet and a more sustainable global economy in the future [1, 2, 4]. The following key changes will accelerate the evolution:

- **The Evolving Role of the Analytical Chemist**: The modern analytical chemist will be a data scientist and a sustainability expert, whose key role is to interpret data from automated intelligent systems, rather than performing manual, solvent-intensive analysis. To predict and optimize methods without extensive experimentation, they use artificial intelligence (AI) which will be essential in designing and implementing holistic green analytical workflows [14, 15].
- **Integration of Intelligent Technologies**: The future labs will be interconnected and smart. AI and ML serve as the "brain" as they enable predictive modeling for method optimization and develop new solvents and reagents by "in silico" designs, thereby reducing the need for extensive trial-and-error experimentation, which saves resources, time, and money. The "body" of the lab will be constructed from the bio-based sorbents and reagents that are the next-generation materials, which serve as a backbone of a truly circular analytical economy [17, 19, 90].
- **Seamless Integration into Industrial Processes**: The "nervous system" of this new approach is the incorporation of GAC principles into PAT. Instead of a reactive approach, where samples are sent to the lab after completion of the process, in-line monitoring systems will give real-time data. By this strategy, a proactive quality control is achieved, through which the fundamental aspects of the manufacturing process are addressed, resulting in preventing errors, reducing waste, and optimizing resources [80, 81].

Redefining analytical chemistry as a tool for sustainability, innovation, and societal well-being, the field can be transformed. By implementing GAC principles in all aspects of scientific practice, from method development to industrial application, analytical science can become not only a tool for measurement but also a catalyst for a greener future [1, 2, 6, 7, 17]. The accuracy is enhanced, and a resource-efficient and ecologically sound green future is established with each advancement [89, 90, 94].

7 Conclusion

Green Analytical Chemistry represents a revolutionary change in analytical science by combining economic efficiency, workplace safety, and environmental sustainability. While the technical challenges, high initial investments, and fragmented regulatory frameworks cause obstacles for the adoption of GAC, the rapid advancements in artificial intelligence (AI) and machine learning (ML), the development of bio-based materials, and zero-waste processes drive toward a future where sustainability and analytical performance can coexist. For achieving this objective, global collaborations among the policymakers, researchers, and industries are necessary. Ultimately, GAC redefines analytical science not just as an evaluation tool but as a catalyst for a greener, resource-efficient and environmentally sustainable future.

References

1. Koel, M. and Kaljurand, M., 2006. Application of the principles of green chemistry in analytical chemistry. *Pure and Applied Chemistry*, *78*(11), pp. 1993-2002.
2. Anastas, P. and Eghbali, N., 2010. Green chemistry: principles and practice. *Chemical society reviews*, *39*(1), pp. 301-312.
3. Brundtland, G.H., 1987. Our common future world commission on environment and developement.
4. Anastas, P.T. and Warner, J.C., 2000. *Green chemistry: theory and practice*. Oxford university press.
5. Capello, C., Fischer, U. and Hungerbühler, K., 2007. What is a green solvent? A comprehensive framework for the environmental assessment of solvents. *Green Chemistry*, *9*(9), pp. 927-934.
6. Armenta, S., Garrigues, S., Esteve-Turrillas, F.A. and de la Guardia, M., 2019. Green extraction techniques in green analytical chemistry. *TrAC Trends in Analytical Chemistry*, *116*, pp. 248-253.
7. Claux, O., Santerre, C., Abert-Vian, M., Touboul, D., Vallet, N. and Chemat, F., 2021. Alternative and sustainable solvents for green analytical chemistry. *Current Opinion in Green and Sustainable Chemistry*, *31*, p. 100510.
8. Fang, Q., Shi, X.T., Du, W.B., He, Q.H., Shen, H. and Fang, Z.L., 2008. High-throughput microfluidic sample-introduction systems. *TrAC Trends in Analytical Chemistry*, *27*(6), pp. 521-532.
9. Muiambo, C.C., Joao, I.M. and Navas, H.V., 2022. Lean waste assessment in a laboratory for training chemical analysts for the pharmaceutical industry. *International Journal of Lean Six Sigma*, *13*(1), pp. 178-202.
10. Al-Shatti, B.J., Alsairafi, Z. and Al-Tannak, N.F., 2023. Green chemistry and its implementation in pharmaceutical analysis. *Reviews in Analytical Chemistry*, *42*(1), p. 20230069.
11. Meher, A.K. and Zarouri, A., 2025. Green analytical chemistry—recent innovations. *Analytica*, *6*(1), p. 10.
12. Valavanidis, A., 2024. Artificial Intelligence and Green Chemistry. *Sci Rev, 1262024*.
13. Mroz, A.M., Basford, A.R., Hastedt, F., Jayasekera, I.S., Mosquera-Lois, I., Sedgwick, R., Ballester, P.J., Bocarsly, J.D., del Río Chanona, E.A., Evans, M.L. and Frost, J.M., 2025. Cross-disciplinary perspectives on the potential for artificial intelligence across chemistry. Chemical Society Reviews., 54, 5433
14. Rahman, S.N.R., Katari, O., Pawde, D.M., Boddeda, G.S.B., Goswami, A., Mutheneni, S.R. and Shunmugaperumal, T., 2021. Application of design of experiments® approach-driven artificial intelligence and machine learning for systematic optimization of reverse phase high performance liquid chromatography method to analyze simultaneously two drugs (cyclosporin A and etodolac) in solution, human plasma, nanocapsules, and emulsions. *AAPS PharmSciTech*, *22*(4), p. 155.
15. Das, R.N. and Roy, K., 2013. Advances in QSPR/QSTR models of ionic liquids for the design of greener solvents of the future. *Molecular diversity*, *17*(1), pp. 151-196.
16. Hussain, C.G., Keçili, R. and Hussain, C.M., 2025. Future of analytical chemistry with green miniaturized technologies. In *Comprehensive Analytical Chemistry* (Vol. 109, pp. 507-537). Elsevier.
17. Qin, G., Song, H., Wu, D., Zhang, Y., Li, P., Zhang, K., Zheng, Y. and Ji, S., 2024. Design of a novel green bio-based organic–inorganic hybrid material for cost-effective and sustainable monitoring of antibiotic residues. Green Chemistry, 26(5), pp. 2807-2824.
18. Kumar, R., Gupta, G., Hussain, A., Rani, A., Thapliyal, A., Gunsola, D., Chattaraj, S., Ganguly, A., Panneerselvam, P., Guerra-Sierra, B.E. and Mitra, D., 2025. Pioneering zero-waste technologies utilization and its framework on sustainable management: international, national and state level. *Discover Applied Sciences*, *7*(3), p. 224.
19. Marcinkowska, R., Namieśnik, J. and Tobiszewski, M., 2019. Green and equitable analytical chemistry. *Current Opinion in Green and Sustainable Chemistry*, *19*, pp. 19-23.

20. Kurul, F., Doruk, B. and Topkaya, S.N., 2025. Principles of green chemistry: building a sustainable future. *Discover Chemistry*, *2*(1), p. 68.
21. Bizzi, C.A., Pedrotti, M.F., Silva, J.S., Barin, J.S., Nóbrega, J.A. and Flores, E.M.M., 2017. Microwave-assisted digestion methods: towards greener approaches for plasma-based analytical techniques. *Journal of Analytical Atomic Spectrometry*, *32*(8), pp. 1448-1466.
22. Cullen, J.M., Allwood, J.M. and Borgstein, E.H., 2011. Reducing energy demand: what are the practical limits?. *Environmental science & technology*, *45*(4), pp. 1711-1718.
23. He, Y., Hii, D.J.C., Wong, N.H. and Peck, T.G., 2022. Sustainable laboratory evaluations: Optimized fume-hood-intensive ventilation and energy efficiency without compromising occupational safety and comfort. *Journal of Cleaner Production*, *333*, p. 130147.
24. Green, D., Smith, L.L., Crain, J., Boparai, A., Kiely, J.T. and Yaeger, J.S., 1995. *Waste minimization in analytical methods* (No. ANL—CMT-ACL/CP-85238). Argonne National Lab., IL (United States).
25. Pena-Pereira, F., Wojnowski, W. and Tobiszewski, M., 2020. AGREE—Analytical GREEnness metric approach and software. *Analytical chemistry*, *92*(14), pp. 10076-10082.
26. Trapasso, G. and Aricò, F., 2025. Organic carbonates as green media: from laboratory syntheses to industrial applications. *Green Chemistry*, *27*(24), pp. 6925-6966.
27. Baltussen, E., Sandra, P., David, F. and Cramers, C., 1999. Stir bar sorptive extraction (SBSE), a novel extraction technique for aqueous samples: theory and principles. *Journal of Microcolumn Separations*, *11*(10), pp. 737-747.
28. Yang, Y., Liu, Z., Chen, H. and Li, S., 2019. Low-density solvent-based dispersive liquid-liquid microextraction followed by HPLC-ICP-MS for speciation analysis of phenylarsenics in lake water. *International Journal of Environmental Analytical Chemistry*, *99*(1), pp. 87-100.
29. Werner, J., Frankowski, R., Grześkowiak, T. and Zgoła-Grześkowiak, A., 2024. Green sorbents in sample preparation techniques–naturally occurring materials and biowastes. *TrAC Trends in Analytical Chemistry*, *176*, p. 117772.
30. de la Guardia, M. and Garrigues, S. eds., 2012. *Handbook of green analytical chemistry* (Vol. 794). Chichester: John Wiley & Sons.
31. Câmara, J.S., Perestrelo, R., Berenguer, C.V., Andrade, C.F., Gomes, T.M., Olayanju, B., Kabir, A., MR Rocha, C., Teixeira, J.A. and Pereira, J.A., 2022. Green extraction techniques as advanced sample preparation approaches in biological, food, and environmental matrices: a review. *Molecules*, *27*(9), p. 2953.
32. Sajid, M. and Płotka-Wasylka, J., 2022. Green analytical chemistry metrics: A review. *Talanta*, *238*, p. 123046.
33. Olives, A.I., Gonzalez-Ruiz, V. and Martín, M.A., 2017. Sustainable and eco-friendly alternatives for liquid chromatographic analysis. *ACS sustainable chemistry & engineering*, *5*(7), pp. 5618-5634.
34. de Godoi Passerine, B.F. and Breitkreitz, M.C., 2023. Recent Developments in Green Chromatography. *Brazilian Journal of Analytical Chemistry*, *10*(40), pp. 17-34.
35. Camel, V., 2000. Microwave-assisted solvent extraction of environmental samples. TrAC Trends in Analytical Chemistry, 19(4), pp. 229-248.
36. Carasek, E., Bernardi, G., do Carmo, S.N. and Vieira, C.M., 2019. Alternative green extraction phases applied to microextraction techniques for organic compound determination. *Separations*, *6*(3), p. 35.
37. Tobiszewski, M. and Namieśnik, J., 2012. Direct chromatographic methods in the context of green analytical chemistry. *TrAC Trends in Analytical Chemistry*, *35*, pp. 67-73.
38. Larsen, C., Lundberg, P., Tang, S., Ràfols-Ribé, J., Sandström, A., Mattias Lindh, E., Wang, J. and Edman, L., 2021. A tool for identifying green solvents for printed electronics. *Nature Communications*, *12*(1), p. 4510.
39. Locatelli, M., Kabir, A., Perrucci, M., Ulusoy, S., Ulusoy, H.I. and Ali, I., 2023. Green profile tools: current status and future perspectives. *Advances in Sample Preparation*, *6*, p. 100068.
40. Tobiszewski, M., Marć, M., Gałuszka, A. and Namieśnik, J., 2015. Green chemistry metrics with special reference to green analytical chemistry. *Molecules*, *20*(6), pp. 10928-10946.

41. Płotka-Wasylka, J., 2018. A new tool for the evaluation of the analytical procedure: Green Analytical Procedure Index. *Talanta, 181*, pp. 204-209.
42. Marteel, A.E., Davies, J.A., Olson, W.W. and Abraham, M.A., 2003. Green chemistry and engineering: drivers, metrics, and reduction to practice. *Annual Review of Environment and Resources, 28*(1), pp. 401-428.
43. Lk, T., NT, R. and UN, M., 2016. A review on bioanalytical method development and validation. *Asian J Pharm Clin Res, 9*(3), pp. 6-10.
44. Borman, P. and Elder, D., 2017. Q2 (R1) validation of analytical procedures: text and methodology. *ICH quality guidelines: an implementation guide*, pp. 127-166.
45. Pharmacopeia, U.S., 2020. The United States Pharmacopeia, USP 43-National Formulary, NF 38. Rockville, MD: US Pharmacopeial Convention.
46. European Pharmacopoeia Commission and European Directorate for the Quality of Medicines & Healthcare, 2013. *European pharmacopoeia* (Vol. 2). Council of Europe.
47. Cinelli, M., Coles, S.R. and Kirwan, K., 2014. Analysis of the potentials of multi criteria decision analysis methods to conduct sustainability assessment. *Ecological indicators, 46*, pp. 138-148.
48. Hoffman, K.C. and Dicks, A.P., 2023. Incorporating the United Nations Sustainable Development Goals and green chemistry principles into high school curricula. *Green Chemistry Letters and Reviews, 16*(1), p. 2185108.
49. Dey, S. and Majumdar, A., 2024. Current status of pollution in Major Rivers and Tributaries of India and protection-restoration strategies. In *Rivers of India: Past, Present and Future* (pp. 69-93). Cham: Springer International Publishing.
50. Yang, H., 2019. Food safety in India: status and challenges. *Gates Open Res, 3*(1043), p. 1043.
51. Singh, A., Singh, B.P., Raghav, Y. and Mishra, V., 2024. Regulatory Frameworks and Guidelines Aimed at Controlling and Minimizing Heavy Metals Exposure. In *Heavy Metal Contamination in the Environment* (pp. 171-179). CRC Press.
52. Stefanache, A., Marcinschi, A., Marin, G.A., Mitran, A.M., Lungu, I.I., Miftode, A.M., Crivoi, F., Lacatusu, D., Baican, M., Cioanca, O. and Hancianu, M., 2025. Green Chemistry Approaches in Pharmaceutical Synthesis: Sustainable Methods for Drug Development. *AppliedChem, 5*(2), p. 13.
53. Hu, J., Zhang, J., Zhao, Y. and Yang, Y., 2024. Green solvent systems for material syntheses and chemical reactions. *Chemical Communications, 60*(21), pp. 2887-2897.
54. Pešić, J., Watson, M., Papović, S. and Vraneš, M., 2021. Ionic liquids: review of their current and future industrial applications and their potential environmental impact. *Recent Patents on Nanotechnology, 15*(3), pp. 225-244.
55. Mashile, P.P., 2023. *Biopolymer-based nanocomposite as recyclable adsorbents for removal of pollutants in wastewater treatment*. University of Johannesburg (South Africa).
56. Fornstedt, T., Enmark, M. and Samuelsson, J., 2022. Method transfer in SFC from a fundamental perspective. *TrAC. Trends in analytical chemistry, 149, 116551.*
57. Babu, S. and Kumar, S., 2025. Potential of deep eutectic solvents as green and sustainable solvents for the recovery of carboxylic acids from aqueous solution: a review. *Journal of Chemical Technology & Biotechnology*., 100, 1541
58. Stephens, N.M. and Smith, E.A., 2022. Structure of deep eutectic solvents (DESs): What we know, what we want to know, and why we need to know it. *Langmuir, 38*(46), pp. 14017-14024.
59. Poojary, H., Selvamuthukumar, K., Parveen, R., Illanad, G., Mandal, D., Koo, S. and Ghosh, C., 2025. Current advances in solid-phase microextraction technique as a green analytical sample preparation approach. *Green Chemistry Letters and Reviews, 18*(1), p. 2486135.
60. Kataoka, H., 2023. Solid-phase microextraction and related techniques in bioanalysis. *Molecules, 28*(6), p. 2467.
61. Kasperkiewicz, A., Lendor, S. and Gionfriddo, E., 2023. Practical aspects for SPME method development in complex samples.
62. Dietz, C., Sanz, J. and Cámara, C., 2006. Recent developments in solid-phase microextraction coatings and related techniques. *Journal of Chromatography A, 1103*(2), pp. 183-192.

63. Cortese, M., Gigliobianco, M.R., Magnoni, F., Censi, R. and Di Martino, P., 2020. Compensate for or minimize matrix effects? Strategies for overcoming matrix effects in liquid chromatography-mass spectrometry technique: a tutorial review. *Molecules*, *25*(13), p. 3047.
64. FDA, F., 2001. Guidance for industry: bioanalytical method validation. *http://www. fda. gov/cder/Guidance/4252fnl. pdf*.
65. Guideline, I.H.T., 2005. Validation of analytical procedures: text and methodology. *Q2 (R1)*, *1*(20), p. 05.
66. Pawar, S.V. and Gaware, V.M., 2025. A Review on: Bioanayltical Method Development and Validation. *Asian Journal of Pharmaceutical Research and Development*, *13*(3), pp. 167-179.
67. Taverniers, I., De Loose, M. and Van Bockstaele, E., 2004. Trends in quality in the analytical laboratory. II. Analytical method validation and quality assurance. *TrAC Trends in Analytical Chemistry*, *23*(8), pp. 535-552.
68. Vander Heyden, Y. and Smeyers-Verbeke, J., 2007. Set-up and evaluation of interlaboratory studies. *Journal of Chromatography A*, *1158*(1-2), pp. 158-167.
69. Freese, T., 2024. Integrating sustainable chemistry into research and industry for a circular economy.
70. Araujo, P., 2009. Key aspects of analytical method validation and linearity evaluation. *Journal of chromatography B*, *877*(23), pp. 2224-2234..
71. Bystrzanowska, M. and Tobiszewski, M., 2020. Chemometrics for selection, prediction, and classification of sustainable solutions for green chemistry—A review. *Symmetry*, *12*(12), p. 2055.
72. Lesellier, E. and West, C., 2015. The many faces of packed column supercritical fluid chromatography–A critical review. *Journal of Chromatography A*, *1382*, pp. 2-46.
73. Rogers, L. and Jensen, K.F., 2019. Continuous manufacturing–the Green Chemistry promise?. *Green chemistry*, *21*(13), pp. 3481-3498.
74. Cavanagh, E.J., Savelski, M.J. and Slater, C.S., 2014. Optimization of environmental impact reduction and economic feasibility of solvent waste recovery using a new software tool. *Chemical Engineering Research and Design*, *92*(10), pp. 1942-1954.
75. Losacco, G.L., Veuthey, J.L. and Guillarme, D., 2019. Supercritical fluid chromatography–Mass spectrometry: Recent evolution and current trends. *TrAC Trends in Analytical Chemistry*, *118*, pp. 731-738.
76. Ayre, A., Ghude, K., Mane, P., Nemade, M., Gosavi, S., Pathare, A. and Lad, A., 2013. Supercritical fluid extraction-a green paradigm in the area of separation science. *Asian J Biomed Pharm Sci*, *3*(23), pp. 1-7.
77. Hossain, M.I., Ong, T.S., Tabash, M.I. and Teh, B.H., 2024. The panorama of corporate environmental sustainability and green values: evidence of Bangladesh. *Environment, Development and Sustainability*, *26*(1), pp. 1033-1059.
78. Agrawal, A., Yıldız, Ü.Y., Hussain, C.G., Kailasa, S.K., Keçili, R. and Hussain, C.M., 2022. Greenness of lab-on-a-chip devices for analytical processes: advances & future prospects. *Journal of Pharmaceutical and Biomedical Analysis*, *219*, p. 114914.
79. Boysen, R.I. and Hearn, M.T., 2008. Role of green analytical chemistry in process analytical technology. *Chemistry in Australia*, *75*(9).
80. Simon, L.L., Pataki, H., Marosi, G., Meemken, F., Hungerbühler, K., Baiker, A., Tummala, S., Glennon, B., Kuentz, M., Steele, G. and Kramer, H.J., 2015. Assessment of recent process analytical technology (PAT) trends: a multiauthor review. *Organic Process Research & Development*, *19*(1), pp. 3-62.
81. Chanda, A., Daly, A.M., Foley, D.A., LaPack, M.A., Mukherjee, S., Orr, J.D., Reid III, G.L., Thompson, D.R. and Ward, H.W., 2015. Industry perspectives on process analytical technology: tools and applications in API development. *Organic Process Research & Development*, *19*(1), pp. 63-83.
82. Cop, G., 2025. A Systematic Review: Exploration of Process Analytical Technology Techniques (Pat) and Their Multifaceted Advantages in Industrial Processes. *International Journal of Applied Pharmaceutics*.

83. Kourti, T., 2006. The process analytical technology initiative and multivariate process analysis, monitoring and control. *Analytical and bioanalytical chemistry, 384*(5), pp. 1043-1048.
84. Kataoka, H., Lord, H.L. and Pawliszyn, J., 2000. Applications of solid-phase microextraction in food analysis. *Journal of chromatography A, 880*(1-2), pp. 35-62.
85. Sengupta, J., 2024. Natural biodegradable polymers transforming lab on a chip technology: A mini review. *Green Analytical Chemistry, 10*, p. 100119.
86. Pol, R., Céspedes, F., Gabriel, D. and Baeza, M., 2017. Microfluidic lab-on-a-chip platforms for environmental monitoring. *TrAC Trends in Analytical Chemistry, 95*, pp. 62-68.
87. Van der Vorst, G., Van Langenhove, H., De Paep, F., Aelterman, W., Dingenen, J. and Dewulf, J., 2009. Exergetic life cycle analysis for the selection of chromatographic separation processes in the pharmaceutical industry: preparative HPLC versus preparative SFC. *Green Chemistry, 11*(7), pp. 1007-1012.
88. Ali, A. and Sinha, K., 2024. Regulatory Landscape for Generic Drugs in India: Assessing Quality, Safety, and Efficacy Standards. Biotechnology Law Report, 43(3), pp. 131-141.
89. Werner, J., Zgoła-Grześkowiak, A., Grześkowiak, T. and Frankowski, R., 2024. Biopolymers-based sorbents as a future green direction for solid phase (micro) extraction techniques. *TrAC Trends in Analytical Chemistry, 173*, p. 117659.
90. Wongkaew, N., Simsek, M., Griesche, C. and Baeumner, A.J., 2018. Functional nanomaterials and nanostructures enhancing electrochemical biosensors and lab-on-a-chip performances: recent progress, applications, and future perspective. *Chemical reviews, 119*(1), pp. 120-194.
91. Tiwari, P.K. and TG, J.C., 2023. Policy Intervention and Legislative Framework Relating to the Sustainability of Built Structures in India. *GLS Law Journal, 5*(2), pp. 1-13.
92. Sharma, A., 2024. 31 Water Quality and Water Pollution.
93. Gałuszka, A., Migaszewski, Z. and Namieśnik, J., 2013. The 12 principles of green analytical chemistry and the SIGNIFICANCE mnemonic of green analytical practices. *TrAC Trends in Analytical Chemistry, 50*, pp. 78-84.
94. Clarke, C.J., Tu, W.C., Levers, O., Brohl, A. and Hallett, J.P., 2018. Green and sustainable solvents in chemical processes. *Chemical reviews, 118*(2), pp. 747-800.

Green Chromatographic Techniques

Vivekanand Arvind Chatpalliwar

Abstract Green chromatographic techniques represent a transformative approach in analytical chemistry, aiming to minimize environmental impact while maintaining high efficiency and accuracy in separation science. This chapter explores the integration of sustainable practices into chromatographic methodologies, highlighting innovations that reduce solvent consumption, energy usage, and chemical waste. A central focus is placed on solvent-free extraction methods, which eliminate or drastically reduce the reliance on hazardous organic solvents, thereby enhancing safety and sustainability. These approaches not only align with green chemistry principles but also improve sample integrity and reduce operational costs. Miniaturized chromatography is discussed as another pivotal advancement, offering reduced sample and reagent requirements while maintaining analytical precision. By scaling down instrumentation and workflows, miniaturization contributes to lower energy demands and waste generation, making it a practical solution for laboratories committed to eco-friendly practices. Complementary to these techniques are microwave-assisted and ultrasound-assisted methods, which accelerate extraction and separation processes. These energy-efficient approaches shorten analysis times, enhance reproducibility, and reduce the need for excessive solvent volumes, thereby reinforcing the green credentials of chromatographic applications. The chapter also emphasizes the role of supercritical fluid chromatography (SFC), which utilizes supercritical carbon dioxide as a mobile phase. SFC offers a non-toxic, non-flammable, and readily available alternative to conventional solvents, delivering high-resolution separations with minimal environmental burden. Its versatility across diverse analytes underscores its growing importance in sustainable analytical workflows. Collectively, these techniques illustrate the paradigm shift toward greener chromatographic practices. By integrating solvent-free extraction, miniaturization, energy-assisted methods, and supercritical fluid chromatography, researchers can achieve robust analytical outcomes while adhering to environmental stewardship. This chapter underscores the necessity of adopting such innovations to meet the dual goals of scientific excellence and ecological responsibility, positioning green chromatography as a cornerstone of modern analytical science.

V. A. Chatpalliwar (✉)
SNJB's S. S. D. J. College of Pharmacy, Nashik, Maharashtra, India

V. Bhusari and M. Ghante (eds.), *Green Analytical Chemistry in India*,
https://doi.org/10.1007/978-3-032-24564-9_4

Keywords Miniaturized chromatography · Solvent-free extraction · Microwave-assisted and ultrasound-assisted methods · Supercritical fluids

Abbreviations

1, 25-$(OH)_2$-D_3	A diol of cholecalciferol with first and 25th position carrying one hydroxy group each
1-AA	*Torus 1-aminoanthracene* a column
1-OH-D_2	A sterol, cholecalciferol with first position carrying a hydroxy group
24, 25-$(OH)_2$-D3	A diol of cholecalciferol with 24th and 25th position carrying one hydroxy group each
ACN	Acetonitrile
ADME	Absorption, Distribution, Metabolism and Excretion
AGREE	Analytical GREEnness, a metric/software tool designed to evaluate the friendliness of the analytical technique with environment
APCI	Atmospheric Pressure Chemical Ionization
APPI	Atmospheric Pressure Photoionization
ASE	Accelerated Solvent Extraction
BEH 2-EP	Bridged Ethyl Hybrid-2-Ethylpyridine
BPR pressure	Back Pressure Regulator pressure
D_3	Cholecalciferol
DLLME	Dispersive Liquid–Liquid Microextraction
EI	Electron spray
EMA	European Medical Association
ESI	Electron spray ionization
FDA	Food and Drug Administration
FTIR	Fourier Transform Infra-Red
GAC	Green Analytical Chemistry
GAPI	Green Analytical Procedure Index
GC	Gas Chromatography
GC-MS	Gas Chromatography-Mass Spectrometry
HPLC	High Performance Liquid Chromatography
IPA	*Iso*-Propanol
K	Kelvin
LC-MS	Liquid Chromatography-Mass Spectrometry
LC-PB/EIMS	Liquid Chromatography-Particle Beam/Electron Ionization Mass Spectrometry
MAO	*Monoamino oxidase*
MEKC	Micellar electrokinetic chromatography
MeOH	Methanol
MPa	Megapascal

MS/MS	Tandem Mass Spectrometry
MSD	Microwave-Assisted Steam Distillation
MSDf	Microwave-Assisted Steam Diffusion
NDMA	*N*-Nitrosodimethylamine
NDSRI	Nitrosamine drug substance-related impurities
NEMI	National Environmental Methods Index
PAH	Poly Aromatic Hydrocarbon
Palmitate-D3	A preparation combining **vitamin A palmitate** and **vitamin D3**
PB-EIMS	**Particle Beam-Electron Ionization Mass Spectrometry**
PCB	Polychlorinated biphenyls
PLOT	Polymeric Layer-Open Tubular
POP	**Porous Organic Polymer**
psi	Per square inch
PTFE	Polytetrafluoroethylene (Teflon)
QbD	Quality by design
QqQ	Triple-quadrupole mass spectrometer
QuEChERS	Quick, Easy, Cheap, Effective, Rugged, and Safe analytical sample preparation method
R	*Rectus*; one of the two configurations of enantiomers
S	*Sinister*; one of the two configurations of enantiomers
SFC	Supercritical Fluid Chromatography
SFC-DAD	Supercritical Fluid Chromatography-Diode Array Detection
SFC-MS/MS	Combination of supercritical fluid chromatography and tandem mass spectrometry
SF-CO_2	Supercritical carbon dioxide
SFE	Solid Phase Microextraction
SFME	Solvent-free Microwave Extraction
SPME	Solid-Phase Microextraction
TOF-MS	Time-of-flight mass spectrometry
UHPSFC	Ultra-High Performance Supercritical Fluid Chromatography
US	*UniSpray Ionization*
UV	Ultra-Violet spectroscopy
VMHD	Vacuum-assisted microwave hydro-distillation
WCOT	Wall-Coated Open Tubular

Green chromatographic techniques represent a modern evolution in analytical chemistry, designed to minimize environmental impact while maintaining high performance.

Over the past two decades, the development of green chromatography has been driven by the principles of green analytical chemistry, which emphasize sustainability, reduced solvent consumption, and safer laboratory practices. Traditional chromatographic methods often rely on large volumes of toxic organic solvents, generating significant waste and posing health risks. In response, researchers have advanced techniques such as supercritical fluid chromatography (SFC), micellar electrokinetic chromatography (MEKC), and ionic liquid-based chromatography, all of which aim to replace hazardous solvents with eco-friendly alternatives. Additionally, innovations in miniaturized systems (like micro- and nano-LC) and solid-phase microextraction (SPME) have reduced sample and reagent requirements, further lowering the ecological footprint. These approaches not only align with global sustainability goals but also enhance efficiency and cost-effectiveness in industrial and research applications.

Automation and digitalization in GAC shift the paradigm from manual, resource-heavy workflows to smart, optimized, and sustainable systems. By combining robotics, AI, and IoT, laboratories achieve precision, reproducibility, and eco-efficiency—reducing both chemical waste and energy demand.

The present article describes such processes developed over a span of decades. A discussion regarding their suitability for extracting natural chemical constituents of interests is also included in the following description of advanced techniques.

- **Miniaturized Chromatography**: Micro-HPLC and capillary electrophoresis reduce solvent volumes drastically.
- **Solvent-Free Extraction**: Techniques like solid-phase microextraction (SPME) and stir-bar sorptive extraction (SBSE) eliminate organic solvents.
- **Microwave-Assisted and Ultrasound-Assisted Methods**: Accelerate reactions and extractions with less energy and fewer reagents.
- **Supercritical Fluids**: CO_2-based chromatography and extraction replace hazardous solvents with recyclable, non-toxic alternatives.

1 Miniaturized Chromatography

Ever since the first presentation of liquid chromatography by Mikhail Tswett [1], chromatographic tools have been constantly evolving, owing to their versatility in qualitatively and quantitatively separating mixtures of several diverse nature into their individual components. The status of the same separating technique is much more advanced in terms of efficiency, instrumental set-up, decrement in required quantities of sample, stationary phase material, and the solvents that compose the mobile phase. Emergence of Ultra High Pressure-Liquid Chromatography (UHP-LC), which is devised using a 30 μm *i. d.* capillary of fused silica filled with nonporous octadecylsilane-modified silica of particle size 1.5–1.8 μm diameter could withstand pressure at 20000 psi at which the mobile phase could flow through it without causing back pressure, achieved extreme efficiency in chromatographic separations spanning short-analysis time [2]. Success achieved in chromatographic analysis by UHP-LC led to the advent of miniaturized liquid chromatography—a system with much reduced length of analytical column thereby requiring further less quantity of solvents and sample size.

The technique proved too beneficial to the fields of science where sample volume is often limited, viz., medicine chiral separation, and forensic. The system resulted in enhanced separation, resolution, and sensitivity of the analytical methods. The reduced volume of solvents and flow rates (few *mL/min* in HPLC to few *nL/min*), reduction in number of replicate analyses required due to the high sensitivity and high resolution could prove economic in terms of time and money. Hence, miniaturized chromatography consequently lowered solvent consumption and waste generation. This scores a point over the older HPLC system such that the majority of solvents, that were conventionally used from the eluotropic series of solvents, to constitute the mobile phases in majority of analyses posit potential environmental hazards.

The miniaturized chromatographic technique when hyphenated with other sensitive physical sophistication, like the MS, adds more value to the chromatographic system in bioanalysis. The separating and characterizing instrument has proved seminal in identifying biomarkers appearing in breath-condensate [3], blood, urine [4] and other biological fluids, stem cell and tumour analyses [5]. It has also been utilized for monitoring the success of a drug therapy regimen [6].

In the food industry, miniaturized liquid chromatography has been applied to exercise food safety, food authenticity, and to evaluate nutritional and toxicological characteristics of edible products by obtaining informative chromatograms. Milk samples were tested to separate and quantify 16 benzimidazole derivatives and metabolites with proper resolution in less than an hour.

Extreme sensitivity promised by the technique makes it applicable to methods requiring detection of several toxic compounds at trace levels, where the normal chromatographic methods are compromised very often on sensitive issue. Accordingly, as many as 13 banned colorants used very often in cosmetics for aesthetic enhancement were separated and analyzed to find them contaminating the products beyond the individual toxic limits. The objectionable colorants, Basic Blue 26, Basic Red 2, and Basic Yellow 28, were quantified to be between 0.02 and 3.60 g/mL, with 80.23–108.28% accuracy [7]. Figure 1 depicts the workflow of the technique applied to separate and identify banned colorants from branded cosmetics.

The instrumental modifications of the system require manifold reduction in dimensions. The internal diameters in the miniatured columns range from 100 μm for capillary LC, 10 μm for nano LC to 5 μm in open tubular LC. The flow rate of mobile phases varies from 1 μL/min in capillary LC, 10 nL/min in nano LC, and < 1 nL/min in open tubular LC.

As such, there are four types of columns that have been utilized in miniaturized chromatography; (i) packed column [8], (ii) monolithic column [8], (iii) open tubular [9], and iv) chip-based microfluidic system [10]. The packing material reported to have been used in column (i) are the bonded C8, C18, Cyano, Phenyl-hexyl, Cholesterol-hydride, NH_2, and diol; column (ii) has utilized silica based monoliths, hybrid monoliths, certain organic nanoparticles, aptamer, and boronated material; column (iii) is reported to have packed with nanoparticles adsorbed physically on polymeric phases (WCOT) and certain polymeric phases that are chemical bonded to the column wall (PLOT); column (iv) employs polymeric pillar arrays as stationary

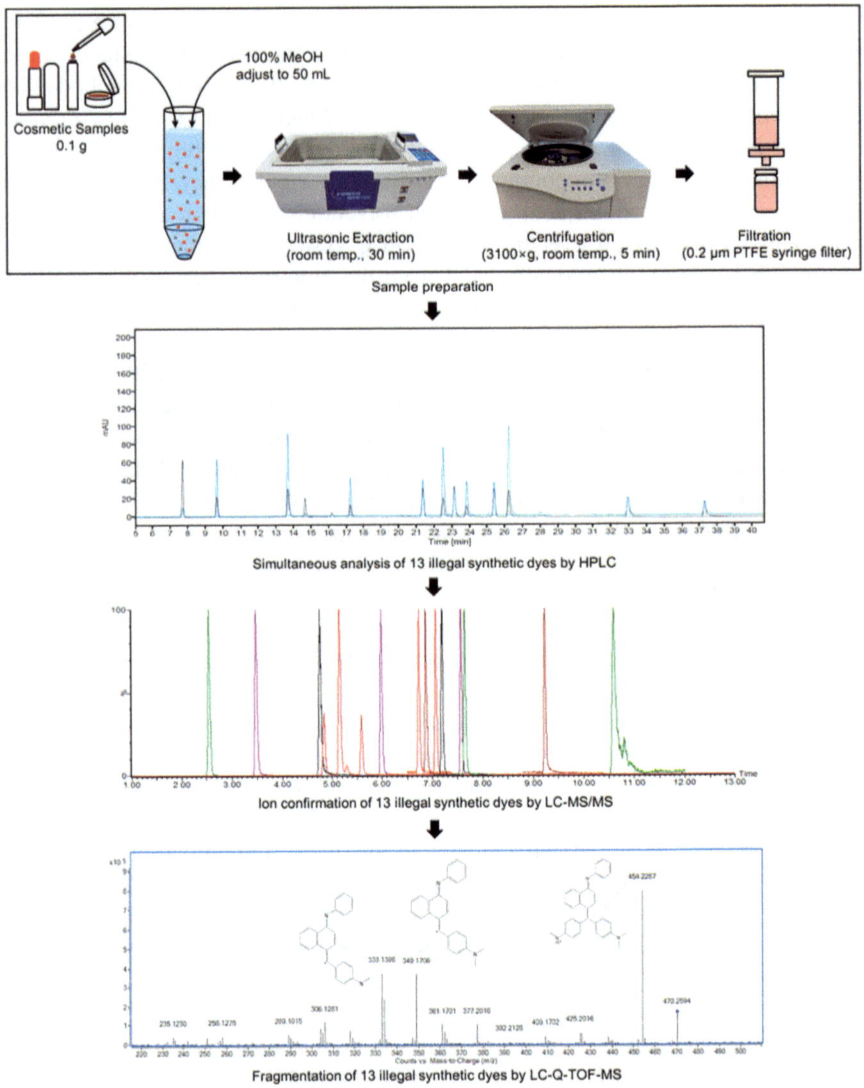

Fig. 1 Workflow chart depicting separation, identification, and quantification of banned colorants in branded cosmetics using miniaturized and hyphenated chromatographic techniques. Reproduced from Ref. [7]

phase in respective columns. Most applications of column (i) have been found applied to pharmaceutical analysis, whereas column (ii) has been specifically popular to achieve separations of chiral compounds. They have also been applied in bioanalysis, in that the column suits for the principles of affinity chromatography. The third type, the open tubular, is a favorite column to achieve separation, identification, and quantification in fields where sample quantity is sparse, such as in forensics, natural products, and biomedicine. Separations of macromolecules, whether sample quantity is large or sparse, have been achieved by column (iv). The portability provided by this advancement have allowed the analysts to work *on-site* carrying *real-time analysis*, making it an ideal technique for forensic, genetics, and *metabolic* studies.

One such development has been a very versatile, hyphenated technique popularized as *Mini 12* [11], which is a portable, table-top mass spectrometer that utilizes dried-blood spot on paper, an extraction spray ionization method, and a paper spray ionization technique for sampling. Figures 2 and 3 depicts the workflow, instrumentation and MS/MS spectra of separated, and quantified constituents.

Some other techniques of miniaturized liquid chromatography have found applications in various fields including bioanalysis, environmental analysis, and pharmaceutical analysis. Some highlight specifications of the outcomes are mentioned in Table 5.

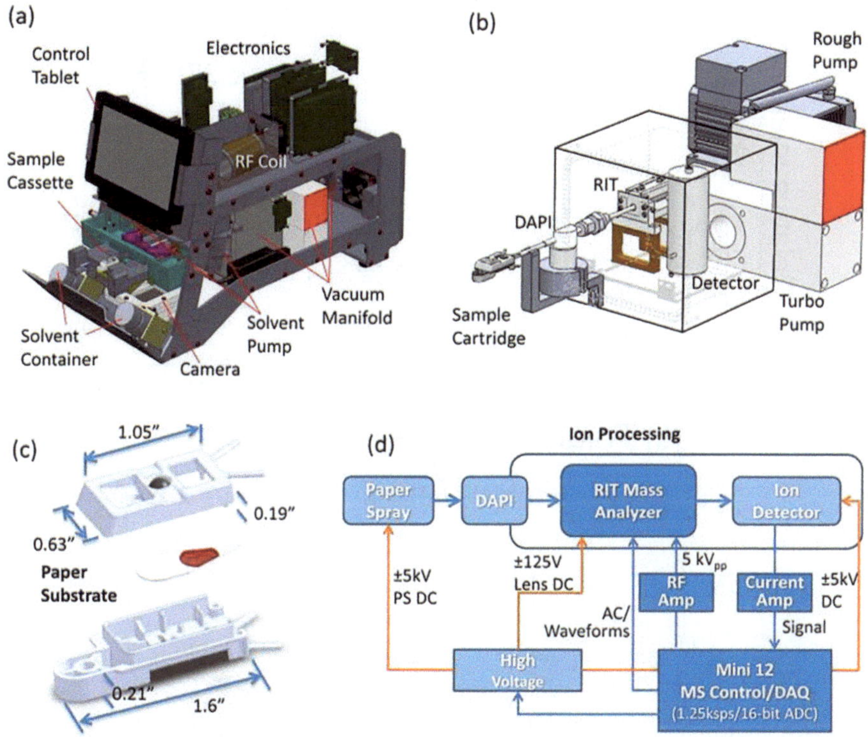

Fig. 2 **a** Overview of Mini 12 system and **b** selected subsystems showing ion generation, introduction, and analysis. **c** Design of paper spray sample cartridge used with the Mini 12 system. **d** Schematic diagram of the control electronics in the Mini 12 system. Reproduced from [11] with permission from ACS, licensed CC-BY-NC-ND license

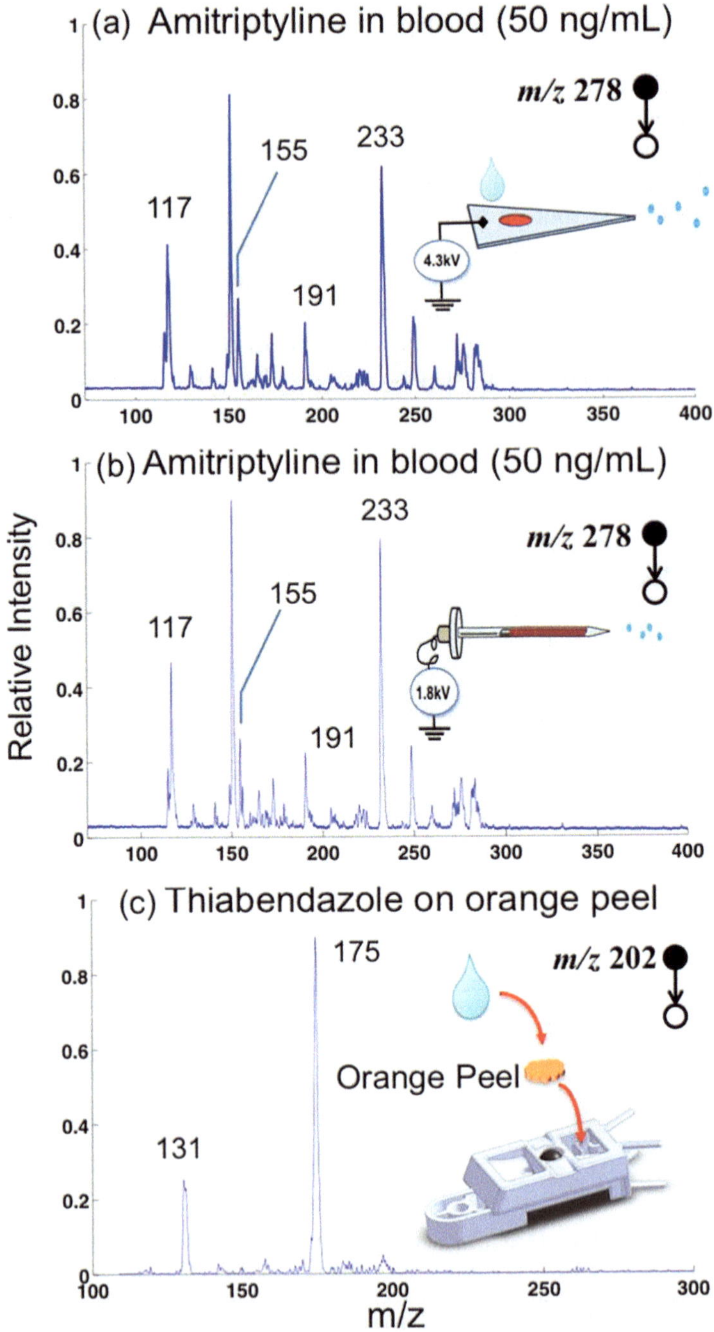

Fig. 3 MS/MS spectra of **a** 50 ng of amitriptyline in blood recorded with paper spray ionization and **b** extraction spray ionization. MS/MS spectra of **c** thiabendazole on an orange peel obtained using paper spray ionization. Reproduced from [11] with permission from ACS, licensed CC-BY-NC-ND license

2 Accelerated Solvent Extraction (ASE)

It is an efficient method for liquid-liquid extraction, requiring optimal combination of sample preparation technique and operational parameters. Some of the important steps involved in sample preparation are grinding, dispersing, drying and extracting the sample with suitable solvent, the selection of which depends on the nature of analyte and the method of analysis being employed. The nature of samples which can be extracted by the method are soil collected from deep oceanic beds [12], tissues of marine organisms like algae [13], parts of fruits and vegetables and pharmaceuticals [14].

Below given operational parameters hold the key to succeed with concentrated analyte extracts.

(i) Selection of solvent for extracting specific active ingredients from biological matrices plays crucial role in this method of extraction that later critically determines the resolution in analyses. Additionally, the literature concerning the method recommends solvent/solvent mixtures for target analyte classes. Worth mentioning is that a solvent performing adequately under normal laboratory conditions show enhanced solubility characteristics under conditions of ASE.
(ii) Temperature elevation impacts the solubility power beneficially, in that, it reduces the viscosity of solvent/solvent mixture thereby enhancing wetting, penetrating the biological matrix and solubilizing the target analytes.
(iii) Pressure maintains the solvent/solvent mixture in liquid state even at higher temperatures. Most ASE operate at pressures of 1000–2000 psi, with 1500 psi being the optimum condition.
(iv) Static Extraction Cycles maintains positive extraction equilibrium by intermittently introducing fresh solvent/solvent mixture. This parameter allows static time for the freshly introduced volume of solvent/solvent mixture with the sample matrix, while the previous volume is drained in the collection vial. This ensures high concentration of analyte being extracted even with matrices posing difficulty to penetrate.

The extracts can then be subjected to analysis using any of the miniaturized chromatographic technique described above. The results of extraction of certain pharmaceuticals from samples of few vegetables using ASE coupled with miniaturized chromatography are tabulated in the Table 5 at the end.

Extraction of bioactive compounds is time-consuming, viz., 4–72 h requirement, consuming large volumes of organic solvents, with a threat to environment of unnecessary flammability, and unwarranted occupational hazards on exposure. These solvents also posit several disposal concerns. ASE, an extraction employing higher pressure and temperature combination produce high-diffusion liquids, resulting faster extraction, reduced exposure of human to organic solvents, consuming lessened volumes of solvents, and increased yields.

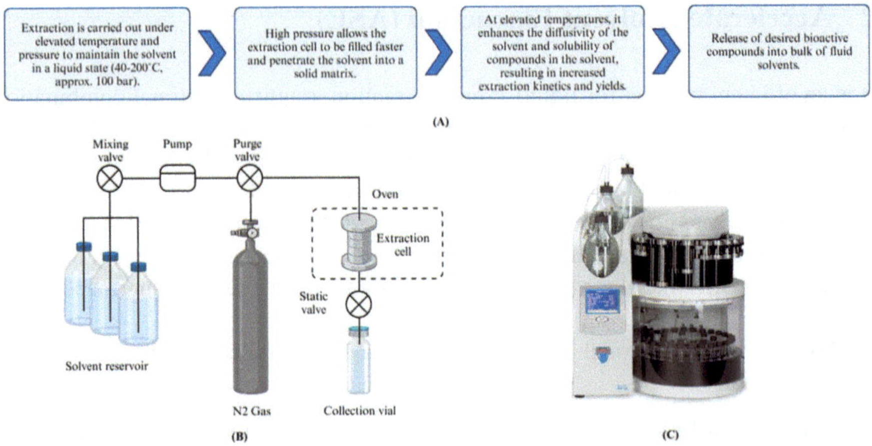

Fig. 4 Mechanism of ASE technique. **b** Schematic of the ASE process. **c** Commercial automated accelerated solvent extractor; Reproduced from "Extraction of bioactive compounds from beach-cast brown algae: a review on accelerated solvent extraction and subcritical water extraction," RSC sustainability, 2024, 2, 2069, under CC BY NC license

Certain bioactive compounds, viz., Alginate, Fucoidan, Laminarin, Phenolics, and Fucoxanthin, with wide applications in food, cosmetic, and pharmaceutical industries, were extracted using ASE techniques from beach cast brown algae. Details of the experiment are tabulated in Table 1.

The schematic Fig. 4 illustrates the working of the Accelerated Extraction System, depicts the workflow, and the actual image of the machinery.

3 Solvent-Free Extraction

Complexity of biological material makes them unsuitable for subjecting directly to any sort of analysis. Every biological specimen requires pre-analytical treatment to make it suitable for the desired analytical process. The pre-treatment process/s convert the sample into forms required by specific analytical technique, without making any slightest difference in initial composition of sample, or if it is to change the composition, it is in a pre-determined way.

Gradual development in extracting biological material over the years has been accomplished using organic solvents [15], however, owing to the toxicity they cause to living beings and the risk posit by them to the environment have been a constant concern for several global regulatory authorities which have off-late laid a strict and stringent guidelines for the limits of residual solvents in the biological products meant for human consumption, may it be food, supplement, nutraceutical, or a medicament formulation.

Owing to the toxicity concerns associated with organic solvents, developing process research for alternative methods of extraction without using toxic solvents became imperative. In the quest, a few techniques of extraction have been developed which are comparatively safe, environment-friendly, easy to process, economic, reproducible, and scalable to produce large quantities of extracted products.

Microwave-Assisted Extraction A rapidly developed, modern extraction technique and a sustainable technology which has a major share in popularizing green analytical chemistry, is microwave assisted extraction. This uses a non-contact heat source for extracting, there-by reduce the chances of decomposition of solid material and change of any sort in chemical nature of vulnerable or thermolabile compounds. It fastens the extraction processes by effective and selective heating, and helps to accelerate energy transfer, reduce thermal gradient, required equipment size and operation units. Advantages provided by the technique are (i) lower time span required than the conventional extractive procedures. The process is completed withing minutes than hours, (ii) lesser volumes of solvent reduce energy consumption, (iii) high reproducibility along with greater purity of the final product, (iv) and lesser tedious process.

A risk-free process, an economically and environmentally viable and a greener process, is the Solvent-free Microwave-Assisted Extraction (SFME), which is known to extract products without using organic—hydrocarbon—solvents thereby ruling out the concerns of residual solvents and the toxicity associated with them. Moreover, the problems of air-pollution caused due to organic vapors are excluded and the required labor have been greatly reduced.

Principle Opposing the principle in conventional extraction technique, wherein the heat is transferred from exterior surroundings to the interior cellular structure of plant material, the heat transfer process initiates from the interior of cells towards their exterior. This happens as the material is irradiated with microwave (MW), water molecules trapped in the cells are subjected to rapid rotations thereby generating heat which is transferred towards the exterior breaking the cell-wall structure and releasing the contents. The generated heat facilitates and fastens solubilizing components of interest. This in situ phenomenon has been extensively studied using SME, *i. e.*, Scanning Electron Microscopy [16].

Advantages provided by the process over conventional extraction methods:

This principle guarantees the reduction in process-cost due to

1. Ruling out the use of organic solvents (most of them are potentially carcinogenic and are hazardous to environment apart from being inflammable).
2. Reduction in hiring trained labor specialized in handling organic hazardous chemicals.
3. Apart from extraction of polar components from plant material, some volatile components and oxygenated hydrocarbons also are extracted at the same temperature and MW-input without inducing degradation and chemical changes in thermolabile material. This substantiates increased efficiency in extraction provided by the process.

4. Complexity of extracted material makes sample preparation imperative prior subjecting to analysis. These preparation methods present risks of loss of sample material either to (i) contamination or (ii) unwarranted chemical change [17]. The risk can be overruled by SFME as material is directly subjected to extraction without processing many preparative steps.

Methods developed for extraction of phytochemicals employing microwaves are described below.

Method-1 SFME The sample of material which is to be subjected to extraction process is placed in a microwave reactor without addition of any solvent or water. A syphon-reflux system is provided above the extractor, which facilitates continuous condensation of the distillate comprising of water and the extracted volatile oils. The syphon continuously separates the oils owing to their lesser densities than water, which is circulated continuously in and out of the reactor ensuring proper hydration of material and humidity in the reactor thereby protecting decomposition of material. Separated volatile oils can be drained in other apparatus and dried using appropriate desiccants prior subjecting to analysis by hyphenated techniques [18].

Method-2 Matrix-Assisted SFME This method is an advanced arrangement to the previously described extraction process of SFME. It employs dry matrix, consisting of any of the material mentioned below, to absorb microwaves and transfers them rapidly to the sample, increasing extraction efficiency. This matrix is placed at the bottom of the flask, and the process of SFME is applied.

Matrix Material: metallic nanoparticles; ceramics; and conductive polymers with heteroatoms such as polyaniline, polythiophene, polypyrrole, and polydopamine, metamaterials, carbon nanotubes anchored with metals, metal oxides, metal sulfides, and other magnetic and dielectric materials, carbon fibers, carbon spheres, carbon micro-tubes, activated carbon, graphene and ultrathin graphene, carbon net-like morphology, and carbon black, a ternary-phased CFC (Complex Forming Compound or Ternary Inclusion Complex), Ni-carbon nano-onions, and Ni/carbon nanotubes anchored to graphene surfaces wrinkled nickel nanosheets covering carbon microspheres, Bi2Te2.7Se0.3/graphene foam (BTS/GF) composites, rGO-based composite aerogel absorbers containing combination of reduced graphene oxide/copper sulfide. These provide excellent microwave-absorption properties and good chemical stability of extractive and thermolabile material ensuring extractive efficiency [17, 19].

Method-3 Vacuum-Assisted Microwave Hydro Distillation (VMHD) In this method, the crude plant material is placed in a flask with water to hydrate the material, which is equipped with a vacuum vent, and the arrangement is then placed in a microwave reactor. The mixture is irradiated with microwave to facilitate extraction of chemical constituents. Vacuum is applied intermittently to evaporate the azeotropic (water-volatile oil) mixture from the biological matrix. Compared to the conventional hydro-distillation, the VMHD method drastically reduces the extraction time. Additionally, the chemical nature of thermolabile material is preserved [20].

Method-4 Microwave-Assisted Steam Distillation (MSD) In this modification, the sample material is adsorbed in a matrix (SFME) and filled in cartridges, which are placed in a microwave and is connected to a condenser in distilling position. Irradiation of the cartridge heats the material. A stream of steam passes through the cartridge and enters the condenser, to the other end of which is a separating funnel, which receives the condensate. Time is allowed for condensed water and extracted oil to separate and is drained down in respective receiver vessel. The extraction proceeds continuously till complete exhaustion of the extractable material. The extract is then dried and analyzed is then dried and analyzed.

Method-5 Microwave-Assisted Steam Diffusion (MSDf) In this method, the cartridges filled with mixture of plant material and dry-matrix (described above) are stacked one-above other, and the stack is placed in a microwave oven. The stacked material is continuously irradiated with MW, and electrically generated steam is then allowed to flow through the stacked cartridges, and the extract is allowed to flow down the condenser attached outside the MW oven, ending low in height re beneath the stack. This process has proved a better extraction method than conventional steam diffusion [21].

Method-6 Pressurized-SFME This method employs the in situ pressure generated within the cavity. The shutters of the oven are fastened to be leak-proof so that it contains the pressure. This pressure is the result of heat generated by irradiating plant material in the cavity, which is placed without adding any water, along with the microwave absorbing matrix (as in Matrix-assisted SFME). The built up-pressure results in rupture of cellular structure/s thereby leaking the aqueous solution of chemical constituents contained in it. The concentrate extract is filtered, and the marc is cooled. This cold marc is re-subjected to extraction process as described, and the filtrate is dried and analyzed for its contents. Cooling and re-extraction is continued till the plant material is exhausted of either water or extractable material [22].

Microwave Dry Diffusion and Gravity The process is identical to the methods described above for 4 to 6. However, this method does not require pressure, and the process is applied to dry plant material [23].

4 Precautions

Every experimental procedure comes with technical process that are performed utilizing scientific tools and skills to perform. Skilled and seasoned scientists proficiently handle machineries, apparatus, equipment, and chemicals including sample material required for a given experiment. However, no scientific experiment can be carried without the threat of an accident, unfortunate though, if due cautions are not exercised. Degree of impact may vary depending on the source and nature of accident. Hence, it becomes imperative to exercise certain precautions while experimenting for extractions. Those mentioned below are specific precautions ought to be implemented while using the individual extraction methods described above.

4.1 Solvent-Free Microwave Extraction

(a) Plant material should be well hydrated before subjecting to the SFME: The plant material is subjected to few preparative steps as pre-washing, drying or freeze-drying, grinding for homogenization, as is usually required. This facilitates internal heating via microwave absorption. Microwaves primarily heat polar molecules, especially water. Well-hydrated plant tissues absorb microwave energy more efficiently, generating internal heat that ruptures glandular structures and releases volatile compounds. It enhances cell wall disruption so that the water inside plant-cells expands rapidly under microwave irradiation. This expansion creates pressure that helps rupture cell walls, improving the release of essential oils and secondary metabolites. Here, charring and thermal degradation is prevented by avoiding overheating or burning the plant material. Adequate hydration acts as a thermal buffer, maintaining temperature control and preserving sensitive compounds, allows uniform heating leading to better extraction efficiency, and reduces the risk of extracting unwanted degradation products or non-volatile impurities. SFME relies on the intrinsic moisture of the plant material to generate steam internally which acts as the carrier for volatile compounds—like hydro-distillation but without external water. If the plant material is naturally dried (e.g., stored herbs), it is often rehydrated with a controlled amount of water, typically 10–30% *w/w* before SFME. This ensures optimal microwave interaction and extraction performance.

(b) Power utilized by MW extractor should be sufficient to provide heat equivalent to 100 °C: This is a critical design requirement for SFME, wherein the plant moisture is internally heated to generate in situ steam as water boils at 100 °C at the atmospheric pressure, and it acts as the carrier for volatile compounds. This ensures efficient rupture of glandular cells and releasing essential oils. Here, it is necessary for the microwave extractor to deliver enough energy per unit time (watts) to raise the temperature of the hydrated plant matrix to ~100 °C. Typical systems operate at ≥750 W, with some industrial setups going up to 1000–1800 W depending on load volume and extraction kinetics [17]. It is worth mentioning here is that in closed vessels, pressure builds up allowing temperatures to exceed 100 °C up to 260 °C in some systems. In the open systems like SFME, the goal is to reach and maintain ~100 °C without external solvents or pressure buildup.

4.2 *Matrix-Assisted SFME*

(a) In this arrangement too, the moisture content of plant matrix is crucial as too little water posit the risk of charring the material, whereas higher percentages dilute the volatile contents and reduce the yield of extraction. Typically, the sample matrix must be hydrated to have 10–30 % *w/w* moisture.
(b) The power system must be controlled to 500–1000 W for lab-scale, as excessive power or prolonged exposure can cause thermal degradation of sensitive compounds. Employing pulsed irradiation if needed to avoid overheating helps a great deal.
(c) Monitoring the employed operating temperature ~ 100 °C to mimic hydro-distillation. Use of infra-red sensors or thermocouples assists to great extent in preventing generation of hot spots that degrade thermolabile constituents.
(d) Particle size uniformity in the sample matrix ensures consistent heating, however, overly fine powder shall be avoided as they may clump or burn. Coarse, evenly distributed material promotes better steam generation and oil release.
(e) Proper ventilation provided in the set-up avoids build-up of undue pressures ensuring safety to avoid accidents during procedural operations.
(f) Residual solvents entrapped in the plant specimen from any previously carried experimental steps can react under microwave conditions, posing safety risks. It must be ensured that the specimen is free from such contaminants.
(g) In case, the purpose of extraction is pharmacognostic or is an analytical profiling, validation of the composition of the extracted material is advisable. This is achieved by hyphenated techniques like GC-MS or LC-MS.

4.3 *Vacuum-Assisted Microwave Hydro-Distillation (VMHD) and Microwave-Assisted Steam Distillation (MSD)*

(a) Here, controlled vacuum is applied to maintain a stable and appropriate vacuum, typically 50–150 mbar. Too low a pressure may cause premature boiling and splashing; too high pressure may not sufficiently reduce boiling points. Use a vacuum regulator and a manometer is helpful to monitor the applied pressures/ vacuum.
(b) Microwave power regulation is kept moderate, ca. 300–800 W, to avoid overheating under vacuum. Sudden power surges can cause bumping or frothing due to rapid vaporization.
(c) Temperature monitoring is a must precaution to avoid generation of localized hotspots; infrared sensors or fiber-optic thermocouples are employed to monitor internal temperature.
(d) Ensure airtight sealing the extraction vessel properly avoids accidental hazards due to built-up of pressure. Use microwave-transparent material, e.g., borosilicate glass, PTFE) that can withstand pressure differentials.

(e) Use of an efficient coolant, e.g., cold finger or recirculating chiller, helps safe and effective condensation and cooling to collect volatile extractives. Under vacuum, vapors condense at lower temperatures—ensure coolant is sufficiently cold (e.g., 0–5 °C).
(f) Preventing Solvent Entrapment or Backflow: Appropriate cold traps are fastened in position between the condenser and vacuum pump to prevent essential oil or water vapor from entering the pump.
(g) The whole assembly should be scrupulously and regularly cleaned and inspected to avoid clogging during operations and contaminating the products.
(h) Sample Preparation: the sample can be in any form except fine powder, which aerosolize under vacuum and contaminate the products.
(i) Certain safety protocols pertaining to experimentation with vacuum are to be observed for monitoring, such as efficient fastening using appropriate valves and interlocks. The microwave shielding should be strong and intact to avoid leaking of the radiation.
(j) Post-extraction, the product requires analytical testing to ensure integrity by performing GC-MS or FTIR to confirm absence of degradation.

4.4 *Microwave-Assisted Steam Diffusion (MSDf)*

(a) Microwave Safety: This is observed by using microwave-friendly apparatus as microwave-transparent materials (e.g., borosilicate glass, PTFE) to avoid arcing or overheating. Any metal parts are avoided inside the microwave cavity unless specifically designed for hybrid systems. Proper shielding is ensured to prevent microwave leakage by fastening the doors and septum properly.
(b) Steam Pressure Control: The steam pressure is continuously monitored to avoid over-pressurization of the extraction chamber, suitable pressure valves and manometers are used to maintain safe operating limits. Sudden release of pressure is avoided to prevent splashing or thermal shock.
(c) Temperature Monitoring: Appropriate thermocouples or infrared sensors are installed and are checked periodically for proper functioning in tracking the internal temperatures.
(d) Moisture and Matrix Preparation: The dried plant material is moistened enough to ensure uniform microwave absorption and prevent localized burning. Overloading the extraction chamber is avoided to maintain efficient steam flow and microwave penetration.
(e) Condensation and Collection: Efficient condensers (double surface) are used to rapidly cool and collect volatiles, minimizing loss or degradation. Hydrosol separation is ensured to avoid emulsification or cross-contamination.
(f) Calibration and Validation: Microwave power output and steam flow rate is calibrated before each run. The extraction kinetics and reproducibility across batches are calibrated for consistent yield and composition.

(g) Personal Protective Gears: Efficient heat-resistant gloves, eye-goggles, and protective laboratory-coats are used while operating to ensure personal safety.

4.5 Pressurized-SFME

Pressurized-SFME operates under elevated temperatures and sealed conditions, making it highly efficient but also prone to pressure-related hazards. Following precautions ensure operator safety, extract quality, system longevity, and regulatory compliance in lab- or pilot-scale setups.

(a) Microwave-compatible pressure vessels with certified overpressure valves.
(b) The vessels to be used are validated to withstand vapour pressure buildup.
(c) The valve-seals and gaskets are checked regularly for wear or leakage.
(d) Microwave shielding and interlocks are confirmed for their integrity to prevent radiation leakage.
(e) Placing metallic components are avoided inside the microwave compartment.
(f) Proper, unclogged venting systems in place ensure efficient solvent vapor management.
(g) The extraction procedures are to be operated in well-ventilated areas or under fume hoods to avoid inhalation risks.
(h) Any chances of cross-contamination are prevented by carrying cleaning validation procedures.
(i) The vessels are appropriately loaded with the extractive material, any sort of overloading. Post-extraction, the vessels are allowed to cool sufficiently to avoid pressure-related accidents.

4.6 Microwave Dry Diffusion and Gravity

This process too warrants all the precautions enlisted above to ensure safe working. However, additional precautionary step required is of Gravity Flow Management.

(a) Here, unobstructed vertical flow of extract is ensured by regularly cleaning the tubing and collection vessels before each run.
(b) Any sort of backpressure or siphoning effects are avoided that may disrupt gravity-driven collection.
(c) Graduated receivers are used to monitor extract volume and phase separation.

4.7 Applications

The extractive methods mentioned till now are powerful green extraction techniques with diverse applications across phytochemistry, food science, cosmetics, and pharmaceuticals. Here are some of its most impactful uses:

1. Essential Oil Extraction: It aids in rapid extraction of volatile oils from aromatic herbs like thyme, rosemary, mint, and basil, preserving thermolabile compounds better than conventional steam distillation. These fragrances and bioactive compounds are used for skincare, haircare, and anti-aging products.
2. Phytochemical Profiling: This is utilized in the systematic identification and quantification of bioactive compounds in plants, especially secondary metabolites like alkaloids, flavonoids, terpenes, and phenolics. It is essential for validating medicinal properties and guiding drug discovery.
3. Efficient isolation of monoterpenes, sesquiterpenes, and phenolic volatiles for GC-MS analysis can be achieved as these chemicals are known to be thermolabile and hence, the conventional extraction methods do not yield acceptable results.
4. These methods are applied in ethnopharmacology and drug discovery to study bioactive profiles, in antioxidant and antimicrobial studies, for example, lavender oil, rose oil, chamomile extract.

These methods utilizing principles of green chemistry and sustainability and promote solvent-free extraction thereby reducing environmental impact, low-energy consumption and shorter processing time make it ideal for eco-friendly labs and industrial scale-up.

5 Supercritical Fluid Chromatography (SFC)

Supercritical fluid chromatography (SFC) is a powerful analytical technique used for isolation and analysis of bioactive compounds having importance in food and pharmaceuticals, forensic and environmental industries. It is a hybrid separation technique that combines the strengths of gas chromatography (GC) and liquid chromatography (HPLC); it uses a supercritical fluid, mostly carbon dioxide, as the mobile phase. This advancement of chromatographic principles exploits the unique solvent properties of supercritical fluids, which exist at conditions above their critical temperature and pressure. In this state, fluids exhibit both gas-like diffusivity and liquid-like solvating power, enabling efficient separation of complex mixtures.

The mobile phase solely consists of supercritical CO_2, which is often modified with small amounts of organic solvents, e.g., methanol to adjust polarity. The gas shows enhanced solubility characteristics at a peculiar combination of temperature and pressure [24], which is utilized for extraction/analysis of several components posing technical problems of instability, less solubility and lesser ability to differentiate between stationary phase and the conventional solvents.

The properties of supercritical fluid that make it suitable for the use as mobile phase are its low viscosity and high diffusivity which confer high efficiency and fast separations, its suitability for extracting thermolabile compounds and high molecular weight analytes that degrade under GC conditions. The use of CO_2 is inexpensive, non-toxic, and reduces reliance on large volumes of organic solvents. The operational parameters, viz., pressure, temperature, and co-solvent composition can be manipulated depending on the need for the analyte.

The columns are packed with several stationary phase material, like those described in the section of Miniaturized Chromatography. Very often, capillary GC columns too are used. Several spectrophotometric/metric instruments are equipped in this chromatographic technique as its detector. Accordingly, UV-visible, flame ionization, or MS; selection of the detecting system solely depends on the analyte under investigation. The effluent solvent can be allowed to escape in the environment without bothering of the hazards faced otherwise with the use of many organic solvents (Fig. 5).

Owing to its fast nature, being economic, environmentally friendly, the technique has been widely applied for chiral separations, impurity profiling, assay of drugs (API and Formulations) in pharmaceutical industry, metabolic studies, peptide analysis, isolation of bioactive compounds from plant and algae extracts from natural origin, detecting pollutants and hazardous mixtures of complex organic nature for environmental analysis, separation and analysis for content uniformity of lipids (trans-fat), flavors and colorants, and nutraceuticals in the food industry. It is particularly effective in isolating thermolabile compounds and those with high molecular weight, making it suitable for complex pharmaceutical formulations.

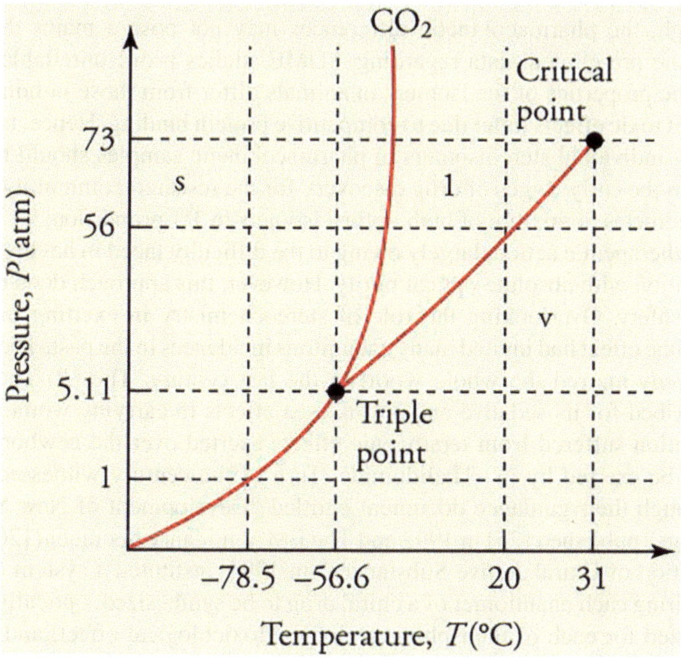

Fig. 5 Phase diagram of carbon dioxide illustrating the phase rule and the influence of temperature and pressure; reproduced from Ref. [24]

Developing SFC methods for extracting/separating/resolving/analyzing constituents/stereoisomers/analyte require considering several of their properties determining the SFC conditions—mobile-phase composition, modifiers, flow rate, mobile phase viscosity, density, and solvation tendency which depend on the combination of pressure and temperature conditions. Additionally, the nature of stationary phase used to retain the passing-by chemicals is one of the important factors to be considered while developing the SFC methods. Plethora of literature is available describing several SFC extracting and analyzing methods developed, the applied conditions from which have been tabulated in Table 2.

Chiral Separations: Existence of stereoisomers having diverse pharmacology is known since ages. The enantiomeric pair of medicinally active stereoisomers are spontaneously distinguished by biological systems, although they are learned to have identical physical, except for optical activity, and similar chemical properties. Both isomers of an optically active pair, *dextro-* and *laevo-*, may exert altogether different pharmacodynamic effect, and face different pharmacokinetic fate as well. Accordingly, the anti-malarial activity of (−)-quinine and anti-arrhythmic property of (+)-quinidine, a synthetic diastereomeric stereoisomer; adrenergic receptor binding action of (−)-ephedrine and inhibition of *MAO* enzyme by (+)-*pseudo-e*phedrine, the latter being the diastereomer of the former, provide apt examples of the concept that prompt to consider these enantiomers to be two different drugs. These are examples of few different stereoisomers having pharmacologically varied action or having synergistic therapeutic effect.

Although, the pharmacokinetic differences may not pose a major therapeutic problem, the pre-clinical data regarding ADME studies prove unreliable as pharmacokinetic properties of the isomers in animals differ from those in humans. The risk to elicit toxic effects loiter due to competitive protein binding. Hence, techniques to quantify individual stereoisomers in pharmacokinetic samples should have been available in the early stages of drug discovery for the research community.

The racemates, a mixture of both optical isomers in 1:1 proportion, were used as drug for a therapeutic action, largely owing to the difficulty faced in having commercial separation with absolute optical purity. However, this approach does not assure complete safety. Overlooking the role of stereochemistry in exerting therapeutic effect or toxic effect had invited many calamitous incidences in the past. The *Thalidomide Tragedy* marred the whole world in the last century. The (*R*)-Thalidomide was prescribed for its sedative and anti-nausea effects to carrying women, instead the population suffered from teratogenic effects exerted over the newborn infants, ensued to be exerted by (*S*)-Thalidomide. Turn of the century witnessed the US-FDA, through their guidance document entitled "Development of New Stereoisomeric Drugs" published [25] in 1992 and The EMA guidance document [26] entitled "Investigation of Chiral Active Substances" in 1994, instituted a system of regulations requiring each enantiomer of a chiral drug to be synthesized, optically purified, and be tested for each of their pharmacological/toxicological effect, and all information be included in the IND and NDA submissions. These compelling prerequisites laid impeding limits for pharmaceutical developers since most of the small

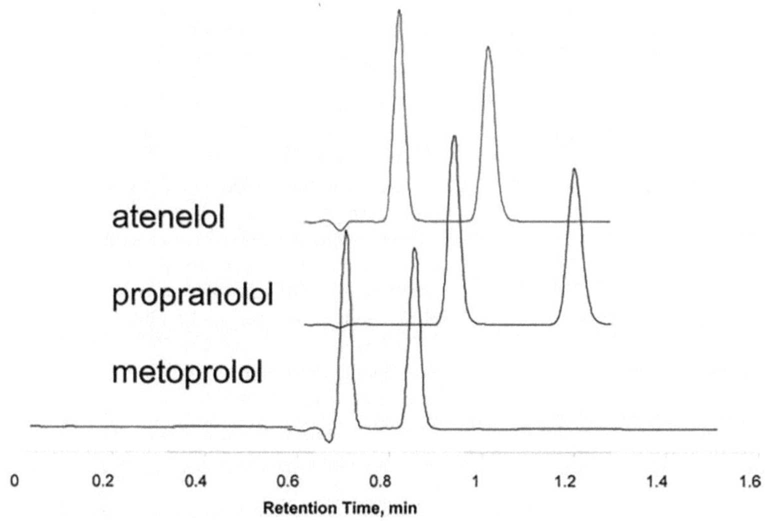

Fig. 6 Chromatogram displaying separation of three β-blockers. Reproduced from Ref. [27]

molecules were chiral, and obtaining all enantiomers in optically pure form lead to the bottle-neck issue.

Advancements in separation technology made it feasible to obtain purified enantiomer. These developments stemmed into concepts of identifying *Eutomer*, *Distomer*, *Dimer*, *Eudismic ratio*, and *Eudismic index*. These terms are used to describe the therapeutic value of stereoisomer NCE with chiral center/s. It is here that SFC becomes effective means of separating enantiomers. It is reported to be remarkably fast, efficient and environment friendly.

Enantiomers are efficiently resolved by SFC which is remarkably fast, efficient. The separation of several beta-blockers achieved in one of the SFC separations, is shown in the chromatogram (Fig. 6).

Racemates of the three adrenergic β-blocker drugs, viz., atenolol, propranolol and metoprolol were collectively resolved into their individual enantiomers by methanol: CO_2 40:60 with 0.2% trimethylamine as modifier, run at 5 mL/min flow rate, maintained at 30 °C with pressure of 150 bar. The separation was achieved using 4.6 × 250 mm column, packed with 5 mm particles of Regispack [27]. The (*S*)-enantiomers of atenolol, propranolol and metoprolol are eudysmic amongst the racemates, (*S*)-atenolol being 100 times more effective than the (*R*)-enantiomer [28], and both, (*S*)-metoprolol and (*S*)-propranolol, being 130 times potent than their respective (*R*)-enantiomers [29].

Similarly, chiral drugs from anti-depressant, and anti-inflammatory category were resolved to get the *eutomer* in optically pure form using SFC. Table 1 summarizes the chromatographic conditions of SFC applied for resolving the eutomers from anti-depressant, anti-histaminic and anti-inflammatory drugs that had been used in their racemate form instead of therapeutically effective eutomer.

Table 1 SFC-chromatographic conditions utilized to resolve racemates into eutomers

Name of the racemate	Eutomer	SFC conditions	Reference
Citaprolam	*S*-(+)-Citaprolam	UHPSFC system inject port: 10 μL; column: Trefoil-CEL2; dimensions: 3.0 mm × 150 mm, filled with 2.5 μm particles; mobile phase: CO_2 mixed with methanol/acetonitrile (70:30, *v/v*) with 10 mM ammonium acetate for gradient elution: 0.3–2.0 min, 30–40%; 2.0–2.8 min, 40% B in 2.8–3.5 min, 40–25% B in 3.5–3.6 min and 25% B in 3.6–4.0 min. Detection: Tandem-quadrupole mass spectrometer	[30]
Amphetamine	*S*-(+)-Amphetamine	UHPSFC system with Chiralpak® AD-3 column; column temp. 10 °C; mobile phase: isocratic elution with 93.5% CO_2 and 6.5% 2-propanol/methanol (1: 1, *v/v*); flow rate: 1 mL/min	[31]
Cetirizine	*R*-(−)-Cetrizine	Column: Chiralpak® (4.6 mm i. d. x 150 mm); particle size: 5 μm; eluent: CO_2/modifier (60: 40); modifier: 1) methanol +0.4% diethylamine, 2) acetonitrile/ethanol (80/20) + 0.4% diethylamine, 3) methyl tert-butyl ether/ ethanol (80/20) + 0.4% diethylamine; flow rate: 3.0 mL/min Column temp.: 40 °C; wavelength: UV: 225 nm, CD: 230 nm; Back pressure.: 15 MPa; injection volume: 5 μL	[32]
Flurbiprofen	*S*-(+)-Flurbiprofen	SFC system with ISCO syringe pump for CO_2 delivery; HPLC pump for modifier delivery; sampling port: autosampler with Rhenodyne valve; column: The RegisWhelk-O1 filled with 10 μm particle; oven temp.: 60 °C; UV-visible detector set at 220 nm	[33]
Ibuprofen	*S*-(+)-Ibuprofen	SFC-SMB; column: 4.6 x 250 mm analytical Rheodyne 6-port valve; column temp.: 40 °C; Mobile phase: supercritical CO_2 with *iso*-propanol 4.5% *w/w*; pressure: 17. 1 MPa	[34]
Naproxen	*S*-(+)-Naproxen	Column: Kromasil CHI-TBB; temp.: 293 K, pressure: 17.8 MPa; Mobile phase: supercritical CO_2 with *iso*-propanol 11% *w/w*	[35]

SFC has been widely applied to carry assay of drugs (API and formulations) in pharmaceutical industry. Pharmaceutical companies now rely on identified supplier to procure synthesized compounds from. In such scenario, it becomes imperative to determine the purity of these compounds, and purify them, if necessary, before testing for biological utility. Here too, SFC has become an important tool in the analysis and purification of such compounds.

Drug development has been immensely contributed by natural products - phytoconstituents, obtained majorly from plants in the last century, while by the turn of the millennium the interest of scientific community has turned towards exploring the zoochemicals isolated from terrestrial and marine animals. Additionally, food and confectionary industry, fragrance and flavour industry, and the cosmetic industry also utilize chemicals obtained from natural sources to broaden the usefulness of their products. However, the difficulty faced by these industrial sectors is the availability of pure phyto/zoo-chemical for their sizeable usage in any of the products, for the benefit of society. These materials are sparsely available in their respective biological sources, moreover, the isolation methods, and subsequently used analytical procedures, used previously were too laborious, arduous, and painstaking; chemicals that were known to be detrimental to the health (inviting occupational hazards to the workers) and hazardous to the environment posing long-term consequences. The same can be said about the analytical techniques existing few decades ago. Advent of green technology—whether it be utilized for extraction of phyto/zoo-chemical or analyzing it—has brought a pleasing and promising modification to provide, faster and safe analytical procedures, which at times can provide provision for on-site analysis too. This has helped overcoming the problems of older extracting and analyzing methods. Application of the technique in the field of lipidomics has been its one keystone achievement. One major aspect in biology is food and food supplements; proteins and lipids form major proportion of human diet to be utilized as source of reserve energy. These perform essential biological functions in the body, and hence their usefulness in living system have made the applicable in fields like food, pharmaceuticals, nanotechnology, floral chemistry, and alternate source for energy production [36–38].

Selectivity and sensitivity of extractive and analytical SFC methods have been raised when it has been hyphenated with mass spectrometry. Machinery working on techniques like electron spray ionization (ESI), atmospheric pressure chemical ionization (APCI), atmospheric pressure photoionization (APPI), and *UniSpray Ionization* (US) has been equipped with SFC to analyze post-chromatographic separation. This development paved way for advent of the promising developments as the Ultrahigh-Performance Supercritical Fluid Chromatography (UHPSFC) [38]. It has established a strong position among the separation techniques in various application fields, and is extensively utilized for the extraction, separation and analyzing of plant extracts, which is often challenging. Figures 7 and 8 displays a chromatogram obtained on separating and analyzing small molecules as terpenes, flavonoids, and phenolic and terpenoic acids, which are present in plant extracts, using UHPSFC hyphenated with the mass spectrometric techniques. Figure 9 given below displays a UHPSFC-(QqQ)-MS/MS chromatogram obtained by chromatographic separation of a lipid mixture containing several isomers of Vitamin D2 and D3 and several hydroxyergocalciferol.

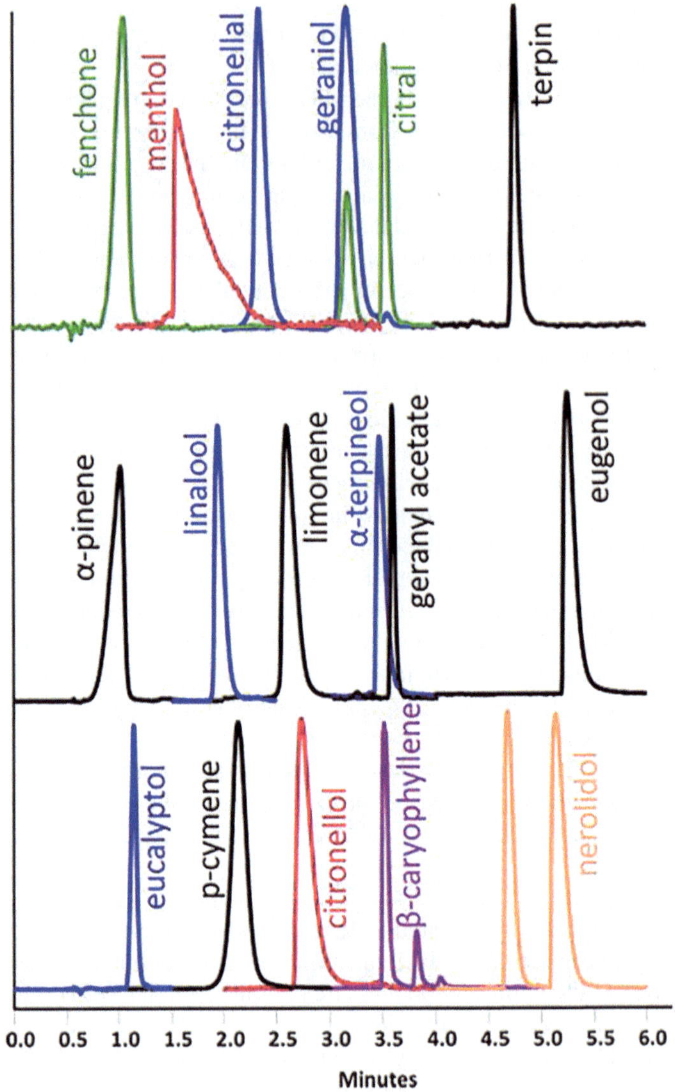

Fig. 7 UV-Chromatogram detected at 211 nm displaying extraction and analysis of terpenes separated by UHPSFC and quantified by MS techniques; with chromatographic conditions: supel carbon LC column (150 mm) at 60 °C, BPR pressure at 3300 psi, flow rate 1.5 mL/min, methanol as organic modifier in gradient elution as following 0% for 1.5 min, 0–40% in 1.5–4.0 min, 40–41% in 4.0–6.0 min, followed by 1.5 min equilibration at starting conditions. Reproduced as-such from Ref. [27] licensed under CC-BY 4.0

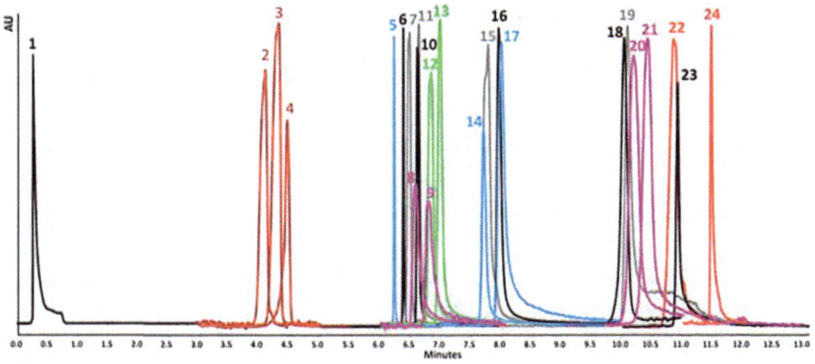

Fig. 8 UV-Chromatogram detected at 211 nm displaying extraction and analysis of flavonoids and phenolic and terpenoic acids separated by UHPSFC and quantified by MS techniques with chromatographic conditions; TorusDiol column (50 mm) at 20 °C and 5% water in MeOH as organic modifier, BPR pressure at 10–13 MPa, flow rate 0.7–1.5 mL/min, methanol as organic modifier in gradient elution as following 0% 0–0.7 min; 3% for 0.7–5 min; 4% 5–5.5 min; 20% for 5.5–8.5 min; 23% for 8.5–12.0 min; 70% for 12.0–13.0 min; 90% for 13.0–14.5 min; 0% for 14.5–15.5 min. Peak of 1. terpenes, 2. betulinic acid, 3. oleanolic acid, 4. ursulic acid, 5. hesperetin, 6. naringenin, 7. apigenin, 8. isorhamnetin, 9. tamarixetin, 10. caffeic acid, 11. protocatechuic acid, 12. kaempferol, 13. luteolin, 14. ellagic acid, 15. taxifolin, 16. gallic acid, 17.quercetin, 18. phloridzin, 19. quercitrin, 20. epicatechin, 21. catechin, 22. hesperidin, 23. hirsutrin, and 24. rutin. Reproduced as such from Ref. [27] licensed under CC-BY 4.0 UHPSFC-(QqQ)-MS/MS is an advanced SFC hyphenated technique allowing simultaneous separation and determination of analytes. Accordingly, vitamins D_2, D_3, with their hydroxylated derivatives and ester analogues were separated and co-analyzed within 8.0 min using the technique [39]. The method was applied to estimate the levels of these micro-nutrients from human blood, proving the technique useful for analyzing other fat-soluble micronutrients too (Fig. 9)

N-Nitrosodimethylamine (NDMA) has emerged as a nitrosamine impurity of major concern in several widely used drugs. It has been detected in medications such as angiotensin receptor blockers (valsartan and losartan), metformin, ranitidine, rifampin, and varenicline, prompting global regulatory reviews. NDMA is classified as a probable human carcinogen based on animal studies, meaning long-term exposure above acceptable intake limits may increase cancer risk. Its presence often arises from manufacturing processes, raw material contamination, or degradation pathways, particularly under certain storage conditions. Regulatory agencies including the FDA and EMA have issued strict guidance on monitoring, controlling, and preventing NDMA and related nitrosamine drug substance-related impurities (NDSRI) to ensure patient safety [40]. SFC has been instrumental in detecting nitrosamines in active pharmaceutical ingredients [41]. This method, which employed a wide-polarity range SFC with a gradient combination of CO_2 with varying proportions of methanol as a modifier, applying a QbD approach proved highly sensitive in detecting carcinogenic impurities of nitrosamine with other monographed impurities, in the sartan-class of drugs at picogram to femtogram range.

The field of forensics require collection of samples and dispatch to a laboratory in a tamper-proof condition, to analyze particularly in the illicit drugs, explosives,

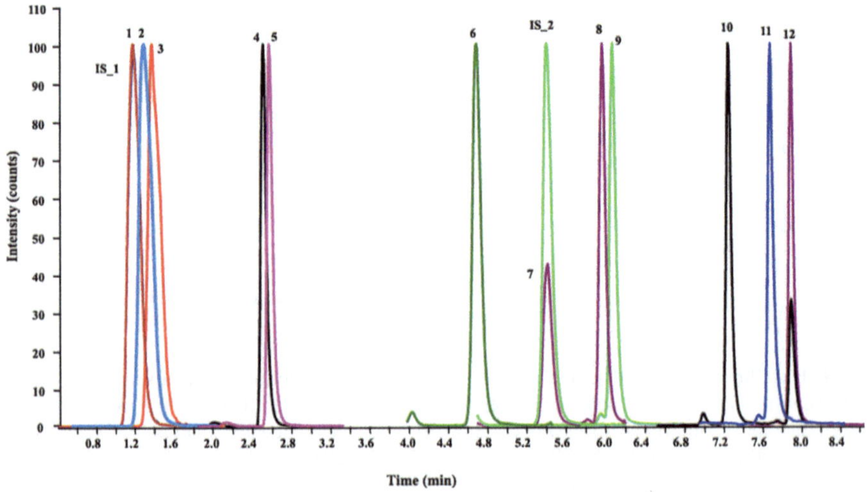

Fig. 9 Normalized UHPSFC-(QqQ)-MS/MS chromatogram with torus 1-aminoanthracene (1-AA) column at 50 °C using a mobile phase consisting of CO_2 and MeOH as the co-solvent in gradient system. Separated and analyzed constituents were (1) palmitate-D3; palmitate-D3-13C16; (2) stearate-D3; (3) linoleate-D3; (4) D3; (5) D2; (6) 25-OH-D2; 25-OH-D3-13C5; (7) 25-OH-D3; (8) 1-OH-D3; (9) 1-OH-D2; (10) 24,25-(OH)2-D3; (11) 1,25-(OH)2-D2; and (12) 1,25-(OH)2-D3

toxic substances, available in trace amount as evidence. In such scenario, SFC and its aid immensely with its speed, sensitivity, and eco-friendly nature. Some of the formative applications of SFC in the field are enlisted in Table 2.

SFC comes very useful in studying the environmental pollutants and determining their percent as pesticides, herbicides, pharmaceutical residues, polycyclic aromatic hydrocarbons (PAHs), polychlorinated biphenyls (PCBs), and other persistent organic pollutants (POPs) that are proven hazardous material for general populace health. It is valued for its eco-friendly nature, reduced solvent use, and ability to handle complex mixtures in soil, water, and air samples (Table 3).

In conclusion, SFC represents a transformative advancement in pharmaceutical analysis, bridging the gap between traditional chromatographic methods and modern demands for eco-friendly, high-performance separations. By employing supercritical carbon dioxide as the primary mobile phase, SFC drastically reduces solvent consumption and energy usage while maintaining excellent resolution and speed of analysis. Its proven utility in chiral separations, impurity profiling, and preparative scale purification highlights its relevance across drug discovery and development pipelines. Although early adoption was limited by instrument robustness and regulatory acceptance, recent technological improvements have enhanced reproducibility and compliance, positioning SFC as a viable alternative or complement to HPLC and GC. Moving forward, the integration of SFC with advanced detectors and hyphenated techniques (e.g., SFC-MS) will further expand its scope, reinforcing its role as a cornerstone of green analytical chemistry and a sustainable solution for the pharmaceutical industry.

Table 2 SFC and hyphenated techniques utilized to identify illicit compounds and their derivatives in biological samples

Name of illicit chemical	Sample matrices which are adulterated	SFC conditions	Reference
Forty-three anabolic agents including Bolasterone, Calusterone, Clenbuterol, Metenolone, Oxandrolone, Tetrahydrogestrinone and their respective metabolites	Urine	UHPSFC; positive ESI; column: Torus DIOL; 100 × 3.0 mm; particle size: 1.7 μm; 40 °C; mobile phase: supercritical CO_2 with MeOH 98%; pressure: 15 MPa; flow rate: 1.3 mL/min	[42]
RS–(±)-Methamphetamines	Drug samples sized from manufacturing sites; in possession of youth	UHPSFC; positive ESI; column: trefoil CEL2; 150 × 3.0 mm; particle size: 2.5 μm; 10 °C; mobile phase: supercritical CO_2 with MeOH (98%) in the proportion 95:5; pressure: 17.2 MPa; flow rate: 1.0 mL/min; detector: PDA	[43]
Sudan dyes I-IV	Food stuffs like turmeric, red paprika, hot chili, etc.	UHPSFC; column: princeton Benzamide; 250 × 4.6 mm; particle size 5.0 μm; temp.: 40 °C; mobile phase: supercritical CO_2 with MeOH; pressure: 20 MPa; flow rate: 4.0 mL/min	[44]
As many as fifty sympathomimetic drugs belonging to β-blockers and stimulant category	Urine	UHPSFC; column: BEH 2-EP; 100 × 3.0 mm; particle size 1.7 μm; temp.: 50 °C; mobile phase: supercritical CO_2 with MeOH 95%; pressure: 13 MPa; flow rate: 2.0 mL/min	[45]

6 Energy Efficiency

Green chemistry improves energy efficiency by replacing high-energy, solvent-intensive methods with miniaturized, faster, and lower-power techniques such as microwave, ultrasound, and room-temperature processes. Traditional methods like distillation, Soxhlet extraction, and long chromatographic runs consume large

Table 3 SFC and hyphenated techniques utilized to identify pesticides dispersed in the environment

Name of the pesticide	Sample matrices which are adulterated	SFC conditions	Reference
Chlorotriazine class of pesticides	Crop, soil	SFC-MS; injector valve: 5 μL; column: 2-Ethylpyridine-bonded silica; gradient mobile phase: SF-CO_2 with MeOH 3% for 0–2 min; 3–15% further 2–7 min; pressure 15 MPa, flow rate: 3 mL/min	[46]
Sulfoxaflor	Vegetables, fruits, crop, soil	SFC-MS/MS; column: Chiralpak® IA-3; mobile phases: SF-CO_2 with ACN, IPA, MeOH, ethanol, *n*-hexane, and 1-butanol in varying proportions; flow rate and depending on the composition of mobile phase	[47]
Organophosphates, carbamates, and triazine pesticides	Fruits like avocado and papaya	SFC-DAD; column: 2-Ethylpyridine-bonded silica; gradient mobile phase: SF-CO_2 with MeOH 5% for 0–0.4 min, 5–60% for 0.4–1.6 min: pressure: 10.34 MPa; temp.: 40 °C; flow rate: 1.5 mL/min	[48]
Pyrethroid insecticide	Vegetables	SFC-UV; column: hypersil APS; mobile phase: CO_2 mixed with MeOH 98: 2; pressure: 13–20 MPa; temp. 60 °C; flow rate: 1–3 mL/min	[49]

amounts of electricity and heat. Energy demand is directly tied to waste generation; requirement of heat energy depends on volume of solvents and their boiling points, causing them to evaporation and higher quanta of energy emission, with causing added wastage. Laboratory sustainability depends not only on reducing chemical waste but also on lowering the carbon footprint of instruments. Using Microwave-assisted extraction reduces the energy demand to heat samples rapidly and uniformly. It cuts the extraction times from hours to minutes, reducing electricity use. The ultrasound-assisted extraction employs acoustic cavitation to disrupt matrices. It requires less heat and solvent compared to conventional reflux or Soxhlet methods. The extractive methods using carried at room temperature and ambient-pressure techniques avoids energy-intensive heating and pressurization. Miniaturization analytical processes reduce sample-size requirement and reagent volumes. Smaller analytical set-up warrants less power for all the equipment.

Supercritical fluids replaced high-boiling, inflammable and carcinogenic (benzene and other hydrocarbons) with carbon dioxide, although some extractions have also been carried using mixtures of few alcohols. The chromatographic run-time has been greatly reduced. This requires less energy for solvent recovery and recycling.

7 Safer Alternatives

Traditional extractive procedures employed harmful volatile solvents like benzene, chloroform, dichloromethane, acetonitrile, carbon tetrachloride, etc. A word of caution was necessarily carried due to flammability of these solvents, and few of those are carcinogenic. Green chemistry allows safer, non-toxic, alternatives. Most of these are renewable, non-toxic, recyclable, excellent for extractions.

Sample preparation steps in the conventional extractive and analytical procedures too utilized large volume of solvents. Modern, green methods like Solid Phase Microextraction (SPME) and Dispersive Liquid–Liquid Microextraction (DLLME) require microliters of safer solvents, whereas miniaturized techniques suggest solvent-free sample preparation.

Traditional chromatographic systems like HPLC and GC consume large volumes of solvents and gases, whereas methods like capillary electrophoresis (CE) and Ultra-High-Performance LC (UHPLC) are engage shorter runs utilizing less solvent, use aqueous buffers, whereas SFC utilizes CO_2.

For detection of the extracted principles sophisticated miniaturized instrumentation are utilized which are portable. These reduce burden of transporting samples from collection sites to analytical laboratory.

8 Waste Minimization

Waste management in Green Analytical Chemistry (GAC) emphasizes designing analytical methods that inherently minimize hazardous byproducts, solvent use, and energy consumption while ensuring accuracy. Guided by the 12 principles of green chemistry, it prioritizes prevention over treatment, atom economy, safer solvents, and energy-efficient processes. Strategies include miniaturization, automation, solvent substitution, microextraction, supercritical fluid use, and solvent-free techniques, complemented by recycling and catalytic conversions. Tools such as NEMI, GAPI, ECO-Scale, and AGREE assess greenness, while neutralization, solvent recovery, and bio-based degradation support waste treatment. Overall, GAC waste minimization enhances environmental protection, safety, cost efficiency, and sustainability, aligning laboratory practices with global ecological goals. Waste minimization in Green Analytical Chemistry (GAC) focuses on reducing hazardous byproducts, solvent use, and energy consumption while maintaining analytical accuracy. It is achieved through miniaturization, solvent substitution, automation, and eco-friendly sample preparation.

In essence, waste minimization in green analytical chemistry is achieved by shrinking scale, substituting safer solvents, eliminating unnecessary reagents, and embedding sustainability into every analytical step.

Table 4 is a comparative chart showing how traditional analytical chemistry and green analytical chemistry differ in their approach to waste management:

Table 4 Waste management in analytical chemistry: traditional versus green (GAC)

Parameter	Traditional	GAC
Sample size	Milliliters or more	Microliters to Nanoliters
Solvent used	Hazardous solvents (e.g., chloroform, benzene)	Non-toxic alternatives (e.g., water, ethanol, ionic liquids)
Waste generation	High volume of hazardous waste	Minimal waste, often non-hazardous
Safety	High-risk due to toxic chemicals	Low-risk, improved lab safety
Disposal strategy	Neutralization, incineration	Prevention, recycling, closed-loop circulation aid recycling
Reagent consumption	Excessive, large scale	Miniaturized, micro-scale methods
Energy demand	High (e.g., heating, vacuum pumps)	Low (e.g., room temperature methods)
Environmental impact	High-risk due to toxic chemicals	Reduced ecological footprint

Several chemicals that are important for pharmaceutical, food and fragrance, and cosmetic industry have been extracted, analyzed for their quality using the green extractive and analytical methods described above and are tabulated in Table 5.

Table 5 Pharmaceuticals, phytoconstituents, zoochemicals that have been isolated and analyzed employing several green technologies

Analyte	Technique employed to analyze	Sample size of matrix required	Limit of quantification	Reference
Amitriptyline, Clenbuterol, Imatinib, Tacrolimus	Miniaturized paper-MS	1 µL, blood	50 ng/mL	[11]
(−)-Menthol and (+)-Menthol	SFE—Achiral liquid chromatography with circular dichroism	20 µL, laboratory solution	90–92 ug/mL	[50]
Polyethylene glycol (PEG) 600 and PEG 1500, Gly-Pro-Arg-Pro tetrapeptide, Methicillin-Resistant Staphylococcus Aureus, Reserpine, Rhodamine B, *N, N*-Diethyl-*m*-toluamide and Cytochrome C	Miniature mass spectrometer with continuous atmospheric pressure Interface	1 µL, urine	1–100 ug/mL	[51]
Catechin, Epicatechin Epigallocatechin epigallocatechin gallate, Epicate chin gallate Gallic acid, Proxyphylline	nanoCapillary-LC-PB/EIMS	50 µL aqueous extracts of several green tea preparation available commercially	5–75 ng/mL	[52]
Sixteen different mycotoxins contained in peanuts, almonds and pistachio	Nano flow liquid chromatography high resolution-mass spectrometry	0.1 µL of sample solution prepared by QuEChERS technique	0.05–5 µg/kg nuts	[53]
Ritonavir, Darunavir, Atazanavir, Tenofovir, Emtricitabine	Nanoflow-Nanospray mass spectrometry	Accurate 0.5 µL extracts from urine samples prepared by NanoFlowUHPLC-nanospray TOF-MS	–	[6]
Several Chlorophenoxybutyric acids, Endosulphans and Aldrin pesticides	Liquid chromatographic-atmospheric pressure ionization mass spectrometry	500 nL of prepared samples originally collected from rives	75–2000 pg	[54]
Arsine Chloride, Dimethyl arsenic acid, Arsenosugar	LC/PB-EIMS	5 µL after sample preparation	1–30 ng/mL	[55]

(continued)

Table 5 (continued)

Analyte	Technique employed to analyze	Sample size of matrix required	Limit of quantification	Reference
Alginate (a) Fucoidan (b) Phenolics (c) Fucoxanthin (d)	ASE using water, methanol, ethanol in varying proportions	Not mentioned	Extractive value in mg/g extract (a) 2.08, (b) 12.52, (d) 51.04, (e) 0.42	[13]
Acetaminophen (a) Caffeine (b), Carbadox (c), Carbamazepine (d), Lincomycin (e), Monensin sodium (f), Oxytetracycline (g), Sulfadiazine (h), Sulfamethoxazole (i), Trimethoprim (j) and Tylosin (k) extracted from celery and lettuce, after the solution of these pharmaceuticals being spiked in vegetable tissue, vortexed and dried to prepare sample.	ASE coupled with LC-MS/MS	500 mg of vegetable sample along with 200 ng/g concentration for each pharmaceutical in the vegetable samples	(a) 4. 4 ng/g, (b) 2. 3 ng/g, (c) 7.4 ng/g, (d) 1.9 ng/g, (e) 6.7 ng/g, (f) 5.4 ng/g, (g) 3.1 ng/g, (h) 11.9 ng/g, (i) 15.8 ng/g, (j) 2.4 ng/g, (k) 4.6 ng/g	[14]
Extraction of (a) terpenes, (b) terpenoic acids, (c) flavonoids, (d) phenolic constituents	UHPSFC with APCI for (a); with US for (b) and (c); and EI for (d)	Accurately, 2–10 μL of samples prepared by dissolving 0.1–1.0 mg/mL	(a) 0.01–1 ng/mL; (b) and (c) 0.5–20 ng/mL (d) 7 ng/mL	[38]
Separation and estimation of (a) D_3, (b) 1-OH-D_2, and (c) 24, 25-$(OH)_2$-D3; (d) 1, 25-$(OH)_2$-D_3; (e) palmitate-D3	UHPSFC-(QqQ)-MS/MS	Human plasma samples 500 ± 0.1 μL	(a) 1.30 μg/L (b) 1.23 μg/L (c) 1.44 μg/L (d) 2.33 μg/L (e) 9.10 μg/L	[39]

References

1. Tswett M. Adsorption analyse und chromatographische methode. Anwendung auf die chemie des Chlorophylls. Ber Dtsch Bot Ges. 1906; 24: 384–93.
2. MacNair JE, Lewis KC, Jorgenson JW. Ultrahigh-Pressure Reversed-Phase Liquid Chromatography in Packed Capillary Columns. Anal Chem. 1997 Mar 1;69(6):983–9.
3. Kononikhin AS, Starodubtseva NL, Chagovets VV, Ryndin AY, Burov AA, Popov IA, et al. Exhaled breath condensate analysis from intubated newborns by nano-HPLC coupled to high resolution MS. Journal of Chromatography B. 2017 Mar;1047:97–105.
4. Chetwynd AJ, David A. A review of nanoscale LC-ESI for metabolomics and its potential to enhance the metabolome coverage. Talanta. 2018 May; 182: 380–90.
5. Luo X, Li L. Metabolomics of Small Numbers of Cells: Metabolomic Profiling of 100, 1000, and 10000 Human Breast Cancer Cells. Anal Chem. 2017 Nov 7;89(21):11664–71.
6. Chetwynd AJ, Samarawickrama A, Vera JH, Bremner SA, Abdul-Sada A, Gilleece Y, et al. Nanoflow-Nanospray Mass Spectrometry Metabolomics Reveals Disruption of the Urinary Metabolite Profiles of HIV-Positive Patients on Combination Antiretroviral Therapy. JAIDS Journal of Acquired Immune Deficiency Syndromes. 2017 Feb 1;74(2): e45–53.
7. Han KM, Kim YK, Seo S, Kim JH, Lee JH, Kim H Il, et al. Analysis of 13 Banned Colorants in Cosmetics via Liquid Chromatographic and Mass Spectrometric Techniques. Applied Sciences. 2023 May 12;13(10):5967.
8. Žuvela P, Skoczylas M, Jay Liu J, Bączek T, Kaliszan R, Wong MW, et al. Column Characterization and Selection Systems in Reversed-Phase High-Performance Liquid Chromatography. Chem Rev. 2019 Mar 27;119(6):3674–729.
9. Qiao X, Chen R, Yan H, Shen S. Polyhedral oligomeric silsesquioxane-based hybrid monolithic columns: Recent advances in their preparation and their applications in capillary liquid chromatography. TrAC Trends in Analytical Chemistry. 2017 Dec;97:50–64.
10. Berthier E, Dostie AM, Lee UN, Berthier J, Theberge AB. Open Microfluidic Capillary Systems. Anal Chem. 2019 Jul 16;91(14):8739–50.
11. Li L, Chen TC, Ren Y, Hendricks PI, Cooks RG, Ouyang Z. Mini 12, Miniature Mass Spectrometer for Clinical and Other Applications—Introduction and Characterization. Anal Chem. 2014 Mar 18;86(6):2909–16.
12. ZHU QZ, SUN Q, SU ZG, XIE MM, SONG JY, SHAN YB, et al. A Soil Water Extraction Method with Accelerated Solvent Extraction Technique for Stable Isotope Analysis. Chinese Journal of Analytical Chemistry. 2014 Sep;42(9):1270–5.
13. Zhang Y, Hawboldt K, MacQuarrie S. Extraction of bioactive compounds from beach-cast brown algae: a review on accelerated solvent extraction and subcritical water extraction. RSC Sustainability. 2024;2(8):2069–91.
14. Chuang YH, Zhang Y, Zhang W, Boyd SA, Li H. Comparison of accelerated solvent extraction and quick, easy, cheap, effective, rugged and safe method for extraction and determination of pharmaceuticals in vegetables. J Chromatogr A. 2015 Jul;1404:1–9.
15. Furniss BS, Hannaford AJ, Smith PWG, Tatchell AR, editors. Vogel's Textbook of Practical Organic Chemistry. 5th ed. London: Pearson Education; 2008. 136–137 p.
16. Ferhat MA, Meklati BY, Smadja J, Chemat F. An improved microwave Clevenger apparatus for distillation of essential oils from orange peel. J Chromatogr A. 2006 Apr;1112(1–2):121–6.
17. Li Y, Fabiano-Tixier AS, Vian MA, Chemat F. Solvent-free microwave extraction of bioactive compounds provides a tool for green analytical chemistry. TrAC Trends in Analytical Chemistry. 2013 Jun; 47: 1–11.
18. Lucchesi ME, Chemat F, Smadja J. Solvent-free microwave extraction of essential oil from aromatic herbs: comparison with conventional hydro-distillation. J Chromatogr A. 2004 Jul;1043(2):323–7.
19. Wang Z, Ding L, Li T, Zhou X, Wang L, Zhang H, et al. Improved solvent-free microwave extraction of essential oil from dried Cuminum cyminum L. and Zanthoxylum bungeanum Maxim. J Chromatogr A. 2006 Jan;1102(1–2):11–7.

20. Passos CP, Coimbra MA. 17. Microwave extraction of bioactive compounds from industrial by-products. In: Microwave Chemistry. De Gruyter; 2017. p. 302–33.
21. Farhat A, Fabiano-Tixier AS, Maataoui M El, Maingonnat JF, Romdhane M, Chemat F. Microwave steam diffusion for extraction of essential oil from orange peel: Kinetic data, extract's global yield and mechanism. Food Chem. 2011 Mar;125(1):255–61.
22. Michel T, Destandau E, Elfakir C. Evaluation of a simple and promising method for extraction of antioxidants from sea buckthorn (Hippophaë rhamnoides L.) berries: Pressurised solvent-free microwave assisted extraction. Food Chem. 2011 Jun;126(3):1380–6.
23. Farhat A, Fabiano-Tixier AS, Visinoni F, Romdhane M, Chemat F. A surprising method for green extraction of essential oil from dry spices: Microwave dry-diffusion and gravity. J Chromatogr A. 2010 Nov;1217(47):7345–50.
24. https://philschatz.com/physics-book/contents/m42218.html. 2025.
25. Food and Drug Administration. CHIRALITY. 1992; 4: 338–40.
26. Investigation of Chiral Active Substances; European Medicines Agency, 1994.
27. Berger TA. Supercritical Fluid Chromatography: Applications. In: Reference Module in Chemistry, Molecular Sciences and Chemical Engineering. Elsevier; 2013.
28. Maia PP, Guimarães L, Nascimento CS. Elucidation of the enantiomer migration order of atenolol by theoretical calculations. Struct Chem. 2024 Oct 7;35(5):1589–94.
29. Seyfinejad B, Jouyban K, Houshyar J, Jabbaripour Sarmadian A, Jouyban A. Profiling metoprolol enantiomers in urine of hypertensive patients. DARU Journal of Pharmaceutical Sciences. 2025 Jul 3;33(2):19.
30. Hegstad S, Havnen H, Helland A, Falch BMH, Spigset O. Enantiomeric separation and quantification of citalopram in serum by ultra-high performance supercritical fluid chromatography-tandem mass spectrometry. Journal of Chromatography B. 2017 Sep;1061–1062:103–9.
31. Havnen H, Hansen M, Spigset O, Hegstad S. Enantiomeric separation and quantification of R/S-amphetamine in serum using semi-automated liquid-liquid extraction and ultra-high performance supercritical fluid chromatography-tandem mass spectrometry. Drug Test Anal. 2020 Sep 16;12(9):1344–53.
32. Chromatography Group. https://www.jasco-global.com/solutions/investigation-of-chiral-separation-conditions-of-cetirizine-by-supercritical-fluid-chromatography-methods-scouting-system/. 2025. Investigation of Chiral Separation Conditions of Cetirizine by Supercritical Fluid Chromatography Methods Scouting System.
33. Mack WN, Sunol SG, Sunol A. Separation of flurbiprofen enantiomers on a chiral stationary phase by supercritical fluid chromatography. AIChE Annual Meeting, Conference Proceedings. 1534-1540.. In: AIChE Annual Meeting. 2005. p. 1534–40.
34. Peper S, Lübbert M, Johannsen M, Brunner G. Separation of ibuprofen enantiomers by supercritical fluid simulated moving bed chromatography. Sep Sci Technol. 2002 Jul 24;37(11):2545–66.
35. Yang Y, Su B, Yan Q, Ren Q. Separation of naproxen enantiomers by supercritical/subcritical fluid chromatography. J Pharm Biomed Anal. 2005 Sep;39(3–4):815–8.
36. Santos AL, Preta G. Lipids in the cell: organisation regulates function. Cellular and Molecular Life Sciences. 2018 Jun 9;75(11):1909–27.
37. Robbins AL, Savage DB. The genetics of lipid storage and human lipodystrophies. Trends Mol Med. 2015 Jul;21(7):433–8.
38. Plachká K, Pilařová V, Kosturko Š, Škop J, Svec F, Nováková L. Ultrahigh-Performance Supercritical Fluid Chromatography–Multimodal Ionization–Tandem Mass Spectrometry as a Universal Tool for the Analysis of Small Molecules in Complex Plant Extracts. Anal Chem. 2024 Feb 1
39. Socas-Rodríguez B, Pilařová V, Sandahl M, Holm C, Turner C. Simultaneous Determination of Vitamin D and Its Hydroxylated and Esterified Metabolites by Ultrahigh-Performance Supercritical Fluid Chromatography–Tandem Mass Spectrometry. Anal Chem. 2022 Feb 22;94(7):3065–73.
40. https://www.fda.gov/drugs/drug-safety-and-availability/information-about-nitrosamine-impurities-medications.

41. Schmidtsdorff S, Schmidt AH. Simultaneous detection of nitrosamines and other sartan-related impurities in active pharmaceutical ingredients by supercritical fluid chromatography. J Pharm Biomed Anal. 2019 Sep; 174: 151–60.
42. Desfontaine V, Nováková L, Ponzetto F, Nicoli R, Saugy M, Veuthey JL, et al. Liquid chromatography and supercritical fluid chromatography as alternative techniques to gas chromatography for the rapid screening of anabolic agents in urine. J Chromatogr A. 2016 Jun; 1451: 145–55.
43. Segawa H, Iwata YT, Yamamuro T, Kuwayama K, Tsujikawa K, Kanamori T, et al. Enantioseparation of methamphetamine by supercritical fluid chromatography with cellulose-based packed column. Forensic Sci Int. 2017 Apr; 273: 39–44.
44. Nie X, Xie Y, Wang Q, Wei H, Xie C, Li Y, et al. Rapid Determination of Sudan Dyes in chilli products using ultra high-performance supercritical fluid chromatography-photodiode array detection. CyTA - Journal of Food. 2021 Jan 1;19(1):560–70.
45. Parr MK, Wuest B, Naegele E, Joseph JF, Wenzel M, Schmidt AH, et al. SFC-MS/MS as an orthogonal technique for improved screening of polar analytes in anti-doping control. Anal Bioanal Chem. 2016 Sep 23;408(24):6789–97.
46. West C, Ogden J, Lesellier E. Possibility of predicting separations in supercritical fluid chromatography with the solvation parameter model. J Chromatogr A. 2009 Jul;1216(29):5600–7.
47. Chen Z, Dong F, Xu J, Liu X, Cheng Y, Liu N, et al. Stereoselective separation and pharmacokinetic dissipation of the chiral neonicotinoid sulfoxaflor in soil by ultraperformance convergence chromatography/tandem mass spectrometry. Anal Bioanal Chem. 2014 Oct 29;406(26):6677–90.
48. Pano-Farias NS, Ceballos-Magaña SG, Gonzalez J, Jurado JM, Muñiz-Valencia R. Supercritical fluid chromatography with photodiode array detection for pesticide analysis in papaya and avocado samples. J Sep Sci. 2015 Apr 14;38(7):1240–7.
49. El-Saeid MH, Khan HA. Determination of Pyrethroid Insecticides in Crude and Canned Vegetable Samples by Supercritical Fluid Chromatography. Int J Food Prop. 2015 May 4;18(5):1119–27.
50. Zougagh M, Aranda P, Castañeda G, Ríos Á. Supercritical fluid extraction—Achiral liquid chromatography with circular dichroism detection for the determination of menthone enantiomers in natural peppermint oil samples. Talanta. 2009 Jul;79(2):284–8.
51. Zhai Y, Feng Y, Wei Y, Wang Y, Xu W. Development of a miniature mass spectrometer with continuous atmospheric pressure interface. Analyst. 2015;140(10):3406–14.
52. Castro J, Pregibon T, Chumanov K, Marcus RK. Determination of catechins and caffeine in proposed green tea standard reference materials by liquid chromatography-particle beam/electron ionization mass spectrometry (LC-PB/EIMS). Talanta. 2010 Oct;82(5):1687–95.
53. Alcántara-Durán J, Moreno-González D, García-Reyes JF, Molina-Díaz A. Use of a modified QuEChERS method for the determination of mycotoxin residues in edible nuts by nano flow liquid chromatography high resolution mass spectrometry. Food Chem. 2019 May; 279: 144–9.
54. Famiglini G, Palma P, Termopoli V, Trufelli H, Cappiello A. Single-Step LC/MS Method for the Simultaneous Determination of GC-Amenable Organochlorine and LC-Amenable Phenoxy Acidic Pesticides. Anal Chem. 2009 Sep 1;81(17):7373–8.
55. Balarama Krishna M V., Castro J, Brewer TM, Marcus RK. On-line separation and identification of inorganic and organic arsenic species in ethanolic kelp and bladderwrack extracts through liquid chromatography/particle beam-electron ionization mass spectrometry (LC/PB-EIMS). J Anal. at Spectrom. 2009; 24(2):199–208.

Green Spectroscopic Techniques

Hemlata M. Nimje, Suchita P. Dhamane, Trushal V. Chorage, and Rajesh E. Jesudasan

Abstract The book written for emphasizes the importance of making spectroscopic techniques more sustainable in line with green chemistry principles using Infrared (IR), Ultraviolet-Visible (UV-Vis) and Fluorescence techniques. It presents recent innovations and techniques such as miniaturized instruments, solvent-free/low-solvent methods, energy-efficient approaches and computational tools that minimize waste of time, solvent and energy. Covering both theoretical study and practical aspects, the book chapter explores uses in biomolecular studies, natural product characterization, environmental monitoring, pharmaceuticals drug product identification, development and nanomaterials. The book is intended for researchers, academicians and students, aiming to inspire environmentally responsible advancements in spectroscopy. By consolidating current knowledge and future prospects, it aims to inspire continued innovation in creating environmentally responsible analytical solutions while maintaining analytical accuracy and sensitivity. We extend our gratitude to the contributors whose expertise and dedication have made this volume possible, and to the readers who share our vision of advancing spectroscopy not only as a tool for discovery, but also as a practice rooted in sustainability. We hope it serves as both a foundation and an inspiration for continued innovation in developing spectroscopic methods and technologies that contribute to a greener and healthier future.

Keywords Green infrared techniques · Ultraviolet-visible spectroscopy · Fluorescence spectroscopy · Eco-friendly sample preparation · Applications

H. M. Nimje (✉) · S. P. Dhamane
JSPM's Jayawantrao Sawant College of Pharmacy and Research, Pune, Maharashtra, India
e-mail: hemanimje@gmail.com

T. V. Chorage
JSPM University's School of Pharmaceutical Sciences, Pune, Maharashtra, India

Present Address:
R. E. Jesudasan
Government Pharmacy College, Sikkim, India

V. Bhusari and M. Ghante (eds.), *Green Analytical Chemistry in India*,
https://doi.org/10.1007/978-3-032-24564-9_5

1 Infrared (IR) Spectroscopy

Green chemistry combines the use of renewable feedstocks, environmentally friendly synthesis methods, and safer solvents to promote sustainability while lowering ecological impact. The twelve principles of green chemistry (illustrated in Fig. 1) provide guidance for developing chemical processes that are both eco-friendly and efficient, aiming to cut down on waste, pollution, expenses, and hazards. Kurul [1] highlighted the application of these principles in nanoparticle production, stressing green routes, sustainable uses, and the ongoing challenges of creating nanomaterials that are both scalable and biocompatible.

1.1 Advances Infrared (IR) Spectroscopy

As per Escobedo [2], Green chemistry (GC) has established itself as a modern scientific field that focuses on preventing pollution and thereby supports sustainable development (SD). Since its emergence in the early 1990s, GC has become a key scientific discipline promoting pollution prevention and supporting SD. Aligned with UNESCO's Decade to Educate for Sustainability (DES, 2005–2014), GC has been increasingly adopted by educational institutions worldwide to integrate SD into education through innovative technologies. A major focus of GC is reducing energy consumption to lessen dependence on fossil fuels. Various alternative energy sources have been explored, including ultrasound (US), microwave (MW), electrochemical

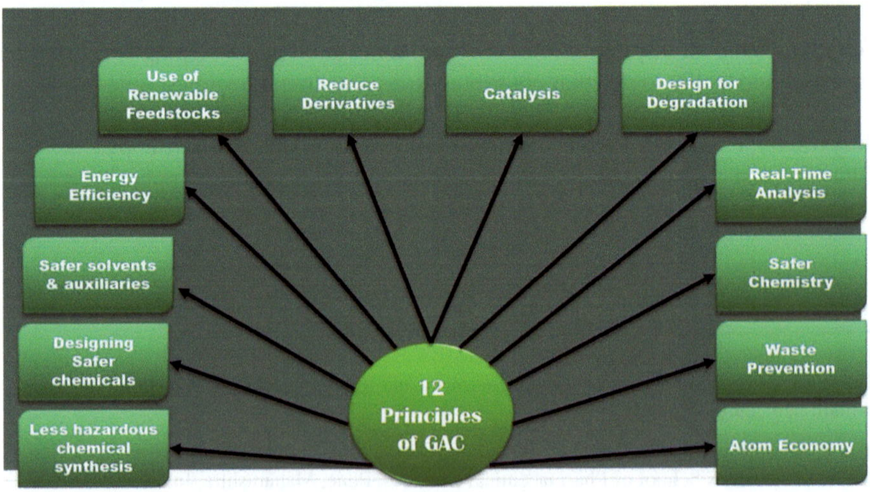

Fig. 1 12 principles of Green Analytical Chemistry (GAC) form a framework promoting sustainability through waste reduction, atom economy, energy efficiency, and safer chemicals in analytical methods

(EC) methods, photochemical (PC), mechanical milling (MM), and more recently, infrared irradiation (IR).

NIR stands out for its fast response time, efficient convection heating, and durable tungsten-halogen light sources, making it especially useful in green chemistry applications.

In response to UNESCO's Decade for Education in Sustainability initiative, this chapter introduces a green chemistry perspective. It highlights recent progress in employing infrared irradiation as an alternative energy source, both to drive organic reactions and to efficiently extract secondary metabolites. The chapter is structured into two main sections to provide a thorough overview: first, an extensive compilation of relevant information gathered from the literature is presented; second, Martínez [3] reports on recent unpublished findings from our laboratory as well as from other research groups.

1.2 Infrared (IR) Heating Principle

Infrared (IR) radiation occupies the region of the electromagnetic spectrum between visible light and microwaves, covering wavelengths from 0.76 to 1000 μm, as illustrated in Fig. 2. According to the electromagnetic spectrum, the IR range is classified into distinct regions.

- Near-IR (NIR): 0.76 μm–2 μm.
- Mid-IR (MIR): 2 μm–4 μm.
- Far-IR (FIR): 4 μm–1000 μm.

Infrared irradiation has shown great potential in promoting organic reactions and enhancing natural product extractions. IR has longer wavelengths than visible light but shorter than microwaves.

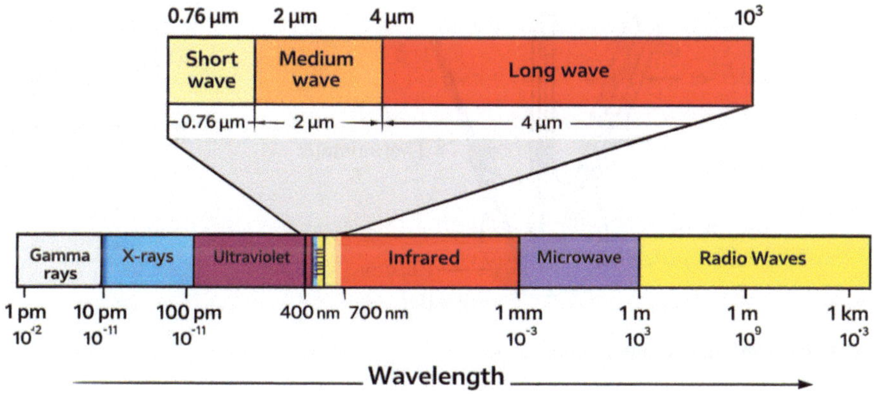

Fig. 2 Infrared irradiation zones

The peak absorption wavelength of an IR emitter is determined by its temperature and emissivity, as explained by blackbody radiation laws (Planck's, Wien's, and Stefan-Boltzmann's) reported by Jofre [4].

Different components in a sample absorb IR radiation at characteristic wavelengths as follows:

- Proteins: 3 to 4 μm.
- Carbohydrates: 3 and 7 to10 μm.
- Lipids: 3 to 4, 6, 9 to 10 μm.
- Water: Strong absorption due to O–H bonds at 2.78 to 4 μm, causing molecular vibration and potential evaporation.

1.3 Reaction Monitoring Using FTIR

Workman [5] demonstrated an experiment on reaction monitoring using FTIR. The reaction was carried out in a 50 mL three-necked flask fitted with a magnetic stirrer. An in situ FTIR monitoring device equipped with a high-temperature ATR head was placed into one of the flask's necks. A background spectrum was collected in air for 1 min. The reaction was monitored using a Remspec fiber-optic probe containing a ZnSe ATR crystal, which was linked to a mid-IR spectrometer. The entire system was operated through a desktop computer with OPUS software, allowing real-time data collection [6]. Both the FTIR monitoring equipment and the software were provided by Remspec Corporation, and the complete setup is depicted in Fig. 3.

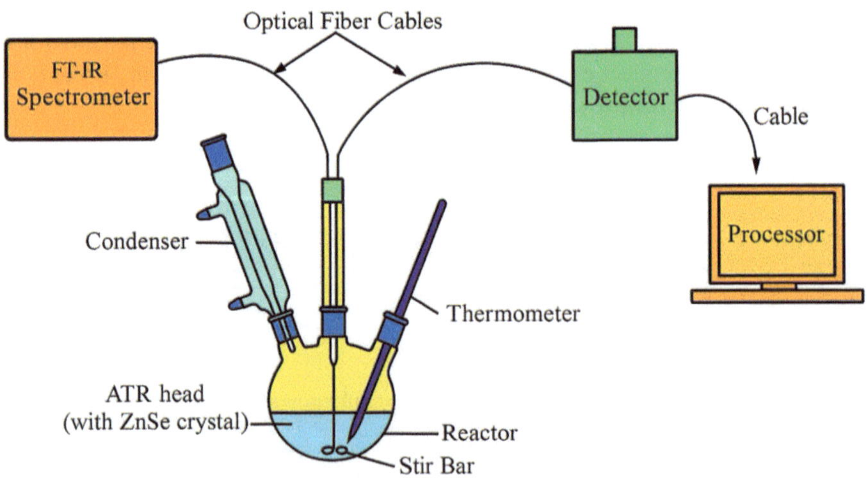

Fig. 3 Real-time monitoring reaction device

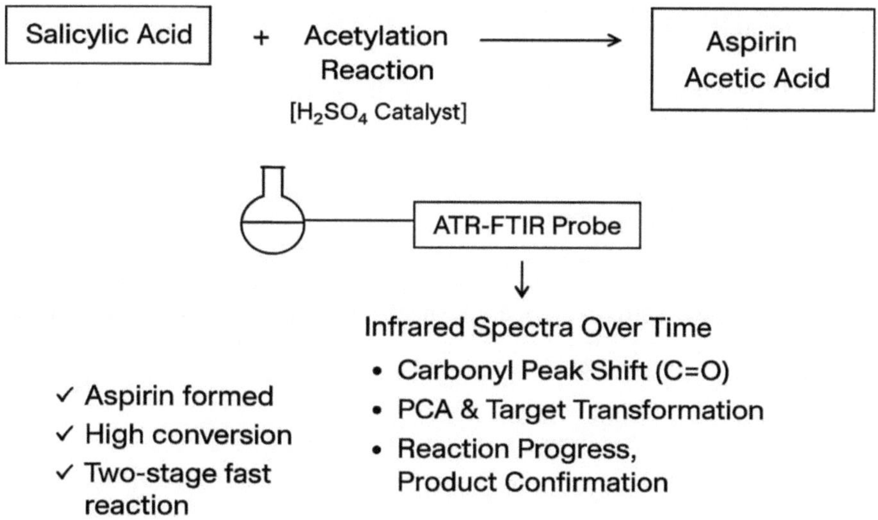

Fig. 4 Real-time monitoring of aspirin synthesis using ATR-FTIR

1.4 Study of Reaction Using Infrared Spectroscopy

1.4.1 Example

Chen [6] presented the use of online ATR-FTIR for real-time tracking of the aspirin esterification process during an acid-catalyzed reaction.

1.4.2 ATR-FTIR Monitoring of Aspirin Synthesis

Aspirin is produced through the acetylation of salicylic acid with acetic anhydride, using sulfuric acid as the catalyst. The reaction is illustrated in Fig. 4.

1.4.3 Objective of Work

To monitor the synthesis of aspirin from salicylic acid and acetic anhydride using ATR-FTIR spectroscopy in real time.

1.4.4 Catalyst Used in the Reaction

Sulfuric acid (H_2SO_4) acted as a catalyst for the acetylation reaction.

1.4.5 Technique Used

ATR-FTIR was used to observe the reaction without sample extraction. In situ, real-time monitoring was performed.

1.4.6 Spectral Observations

It was observe that changes in carbonyl peak positions, shape and intensities were noted. These shifts indicated chemical transformation during the reaction. Time trendlines and concentration changes was helps determine reaction endpoint accurately.

1.4.7 Data Analysis

Principal Component Analysis (PCA) helped simplify and visualize complex spectral data. Target Transformation confirmed the presence and progression of specific compounds.

Reaction Kinetics: The reaction followed a two-stage mechanism in which first reaction is fast, second reaction is slower. It showed a rapid conversion with high yield. Crystallization of aspirin is visible at the end.

1.4.8 Advantages

No need for sample purification or isolation. ATR-FTIR is an effective tool for process monitoring and optimization in organic synthesis. Improved aspirin yield in this process.

1.4.9 Uses of Aspirin

Aspirin is a widely used and versatile drug found in over 100 medications (e.g., Alka-Seltzer, Excedrin, Vanquish). It exhibits antipyretic, analgesic, anti-inflammatory, and anti-rheumatism properties. Extensively used in Prevention and treatment of artery infarction. It is used for Inhibiting platelet aggregation and cardiovascular disease management. Also found effective in treating neurodegenerative diseases and certain cancers.

1.4.10 Knoevenagel Condensation

Table 1 presents three aldol-type reactions carried out using MIR as the activation method, with natural bentonitic clay-Tonsil Actisil FF (TAFF) serving as the catalyst.

Table 1 Knoevenagel condensation reactions

Reactions					Ref.
1. 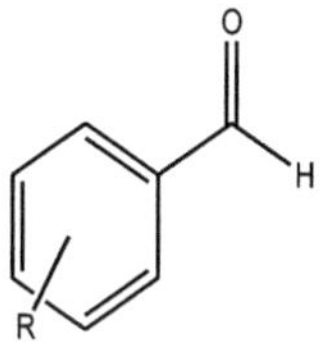	+		MIR 15 min, → TAFF solventless	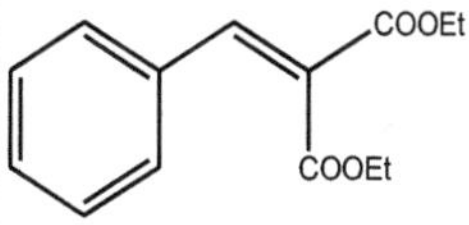	[7]
2. 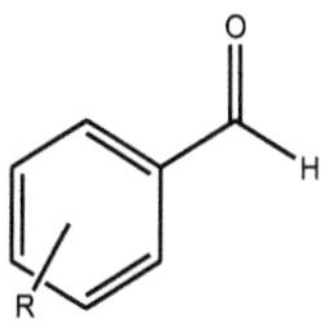	+	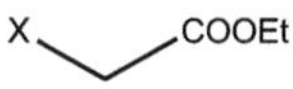	MIR 15 min, → TAFF solventless	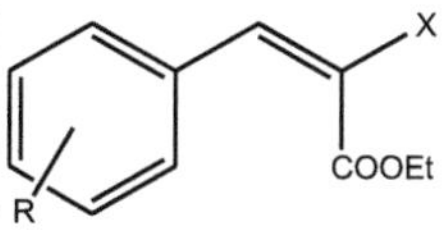	[8]
3. 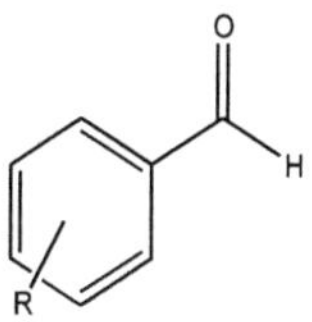	+	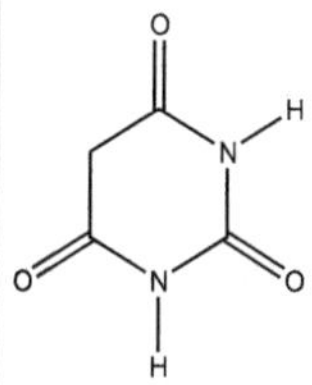	MIR 45 min, → solventless	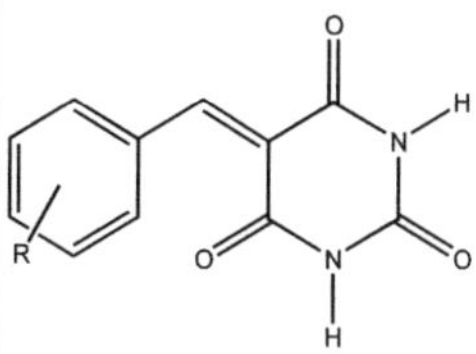	[9]

According to Martínez [3], the desired products were obtained in high yields, with short reaction times, under solvent-free conditions, and with a simple workup process.

1.5 Natural Product Extraction

Recent studies have highlighted the potential of infrared (IR) radiation for the extraction of secondary metabolites from plant materials. Cheaib et al. [7] demonstrated that far-infrared (FIR) assisted extraction from apricot pomace yielded higher levels of polyphenols, flavonoids and tannins compared with traditional solid and liquid extraction. In addition, the FIR-derived pomace extract exhibited stronger antibacterial effects against both gram-positive and gram-negative bacteria. In another study (Entry 5), Martínez et al. [8] used near-infrared (NIR) radiation to extract capsaicin and dihydrocapsaicin from habanero pepper using ethanol, achieving yields of 43.88% and 29.44%, respectively. SEM analysis revealed better preservation of cell structure in NIR-treated samples, indicating the efficiency of nonconventional IR-based methods. Table 2 illustrates the extraction of various natural compounds.

Table 2 Extraction of natural compounds

Entry	Vegetable specimen	Extracted compounds	Ref.
1	Oryza sativa L	4-hydroxybenzaldehyde, p-Coumaric acid, among others	[9, 10]
2	Arachis hypogaea L	2-Methoxyphenol, methoxybenzoic acid, among others	[11]
3	Morus alba L	Benzoic acid, (+)-catechin, p-Coumaric acid, among others	[12]
4	Habanero pepper	Capsaicin and dihydrocapsaicin	[9]
5	Apricot	Polyphenolic, tannins, flavonoids compounds	[13]

1.6 IR Strategies on Sample Preparation

Infrared (IR) radiation holds great potential in analytical chemistry due to its ability to rapidly heat solutions. It is commonly used for drying samples in clean conditions and preparing sub-boiling acids. IR has also been effectively applied for heating liquid or solid samples to determine volatile analytes, such as iodine in milk. Less conventional applications include using a three-tungsten lamp setup for direct analysis of solid samples. Campos [14] used this to burn organic materials, with volatile elements carried by gas to a flame AAS, while Magalhães [15] used it to volatilize mercury from sediment samples.

Despite its recognized heating capabilities, IR's application in this area has been limited. One notable effort was by Nobrega [16] group, who utilized tungsten lamps. Research on this topic remained sparse until 2013. To address this, the manuscript will be organized into two parts: studies before the green analytical chemistry (GAC) era and those after its implementation. Emphasis will be placed on post-2000 developments, focusing on the environmental impact of IR-based sample preparation using green indices.

Nobrega [16] presented a simple, low-cost sample preparation method using infrared (IR) radiation from tungsten lamps for rapid heating of acidic sample suspensions. The heating occurs in under 60 seconds, with total preparation completed in less than 5 min, making it suitable for trace element analysis. The method was utilized to measure Cu, Mn, Fe, and Zn in standard reference materials, such as bovine liver and apple leaves, producing results consistent at a 95% confidence level. Furthermore, copper and manganese were analyzed in both untreated and freeze-dried bovine liver samples, with validation carried out using a conventional microwave-assisted approach.

In this study, three 150 W tungsten lamps arranged in a triangular setup were used to heat a glass flask containing the sample. The procedure was assessed for determining Cu, Mn, Fe, and Zn in botanical as well as biological samples.

1.7 Evaluation of the IR Procedure

1.7.1 Ashing IR Sample Preparation (Post-Green Analytical Principles)

After the introduction of green analytical principles, infrared (IR)-based ashing methods saw major innovations.

1.7.2 Early Development

Strenger [17] introduced the IR-HP-asher, a high-pressure IR incineration device using quartz digestion vessels in a steel autoclave. It achieved complete oxidation of organic matrices and full metal recovery, outperforming traditional techniques (HPA, microwave digestion) in completeness and analyte retention. However, its use declined for over a decade.

1.7.3 Zhang Et al.'s Innovations

From 2017–2018, Zhang et al. [18–20] developed advanced IR-based systems.

Fila (2017)

It is a novel dry incineration method. A compact, multi-channel IR system using gold-coated quartz tubes for efficient heating and oxygen-assisted incineration. Enabled 12-sample parallel ashing within 30 min.

AFIAS and AFIA (2018)

Sample preparation system include Automated Focused IR Ashing sample (AFIAS) system and Fully Automated Focused IR Microashing (AFIA) system. Fully automated versions of FILA integrating cold traps for analyte preservation and computer-controlled operation for sample handling, heating, ash dissolution, and homogenization. These systems boosted multi-element analysis efficiency and minimized analyte loss.

1.7.4 Green Metrics Assessment

Zhang's (2018) methods [18–20] scored higher on green indices such as AGREE, AGREEPrep, and Analytical Eco-scale, particularly for:

- Integration and automation (criterion 7).
- Reduced preparation stages and high sample throughput (criteria 4 and 8).
- Efficient single/multi-analyte analysis (e.g., via ETAAS, ICP).
- In contrast, Strenger and Hirner's method scored lower due to:
- High reagent consumption (criterion 2),
- Longer preparation times (criterion 6).

All methods perform well on the Analytical Eco-scale (Table 3), but fall into category B on the Green Certificate, largely due to deductions for reagent use and waste generation. In conclusion, Zhang's automated IR systems represent a significant advancement in eco-friendly dry sample treatment, offering speed, efficiency, and environmental benefits.

The AGREEPrep indices for various sample preparation methods range from 0.35 to 0.49, indicating varying degrees of environmental friendliness. Zhang et al. [19, 21] stand out for their high eco-friendliness, particularly due to automation and integration, use of hermetic seals, and reduced number of preparation steps, which enhance efficiency in multi-element analysis. Despite focusing on single-analyte determination, Zhang et al. [19] improves throughput and uses ETAAS, supporting its green credentials.

In contrast, Strenger and Hirner's method scores lower due to high reagent consumption and longer preparation time.

AGREE indices across methods vary from 0.47 to 0.53. Among IR-based ashing techniques, Zhang et al. [20] again shows strong performance by enabling multi-analyte determination from a single sample and simplifying analysis steps.

However, Zhang et al. [18, 19] show reduced performance due to more complex preparation processes. Strenger's approach benefits from minimized sample mass and reduced steps.

Table 3 Evaluation of ashing and wet IR sample preparation using summarized green indexes

Analytical eco-scale	Green certificate	AGREE	AGREEPrep	Ref. works
Acceptable	D	0.42	0.31	[22]
Excellent	B	0.55	0.46	[23]
Excellent	B	0.49	0.42	[24]
Excellent	B	0.47	0.41	[19]
Excellent	B	0.52	0.49	[18]
Excellent	B	0.53	0.43	[20]
Acceptable	B	0.48	0.36	[21]

The Infrared (IR) sample preparation procedure offers several advantages-affordability, simplicity, and rapidity. However, limitations include partial decomposition of samples and a balance between heating efficiency and environmental impact. A comprehensive green assessment was conducted using Eco-Scale, Green Certificate, AGREE, and AGREEPrep metrics as shown in Table 3.

AGREE scores ranged from 0.42 to 0.55, with highest performance in fewer steps, no derivatization, small sample size and multi-analyte capability.

AGREEPrep scores ranged from 0.31 to 0.46, benefiting from small sample size, low energy consumption, short preparation time.

Top performers: Helmeczi [22] showed consistently high AGREE scores, while Helmeczi [21] received a zero score in criterion 11 due to use of toxic reagents.

Wang [21] showed varied AGREEPrep results depending on reagent combinations (I–IV), with III and IV excelling in reagent safety.

Energy efficiency was uniformly strong across IR applications.

1.8 Wet IR Sample Preparation

In the study by da Silva [14] infrared (IR) radiation was evaluated as a heat source for preparing shrimp samples to determine trace elements (As, Cd, Co, Cr, Cu, Fe, Mn, Ni, Pb, Se, Sn, V) using inductively coupled plasma atomic emission spectrometry (ICP OES). Using optimized conditions (1000 mg sample, 8 mL HNO_3, 500 W IR lamp for 8 min), the IR-based open sample preparation system demonstrated decomposition efficiency comparable to traditional digestion block (DB) methods. However, IR digestion was significantly faster (8 min vs. 3 h). While DB systems accommodate more samples simultaneously (15 vs. 4), the speed and simplicity of the IR method offset this limitation. Overall, the IR-assisted digestion (IRAD) method offered notable advantages in terms of cost, sensitivity, performance, and operational ease.

In 2020, Jofre et al. [25] developed a simplified soil sample preparation method aligned with Green Analytical Chemistry (GAC) principles. The study emphasized minimizing reagent use and employing a newly designed infrared-assisted digestion (IRAD) prototype using two commercial 250 W IR lamps. This cost-effective setup allowed simultaneous digestion of 18 samples in the reported paper of Cora Jofre [25] Unlike earlier studies that used tungsten lamps and vessels made of quartz or glass, this work introduced polytetrafluoroethylene (PTFE) vials for the first time.

The procedure involved digesting 150 mg of soil samples with 2 mL of HNO_3 under IR radiation until dry, followed by treatment with 2 mL of H_2O_2 under gentle heating. The residue was then redissolved in 15 μL of HNO_3 and diluted to 10 mL with ultrapure water. Elemental analysis (Ca, Cd, Cu, Fe, K, Mg, Mn, P, Pb, Zn) was performed using microwave induced plasma atomic emission spectrometry (MIP OES). This IRAD method ensured efficient digestion, high recovery, and analytical precision. The same method was also successfully applied to swine feed analysis for ten elements, demonstrating its versatility and effectiveness.

Wet IR sample preparation outlining the evolution, innovations, environmental considerations, and comparative performance of infrared-assisted digestion (IRAD) techniques for sample preparation.

1.8.1 Background and Evolution of IR in Wet Digestion

- Traditional wet digestion relied heavily on hot blocks and microwave (MW) ovens.
- In recent years, infrared (IR) radiation, initially limited to use in nebulizers and ashing, has been explored as an alternative kinetic source for accelerated wet digestion.
- IR offers faster heating, simpler handling, and potentially higher energy transfer (2–4 orders of magnitude higher than MW).

1.8.2 Key Developments and Studies

Gouveia et al. [16]—First IR Wet Digestion Study

- Used three-tungsten lamps (15–150 W) for acidic sample heating.
- Sample digestion in <5 min for Cu, Fe, Mn, Zn in botanical and biological samples.
- Method evaluated by GFAAS.

Dantas et al. [26]—IR-MW System

- Combined IR pre-heating and MW digestion.
- Applied to human food samples; improved control and digestion efficiency.
- Lower AGREE/AGREEPrep scores due to more reagents, filtration step, and higher sample use.

Wang et al. [27]—ColdBlock™ System

- Utilized short-wavelength, focused IR radiation in a cold environment.
- Enabled rapid (15–20 min) and efficient digestion of environmental and geological samples.
- Modified designs enhanced energy concentration and analyte recovery, even for refractory materials (Au, REEs, U, Th).
- Later versions (6- and 12-channel systems) improved throughput, temperature control, and energy use.

Helmeczi et al. (2018–2021) [23]

- Improved ColdBlock™ with optimized IR lamp positioning and cooling blocks.
- Achieved high digestion performance for REEs and Au with excellent green metrics in [22], but lower ratings in [28] due to large reagent volumes.

Other Notable IR Digestion Innovations

- Oliveira et al. [28]—IRAD for Flaxseed.
- Used 500 W tungsten lamp + aluminum reflector.
- Digest 2000 mg in 8 min with nitric acid.
- Results comparable to MW digestion.

Souza et al. [29]—Milk Sample Digestion

- Explored IR, IR-MW, and IR with V_2O_5 catalyst.
- Enabled higher sample masses and efficient digestion.

Campos et al. [30]—Soybean Matrices

- IR-assisted digestion (IRAD) with reflux enhancement.
- Rapid (5 min) and precise multi-element analysis.
- da Silva et al. (2019) [30]—Shrimp Samples,
- IR vs. digestion block: IR was faster (8 min vs. 3 h), with comparable efficiency.

Environmental Assessment (Green Metrics)

- AGREE and AGREEPrep: Consider 12 and 11 criteria, respectively (e.g., automation, reagent toxicity, energy use).
- Highest Scores: Helmeczi et al. [22] and Cora Jofre et al. [31].
- Lowest Scores: Helmeczi et al. [22] (due to high reagent use).

Eco-Scale and Green Certificate

- Most ColdBlock™ methods scored excellent, except (acceptable).
- Green Certificate categories ranged from A to D, depending on reagent type/volume.
- IR techniques excel in energy efficiency (criterion 8) and reduced preparation time (criterion 6).
- Use of H_3PO_4 instead of hazardous acids (e.g., HF, $HClO_4$) favored green metrics.

General Conclusions

- IR-assisted digestion (especially ColdBlock™) is a fast, economical, flexible, and environmentally safe alternative to MW and hot block digestion.
- Offers rapid throughput, high analyte recovery, and compatibility with green chemistry principles.
- Commercial and custom-built IR systems continue to evolve, supporting both routine and advanced applications in environmental, biological, food, and geological sample preparation.

1.9 Uses of IR Technology in Analytical Chemistry

Infrared (IR) radiation was first utilized in analytical chemistry in the early 1970s by NIST researchers. In 1972, Jofre [4] introduced sub-boiling distillation using non-contact IR lamps to purify ultrapure acids, such as HNO_3, HCl, HF, $HClO_4$, and H2SO$_4$, with significantly reduced cationic impurities. Unlike traditional distillation, this method gently evaporates liquid from the surface, minimizing contamination by preventing aerosol formation. This innovation revolutionized reagent purification, and various improvements have since been made to optimize the process and adapt it to different applications. Another key use of IR radiation is in drying processes, both industrial and laboratory-based.

IR drying offers several advantages:

- Rapid and efficient heat transfer.
- Reduced drying duration and energy consumption.
- Temperature monitoring for safe heat control.
- Versatility across different materials.

Various IR drying devices (lamps, ovens, panels) are available, emitting different wave IR depending on the material's needs.

1.10 Applications of FTIR in Plant Extracts and Standard Phytochemicals

1.10.1 Screening of Phytochemicals Using FTIR

FTIR spectroscopy, a non-destructive analytical technique, was applied to characterize various groups (functional groups) in plant herbal extracts and reference phytochemicals [32] (including tannic acid, quercetin, rutin, and gallic acid) in the range 4000–400 cm^{-1}.

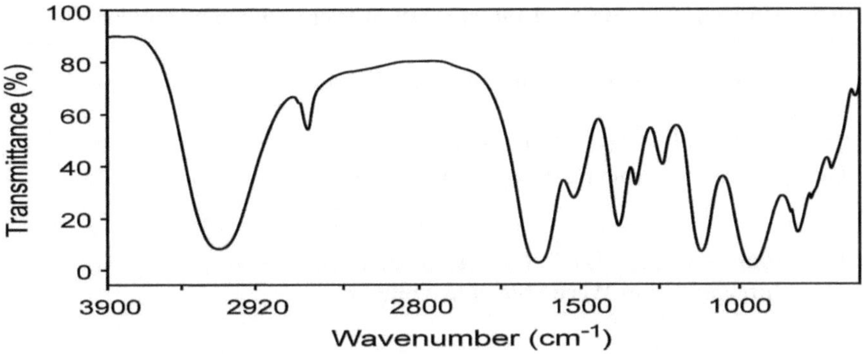

Fig. 5 FTIR spectra of tannic acid

1.10.2 General Observations

- 4000–1500 cm^{-1} → Functional group region.
- 1500–500 cm^{-1} → Fingerprint region.
- 2900–2800 cm^{-1} → C–H stretches.
- 3700–2500 cm^{-1} → O–H and C–H stretches.

1.10.3 Tannic Acid (Fig. 5) Interpretation Were Shown as Follows

- 3495–3280 cm^{-1} → O–H stretch.
- 2929, 2845 cm^{-1} → CH_2 asymmetric/symmetric stretch.
- 1703 cm^{-1} → C=O stretch.
- 1500–1450 cm^{-1} → O–C–O (carboxylic acid).
- 1314–1187 cm^{-1} → O–H bending.
- 1021–1090 cm^{-1} → C–O stretch.
- 750–878 cm^{-1} → meta-substitution in aromatic protons.

1.10.4 Quercetin Interpretation Were Shown as Follows

- 3398–3314 cm^{-1} → O–H stretch (intra-molecular H-bonding).
- 1663, 1606 cm^{-1} → C=O stretch.
- 1520–1500 cm^{-1} → NO_2 bending.
- 1449–1400 cm^{-1} → C–O stretch.
- 1260–1200 cm^{-1} → C–O–C (ester).
- 1164–1100 cm^{-1} → C–O–C stretch.
- 819–800 cm^{-1} → =C–H bending.

1.10.5 Rutin and Gallic Acid Interpretation Were Shown as Follows

- 3550–3400 cm^{-1} → O–H (carboxylic acid, H-bonding).
- 1700–1600 cm^{-1} → C=O stretch.
- 1500–1400 cm^{-1} → C–O, O–H stretch.
- 800–600 cm^{-1} → aromatic proton substitution.

1.10.6 Plant Extracts (Bark, Leaves, Sepals, Fruits, Seeds): Interpretation Were Shown as Follows

- ~3500 cm^{-1} → Alcoholic, carboxylic acids, nitrates, carbohydrates.
- 1748 cm^{-1} → C=O stretch (ester carbonyl).
- 2923 & 2855 cm^{-1} → CH_2 asymmetric/symmetric stretch.
- 1300–1100 cm^{-1} → O–H bending.

FTIR spectroscopy identified the functional groups linked to the biological activities (antidiabetic, antioxidant, antibacterial) of the phytochemicals, with ethanol extraction showing the highest efficiency.

1.11 Utilization of FTIR for the Characterization of Green-Synthesized Nanoparticles

The application of FTIR spectroscopy in the characterization of green synthesized nanoparticles is essential for understanding the role of plant- or microorganism-based agents in reduction, capping, and stabilization processes. The fundamental properties of nanoparticles—such as their size, shape, structure, and stability—are greatly influenced by the synthesis conditions, including factors like the kind and concentration of plant extract used, as well as the pH and temperature of the reaction environment. During green synthesis, the biomolecules in natural extracts act as reducing, capping, and stabilizing agents [33].

1.11.1 Role of Functional Groups in Green Synthesis [34]

- **Reducing Agents**: Amino (-NH_2), Hydroxyl (-OH), carbonyl (-C=O), and methoxide (-OCH_3) groups reduce metal ions to elemental forms.
- **Natural Biomolecules** such as carbohydrates, terpenoids, polyphenols, glycosides, amino acids, flavonoids, acids, alkaloids, enzymes, and proteins serve as effective reducers and stabilizers.
- **Capping Agents** (organic ligands, proteins, surfactants, amino acids, polysaccharides, polymers, dendrimers, bio-extracts, lipids) prevent agglomeration, stabilize colloids, and enable biofunctionalization.

- **Stabilizing Agents** ensure long-term NP suspension stability and prevent aggregation, sometimes overlapping with capping agents.

1.11.2 FTIR Spectroscopy in NP Characterization

FTIR is a key tool to:

- Determine the functional groups responsible for reduction, capping, and stabilization processes.
- Confirm chemical identity of NPs synthesized by green methods.
- Track biomolecules responsible for preventing aggregation.
- Refine the synthesis parameters by analyzing FTIR spectra recorded at various time intervals.

1.11.3 Key Plant Metabolites in NP Formation

- Polyphenols: Strong reducers due to phenolic –OH and –COOH groups; provide antioxidant and metal-chelating action, forming protective coatings.
- Terpenoids: Stabilize NPs via redox reactions.
- Alkaloids: Nitrogen-containing compounds aiding in metal ion reduction.
- Proteins and amino acids: Reduce metals via electron transfer and stabilize through amino/carboxyl groups.
- Carbohydrates: Provide steric stabilization with large molecular structures.

1.11.4 Consistency across Metal Types

FTIR spectra consistently show Chemical groups Amino ($-NH_2$), Hydroxyl (-OH), and carbonyl (-C=O) associated with reduction and stabilization, regardless of the specific metal precursor.

FTIR spectroscopy offers valuable insights into the green synthesis of nanoparticles by detecting the biomolecules and chemical moieties involved in their reduction and stabilization, verifying nanoparticle formation, and aiding in the optimization of synthesis parameters. Its ability to consistently detect active groups across different metals makes it a vital technique in sustainable nanotechnology research.

1.12 Conclusion of IR Spectroscopy

Infrared (IR) radiation has been effectively utilized in several sample preparation applications such as sub-boiling acid distillation, solid sample drying, combustion, volatilization, and desolvation. Its capability for localized and rapid heating makes it advantageous for accelerating and enhancing analytical procedures in fields like environmental analysis and food testing. Despite its proven benefits, IR-based methods are still underutilized in laboratories. Continued research and optimization could improve their adoption and efficiency.

While traditional green assessment metrics focus on analytical performance, the adoption of broader, innovative green metrics is gradually reshaping analytical chemistry. These evolving tools emphasize environmental sustainability, signaling a transformative shift in how analytical methods are developed and evaluated.

2 UV-Visible Spectrometry

2.1 Introduction UV-Visible Spectrometry

The green synthesis of nanomaterials represents an environmentally sustainable method that employs biological sources such as plant extracts, microorganisms, or enzymes for nanoparticle production, as demonstrated by Wanakai [35]. This technique reduces the generation of toxic byproducts, lowers energy usage, and avoids harmful reagents, adhering to the core principles of green chemistry. Among different characterization methods, UV–Visible spectroscopy is essential for tracking nanoparticle formation and analyzing their optical characteristics.

2.2 Role of UV-Visible Spectrometry

This is used for following determinations:

- Verify the formation of nanoparticles by observing surface plasmon resonance peaks.
- Monitor synthesis in real time, providing information about nucleation and growth stages.
- Estimate particle size and distribution, based on peak shape and position.
- Evaluate stability of nanoparticles over time.

2.2.1 Green Synthesis approach

This Method Involves

- Reducing Agents: Natural compounds like flavonoids, terpenoids, alkaloids, and some proteins derived from biological sources.
- Stabilizing Agents: Plant metabolites or extracellular enzymes that prevent aggregation of nanoparticles.
- Solvents: Water or ethanol, which are environmentally benign.

Table 4 Interpretation of nanoparticles using UV and visible spectrometry

S. No.	Nanoparticles type	Peak (Approx.)	Interpretation
1.	Silver (AgNPs)	400–450 nm	Peak indicates spherical nanoparticles
2.	Gold (AuNPs)	500–550 nm	Redshift/blue shift implies size or shape changes
3.	ZnO NPs	350–380 nm	Broad absorption; size-dependent bandgap
4.	Iron oxide NPs	< 400 nm	Broad, less defined peaks

This synthesis is commonly used for metallic nanoparticles like silver, gold, zinc oxide, and iron oxide (Fe_3O_4).

2.2.2 Key Observations in UV-vis Analysis

Interpretation of nanoparticles using UV and Visible Spectrometry was shown in Table 4.

The shift in wavelength or change in peak intensity indicates variation in nanoparticle size, shape, and concentration.

2.2.3 Advantages of Green Synthesis with UV-vis Monitoring

- Non-toxic and sustainable process.
- Simple, cost-effective setup.
- Quick confirmation of synthesis.
- Real-time tracking of reaction progress.
- No need for extensive sample preparation.

2.2.4 Limitations of Green Synthesis with UV-vis Monitoring

- UV-vis spectroscopy provides limited structural information.
- It cannot differentiate between different crystal phases.
- It requires complementary techniques like FTIR, TEM, or XRD for comprehensive characterization.

Green synthesis of nanomaterials offers a sustainable route for nanoparticle production, and UV-Visible spectrometry serves as a rapid, efficient tool for their initial characterization. This approach is consistent with green chemistry principles, promoting environmentally friendly progress in the field of nanotechnology.

2.3 UV-vis Spectroscopy in Nanoparticle Synthesis

UV-Visible spectroscopy is an effective technique for monitoring the bio-reduction of metal NPs in aqueous solutions. It enables indirect analysis of nanoparticle formation by detecting changes in optical properties, such as surface plasmon resonance (SPR) shown by Hosseingholian [36].

For instance, during the environmental friendly synthesis of silver nanoparticles (Ag-NPs), the reduction of Ag + ions to Ag^0 is often accompanied by a visible color change, which can be immediately detected by UV-Vis. The technique also helps assess the stability, particle size and its dimension including shape of the nanoparticles over time.

Additionally, UV-Vis spectra can reveal signs of **agglomeration** or **growth in particle size**, making it a valuable, real-time characterization method in green nanomaterial synthesis.

2.4 Assessment of Analytical Parameters for the Environment Friendly UV–Visible Method

Yisak et al. [3] proposed UV-Vis green technique for determining total sugars level was evaluated for key analytical parameters. The calibration curve for sucrose showed a correlation coefficient between conc. and absorbance (R2 = 0.9996) across a range of 7.91–20.11 mg/L. The method demonstrated good sensitivity, with a LOD of 4.4 mg/L and LOQ of 14.6 mg/L. Specificity was confirmed by overlaying spectra of multiple sugars (maltose, fructose, glucose, sucrose), ensuring selective detection in cereal grain matrices. Precision, assessed through repeat scans, showed an RSD of 6.9%, indicating good repeatability. Accuracy was validated via recovery tests in spiked real samples, yielding recoveries between 85 ± 3 and 105 ± 4%, confirming the method's reliability and reproducibility.

2.5 Application of Green UV-vis Method (Total Sugar Estimation in Grains)

The newly developed Environmental friendly UV-Vis technique was applied to quantify total sugars in the aqueous extracts of white rice, brown tef, red wheat, and white tef. Scanning was conducted in the 182–210 nm range with optimized instrument settings, and sugar content was calculated in % w/w (sucrose equivalent) on a dry basis, accounting for dilution factors.

Tef grains had sugar contents similar to red wheat but larger than white rice. The results aligned with literature values for wheat (7.8% w/w) and were higher than those for maize, sorghum, millet, and rye obtained using phenol-sulfuric acid

Table 5 Amount of total sugars in white and brown tef determined by the newly developed UV-VIS method

Regions	Variety of foods cereals	Amount of sugars (%w/w)
Amhara	Brown tef	8.97 ± 0.19
	White tef	8.30 ± 0.48
Oromia	White tef	9.08 ± 0.30
	Red wheat	9.22 ± 0.20
	Brown tef	8.47 ± 0.50
	White rice	3.19 ± 0.24
SNNPR	Brown tef	7.47 ± 0.24
	White tef	5.48 ± 0.21

UV-Vis methods. The lower values in other methods may be due to monosaccharide degradation caused by hot concentrated acid, which the green method avoids. This demonstrates that the proposed green method is more accurate, non-destructive, and suitable for determining total sugars in cereal food grains.

The measured total sugar contents were:

- Brown tef sugar content (6.17 to 10.32%).
- White rice sugar content (3.19 ± 0.24%).
- White tef sugar content (5.48 to 9.44%).
- Red wheat sugar content (9.22 ± 0.20%).

According to Yisak et al. [3], the total sugar content (mean ± SD) in brown and white tef (Eragrostis tef (Zuccagni) Trotter), red wheat, and white rice grains, as measured by the newly developed UV–Vis method, is presented in Table 5.

Yisak (2022) [37] developed and validated a straightforward, rapid, cost-efficient, and environmentally friendly green UV–Vis technique for the direct quantification of total sugars in aqueous extracts of white rice, brown tef, white tef, and red wheat.

The method avoids hazardous chemicals like phenol and sulfuric acid, using only water for extraction, followed by centrifugation and filtration. It showed reliable and comparable results with literature data and is suitable for routine sugar analysis in food samples, offering environmental and health safety advantages.

2.6 *Application UV–vis Spectrophotometry*

2.6.1 Screening of Phytochemicals Using UV–vis Spectrophotometry

This was applied to detect chromophoric groups, Axochromic groups and aromatic rings in plant extracts by analyzing $\pi \rightarrow \pi^*$, $\sigma \rightarrow \sigma^*$, and $n \rightarrow \pi^*$ electronic transitions. Standard compounds showed characteristic absorption bands published by Patle [32] as follows:

- Gallic acid: 265 nm.

- Quercetin: 250, 370 nm.
- Rutin: 250, 355 nm.
- Tannic acid: 275 nm.

Ethanol, n-hexane and chloroform extracts from various plant parts exhibited bands corresponding to the standards, indicating the presence of phenolic acids, alkaloids, tannins, flavonoids and carotenoids. Figure 6 shows the UV–Vis spectra of bark, fruits, sepals, leaves, and seeds of *D. pentagyna* (ethanol extract A), (chloroform extract B), and n-hexane (extract C), while Fig. 6D presents the UV–Vis spectra of standard compounds such as quercetin, gallic acid, tannic acid and rutin.

- Ethanol extracts (sepals, fruits, seeds): 270–280 nm and 350–500 nm → tannins and flavonoids.
- Bark: 400–450 nm → carotenoids.
- Leaf extracts: Four bands (234–676 nm) → diverse phytochemicals.
- Sepal and fruit extracts (chloroform/n-hexane): 234–676 nm → phenolic compounds, flavonoids, alkaloids.

Two distinct absorption regions for phenolics and flavonoids arise from aromatic rings (280 nm) and other ring systems (300–600 nm), with broader bands partly due to ligand-to-metal charge transfer (LMCT). Overall, UV–Vis method proved effective for identifying chromophoric groups in plant extracts.

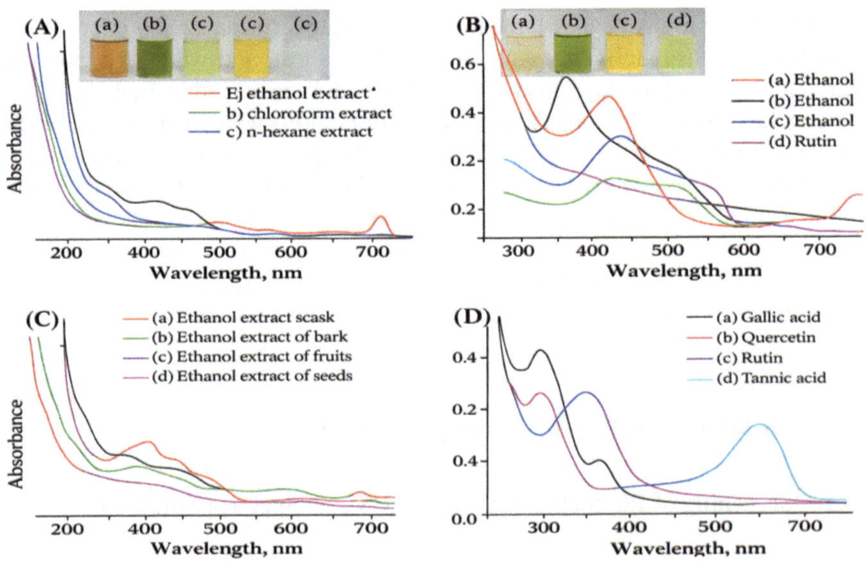

Fig. 6 UV–Vis spectra of different extracts of bark, leaves, sepals, fruits, and seeds of D. pentagyna in ethanol extract solution **A** chloroform extract solution **B** n-hexane extract solution **C** and UV–Vis spectra of standard compound such quercetin, gallic acid, tannic acid, and rutin **D**

3 Fluorescence Spectroscopy

3.1 Introduction of Fluorescence Spectroscopy

Fluorescence and phosphorescence are processes in which photons are emitted as molecules return from electronically excited states to lower energy levels (ground state). These processes involve transitions between the electronic and vibrational states of fluorescent molecules, or fluorophores, typically aromatic compounds like tyrosine, tryptophan, and fluorescein as shown by Naresh [38].

3.1.1 Excitation

Molecules absorb light in the (400–700 nm) visible range, (200–400 nm) UV range, or (700–1100 nm) NIR range, promoting electrons from the ground state (S0) to higher singlet states (S_1, S_2). This excitation happens extremely rapidly (~10–15 s) and is quickly followed by vibrational relaxation to the lowest vibrational level of S1, occurring within femtoseconds to picoseconds.

3.1.2 Fluorescence

- Prompt fluorescence: $S_1 \rightarrow S_0 + h\nu$ (immediate emission from singlet state).
- Delayed fluorescence: $S_1 \rightarrow T_1 \rightarrow S_1 \rightarrow S_0 + h\nu$ (involves intersystem crossing to triplet and back to singlet before emission).
- Phosphorescence: $T_1 \rightarrow S_0 + h\nu$ (delayed emission from triplet state due to spin-forbidden transition).

3.1.3 Types of Fluorescence/Emission Processes

1. Stokes Fluorescence: The photons emitted possess less energy (longer wavelength) than the absorbed photons because of energy loss prior to emission, a phenomenon known as the Stokes shift.
2. Anti-stokes Fluorescence: Emission at shorter wavelengths (higher energy) than absorption, often due to thermal energy or high vibrational level populations.
3. Resonance fluorescence: Emitted photons have the same energy as absorbed photons; observed in gases and crystals, not in solution.

Jablonski diagram illustrates the excited-state structure and these transitions shown by Ladokhin [39] (Fig. 7).

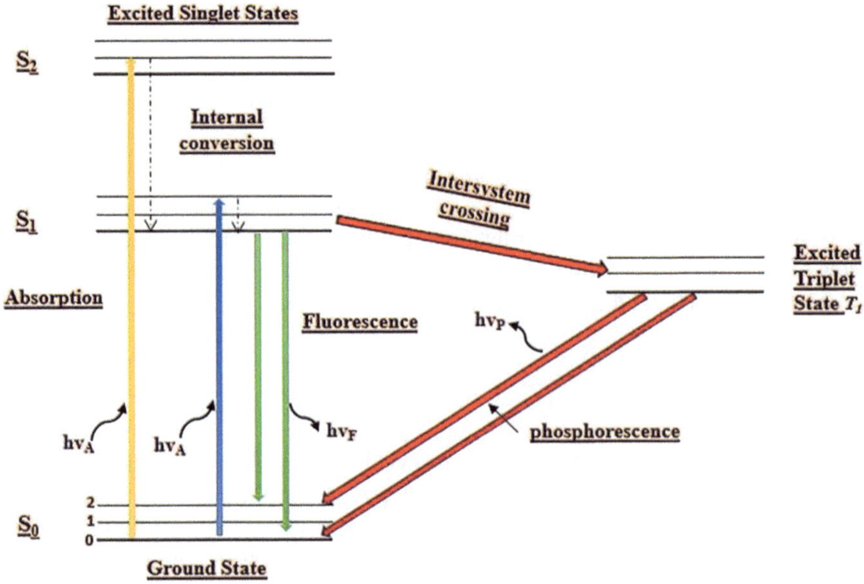

Fig. 7 Jablonski diagram

3.2 Fluorescence Spectroscopy in Protein

3.2.1 Example (Tryptophan)

Weber [40] suggested that the long-wavelength absorption band is made up of two electronic transitions, ^{1}La and ^{1}Lb, whose dipole moments are oriented at a large angle relative to each other. In tryptophan, light absorption occurs via $\pi \rightarrow \pi^*$ transitions in the indole ring, featuring a short-wavelength band near 220 nm (1Bb transition) and a long-wavelength band between 260 and 290 nm that includes overlying ^{1}La and ^{1}Lb transitions. The ^{1}La transition is polarity-sensitive and the primary contributor to fluorescence emission. In aqueous solution, tryptophan shows a broad, structureless fluorescence band peaking near 350 nm (width ~ 60 nm), though small variations in peak position (348–353 nm) occur due to differences in instrument calibration. Tryptophan absorbs light via $\pi \rightarrow \pi^*$ transitions within its indole ring, displaying a short-wavelength band near 220 nm (1Bb transition) and a long-wavelength band in the range of 260–290 nm that results from overlaying ^{1}La and ^{1}Lb transitions. The ^{1}La transition is polarity-sensitive and the primary contributor to fluorescence emission. In aqueous solution, tryptophan shows a broad, structureless fluorescence band peaking near 350 nm (width ~ 60 nm), though small variations in peak position (348–353 nm) occur due to differences in instrument calibration. Excitation increases the dipole moment (~4 D), making emission highly responsive to solvent polarity and mobility, which is useful for probing protein structure and dynamics. The absolute quantum yield in water is about 0.13, but relative yield is often used.

Fluorescence can be quenched by water (high activation energy, 11–12 kcal/mol), polar amino acids, disulfide bonds, peptide groups, and through long-range energy transfer in hemeproteins. Additional quenchers include oxygen, halides, acrylamide, certain metal ions, and various chemicals, with quenching studies providing insight into residue accessibility.

Fluorescence decay is heterogeneous, with lifetimes of ~0.5 and ~ 3.1 ns in aqueous environments, influenced by side-chain rotamers and solvent relaxation.

Excitation anisotropy varies with wavelength (~0.3 at 300 nm, ~0.2 at 270 nm), and the 280–300 nm region is challenging for polarization studies due to solvent-dependent shifts.

Use of tryptophan analogs in protein–protein, protein–nucleic acid interaction, and protein folding research as follows:

- 5-Hydroxytryptophan: Absorbance up to 320 nm, selective excitation possible, high anisotropy—good for polarization studies.
- 4-Fluorotryptophan: Nonfluorescent "silent" analog, useful in non-fluorescence assays.
- 7-Azatryptophan: Large quantum yield change upon water exposure—useful in protein folding studies.

Structure of tryptophan analogs were shown in Fig. 8.

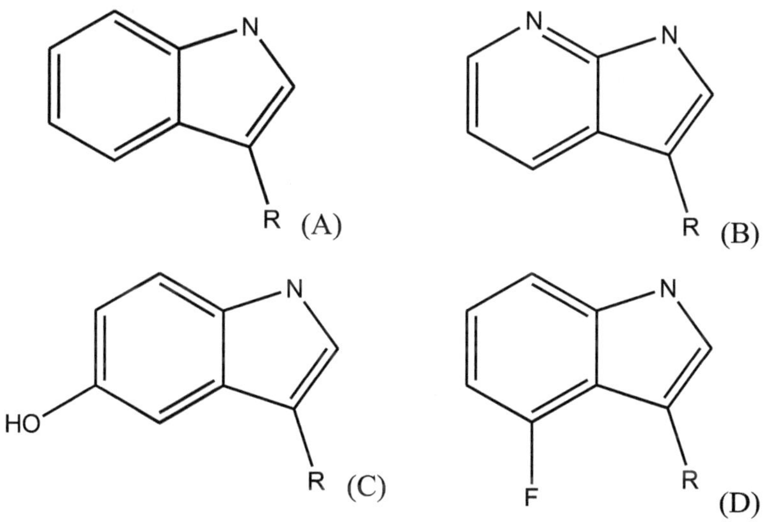

Fig. 8 Structure of **A** Tryptophan ring, **B** 7-azatryptophan, **C** 5-hydroxytryptophan and **D** 4-fluorotryptophan

3.3 Applications of Fluorescence Spectroscopy

Fluorescence spectroscopy provides a sensitive, non-destructive way to monitor protein structure, folding, environment, and interactions in real time[6]. Fluorescence spectroscopy is an easy, fast, and economical method with high sensitivity and specificity, offering linear response over a wide concentration range. More sensitive than absorbance methods, it is widely used for quantitative and qualitative analysis in environmental, industrial, medical, forensic, genetic, and biotechnological applications.

Some applications of fluorescence spectroscopy as follows,

3.3.1 Cancer Diagnosis (Laser-Induced Fluorescence—LIF)

- Used for in situ, real-time detection of tumors without harmful radiation.
- Involves a photosensitizing drug (e.g., hematoporphyrin derivative) that accumulates in tumors and fluoresces under specific light.
- Enables both tumor imaging and photodynamic therapy by generating toxic singlet oxygen to destroy cancer cells.

3.3.2 Marine Petroleum Pollution Detection

- Detects oil slicks, petroleum contaminants, and pollution sources in seawater.
- Petroleum contains a small fraction of fluorescent hydrocarbons ($\approx$10%), mainly emitting in the 270–400 nm range.
- Despite strong UV/blue light absorption, oil fluorescence is a reliable optical detection method.

3.3.3 Glucose Determination

- Measures glucose levels in biological systems; useful for diagnosing conditions like diabetes and hypoglycemia.
- Fluorophotometry is preferred for its simplicity and sensitivity.
- Recent advancements use bio-molecule-stabilized gold nanoclusters as sensitive and selective fluorescence probes.

3.3.4 Studying Protein Folding

Fluorescence spectroscopy, especially using intrinsic fluorophores like tryptophan, helps monitor folding/unfolding of proteins. Changes in fluorescence intensity, wavelength shift, or lifetime indicate exposure of hydrophobic residues or conformational changes during folding.

3.3.5 Structural Dynamics

Fluorescence allows tracking of rapid structural motions in proteins. Time-resolved fluorescence can capture dynamic conformational states, flexibility, or domain movements at nanosecond to millisecond timescales.

3.3.6 Residue Accessibility

By using quenching agents (e.g., acrylamide, iodide), researchers can probe the solvent accessibility of specific residues. If a fluorophore is buried inside the protein, quenching is reduced; if exposed, quenching is stronger. This reveals local structural environments.

3.3.7 Protein–Protein/Protein–Nucleic Acid Linkage (Connections)

Fluorescence resonance energy transfer (FRET) and fluorescence quenching can detect molecular interactions. Changes in fluorescence signal occur when proteins bind to each other or to nucleic acids, allowing determination of binding affinity, conformational rearrangements, and interaction sites.

References

1. Kurul F, Doruk B, Topkaya SN. Principles of green chemistry: building a sustainable future. Discover Chemistry. 2025 Apr 2 (1):68. https://doi.org/10.22034/ijnc.2025.719174.
2. Escobedo R, Miranda R, Martínez J. Infrared irradiation: Toward green chemistry, a review. International Journal of Molecular Sciences. 2016; 17. https://doi.org/10.3390/ijms17040453
3. Martínez J, Miranda R. Infrared irradiation, an excellent, alternative green energy source. Green chemistry applications. 2019 Feb 9: 32–55. doi:https://doi.org/10.5772/intechopen.8380
4. Jofre FC, Savio M. Infrared radiation: A sustainable and promising sample preparation alternative for inorganic analysis. TrAC Trends in Analytical Chemistry. 2024 Jan 1; 170: 117469. https://doi.org/10.1016/j.trac.2023.117469
5. Workman Jr J. Advances in Infrared Spectroscopy: A Peer-Reviewed Special Issue. Spectroscopy Supplements, 2022 August 1; 37: 6.
6. Chen H, Yang H. Online ATR-FT-IR for Real-Time Monitoring of the Aspirin Esterification Process in an Acid-Mediated Reaction, Spectroscopy Supplements, 2022, 7–12.
7. Cheaib D, El Darra N, Rajha HN, Ghazzawi IE, Maroun RG, Louka N. Biological activity of apricot byproducts polyphenols using solid–liquid and infrared-assisted technology. Journal of Food Biochemistry. 2018 Oct; 42(5):e12552. https://doi.org/10.1111/jfbc.12552
8. Martínez J, Rosas J, Pérez J, Saavedra Z, Carranza V, Alonso P. Green approach to the extraction of major capsaicinoids from habanero pepper using near-infrared, microwave, ultrasound and Soxhlet methods, a comparative study. Natural product research. 2019 Feb 1; 33(3):447–52. https://doi.org/10.1080/14786419.2018.1455038
9. Lee SC, Kim JH, Jeong SM, Kim DR, Ha JU, Nam KC, et al. Effect of far infrared radiation on the antioxidant activity of rice hulls. Journal of Agricultural and Food Chemistry. 2003; 51:4400–4403. https://doi.org/10.1021/jf0300285

10. Lee SC, Kim JH, Nam KC, Ahn DU. Antioxidant properties of the far infrared-treated rice hull extract in irradiated raw and cooked Turkey breast. Journal of Food Science. 2003; 68: 1904–1909
11. Lee SC, Jeong SM, Kim SY, Park HR, Nam KC, Ahn DU. Effect of far-infrared radiation and heat treatment on the antioxidant activity of water extract from peanut hulls. Food Chemistry. 2006; 94:489–493. https://doi.org/10.1016/j.foodchem.2004.12.001
12. Wanyo P, Siriamornpun S, Meeso N. Changes in phenolic compounds, antioxidant and physical properties of mulberry tea influenced by intensity of far-infrared radiation. American Eurasian Journal of Agricultural & Environmental Sciences. 2009; 6:470–479.
13. Alcerreca G, Sanabria R, Miranda R, Arroyo G, Tamariz J, Delgado F. Preparation of benzylidene barbituric acids promoted by infrared irradiation in absence of solvent. Synthetic Communications. 2000 Apr 1; 30(7): 1295–301. https://doi.org/10.1080/00397910008087151
14. Campos VM, Dasilva FL, Oliveira JP, Gouveia ST, Ribeiro LP, Lopes GS, Matos WO. Infrared radiation as a heat source in sample preparation of shrimp for trace element analysis. Journal of Food Composition and Analysis. 2019 Jun 1; 79:107–13. https://doi.org/10.1016/j.jfca.2019.03.014.
15. Magalhaes CC, Krug F, Fostier A. Direct determimnation of mercury in sediments by atomic absorption spectrometry. Journal of Analytical Atomic Spectrometry. 1997; 12(10):1231–4. https://doi.org/10.1039/A701870C
16. Nóbrega JA, Gouveia ST, Fatibello-Filho O. Infra-red heating as an alternative technique for fast sample preparation. Journal of the Brazilian Chemical Society. 2000; 11:261–5. https://doi.org/10.1590/S0103-50532000000300011
17. Strenger S, Hirner A. Digestion of organic components in waste materials by high pressure ashing with infrared heating. Fresenius' Journal of Analytical Chemistry. 2001 Nov; 371(6):831–7. https://doi.org/10.1007/s002160101076
18. Zhang N, Zheng J. Rapid sample preparation with multi-channel focused infrared micro-ashing prior to determination of chromium in gelatin capsules by electrothermal atomic absorption spectrometry. Analytical Methods. 2018; 10(8):920–5. https://doi.org/10.1039/C7AY02844J
19. Zhang N, Shen K, Yang X, Li Z, Zhou T, Zhang Y, Sheng Q, Zheng J. Simultaneous determination of arsenic, cadmium and lead in plant foods by ICP-MS combined with automated focused infrared ashing and cold trap. Food chemistry. 2018 Oct 30; 264:462–70. https://doi.org/10.1016/j.foodchem.2018.05.058
20. Zhang N, Zheng J. Fully automated focused infrared microashing combined with use of ICP-based instruments for rapid analysis of multiple elements in biological samples. Analytical and Bioanalytical Chemistry. 2018 Feb; 410(5):1489–99. https://doi.org/10.1007/s00216-017-0790-z
21. Wang Y, Baker LA, Helmeczi E, Brindle ID. Rapid high-performance sample digestion of base metal ores using high-intensity infrared radiation with determination by nitrogen-based microwave plasma optical spectrometry. Analytical chemistry research. 2016 Mar 1; 7:17–22. https://doi.org/10.1016/j.ancr.2016.02.002
22. Helmeczi W, Helmeczi E, Baker LA, Wang Y, Brindle ID. Development of a general acid method for the digestion of gold ore samples together with a comparison of extraction solvents for gold and determination by microwave-induced plasma-atomic emission spectrometry (MIP-AES). Journal of Analytical Atomic Spectrometry. 2018; 33(8):1336–44. https://doi.org/10.1039/C8JA00136G
23. Helmeczi E, Wang Y, Brindle ID. A novel methodology for rapid digestion of rare earth element ores and determination by microwave plasma-atomic emission spectrometry and dynamic reaction cell-inductively coupled plasma-mass spectrometry. Talanta. 2016 Nov 1; 160:521–7. https://doi.org/10.1016/j.talanta.2016.07.067
24. Zhang N, Li Z, Zheng J, Yang X, Shen K, Zhou T, Zhang Y. Multielemental analysis of botanical samples by ICP-OES and ICP-MS with focused infrared lightwave ashing for sample preparation. Microchemical Journal. 2017 Sep 1; 134:68–77.https://doi.org/10.1016/j.microc.2017.05.006

25. Jofre FC, Larregui DN, Murcia VN, Pacheco P, Savio M. Infrared assisted digestion used as a simple green sample preparation method for nutrient analysis of animal feed by microwave induced plasma atomic emission spectrometry. Talanta. 2021 Aug 15; 231:122376. https://doi.org/10.1016/j.talanta.2021.122376
26. Dantas AN, Matos WO, Gouveia ST, Lopes GS. The combination of infrared and microwave radiation to quantify trace elements in organic samples by ICP OES. Talanta. 2013 Mar 30; 107:292–6. https://doi.org/10.1016/j.talanta.2013.01.047
27. Wang Y, Kanipayor R, Brindle ID. Rapid high-performance sample digestion for ICP determination by ColdBlock™ digestion: part 1 environmental samples. Journal of Analytical Atomic Spectrometry. 2014; 29(1):162–8. https://doi.org/10.1039/C3JA50333J
28. Oliveira JP, Silva FL, Monte RJ, Matos WO, Lopes GS. A new approach to mineralization of flaxseed (Linum usitatissimum L.) for trace element analysis by flame atomic absorption spectrometry. Food chemistry. 2017 Jun 1; 224:335–41. https://doi.org/10.1016/j.foodchem.2016.12.096
29. de Sousa V, Da Silva FL, Gouveia ST, Matos WO, Ribeiro LP, Lopes GS. Infrared radiation applied as a heating source in milk sample preparation for the determination of trace elements by inductively coupled plasma-optical emission spectroscopy. Revista Virtual de Química. 2017 Nov; 9(6):2226–36. https://doi.org/10.21577/1984-6835.20170132
30. Campos VM, Silva FL, Oliveira JP, Ribeiro LP, Matos WO, Lopes GS. Investigation of a rapid infrared heating assisted mineralization of soybean matrices for trace element analysis. Food Chemistry. 2019 May 15; 280:96–102. https://doi.org/10.1016/j.foodchem.2018.12.051
31. Jofre FC, Larregui DN, Savio M. An eco-friendly infrared method for rapid soil sample preparation for multielemental determination by microwave induced plasma atomic emission spectrometry. Microchemical Journal. 2020 Dec 1; 159:105448. https://doi.org/10.1016/j.microc.2020.105448
32. Patle TK, Shrivas K, Kurrey R, Upadhyay S, Jangde R, Chauhan R. Phytochemical screening and determination of phenolics and flavonoids in Dillenia pentagyna using UV–vis and FTIR spectroscopy. Spectrochimica Acta Part A: Molecular and Biomolecular Spectroscopy. 2020 Dec; 242:118717. https://doi.org/10.1016/j.saa.2020.118717
33. Pasieczna-Patkowska S, Cichy M, Flieger J. Application of Fourier transform infrared (FTIR) spectroscopy in characterization of green synthesized nanoparticles. Molecules. 2025 Feb 4; 30(3):684. https://doi.org/10.3390/molecules30030684
34. Sankar, R.; Manikandan, P.; Malarvizhi, V.; Fathima, T.; Shivashangari, K.S.; Ravikumar, V. Green synthesis of colloidal copper oxide nanoparticles using *Carica papaya* and its application in photocatalytic dye degradation. *Spectrochim. Acta Part A* 2014, *121*, 746–750. https://doi.org/10.1016/j.saa.2013.12.020
35. Wanakai SI, Kareru PG, Makhanu DS, Madivoli ES. Advances in green nanotechnology: data for green synthesis and characterization of iron nanoparticles synthesized using Galinsoga parviflora, Conyza bonariensis and Bidens pilosa leaf extracts, and their application in degradation of methylene blue dye and rifampicin antibiotic. Data in Brief. 2023 Feb 1; 46:108882. https://doi.org/10.1016/j.dib.2022.108882
36. Hosseingholian A, Gohari SD, Feirahi F, Moammeri F, Mesbahian G, Moghaddam ZS, Ren Q. Recent advances in green synthesized nanoparticles: from production to application. Materials Today Sustainability. 2023 Dec 1; 24:100500. https://doi.org/10.1016/j.mtsust.2023.100500
37. Yisak H, Tukue H, Redi-Abshiro M, Chandravanshi BS, Yaya EE. Development of Green UV-Vis Method for Direct Determination of Total Sugars in the Aqueous Extract of Teff (Eragrostis tef (Zuccagni) Trotter) Grains and Other Cereals. International Journal of Analytical Chemistry. 2022; (1):5129510. https://doi.org/10.1155/2022/5129510
38. Naresh K. Applications of fluorescence spectroscopy. J. Chem. Pharm. Sci. 2014; 974: 2115.
39. Ladokhin AS. Fluorescence spectroscopy in peptide and protein analysis. Encyclopedia of analytical chemistry. 2000 Oct 30:5762–79.
40. Valeur B, Weber G. Resolution of the fluorescence excitation spectrum of indole into the IL, and Ir-., excitation bands. Photochern. Photbiol. 25: 441. 1977; 444.

Green Analytical Chemistry in Nanotechnology

Charmy Kothari, Gautam Sanghavi, and Krishna Bhalodi

Abstract Green Analytical Chemistry (GAC) aims to make chemical analysis safer, cleaner, and more sustainable by reducing the use of toxic chemicals, lowering energy consumption, and minimizing waste generation. In recent years, nanotechnology has emerged as a powerful enabler of the goals by facilitating efficient, high-performance, and environmentally responsible analytical strategies. Nanomaterials possess unique physicochemical properties such as high surface area, enhanced reactivity, tuneable size, and excellent sensitivity that allow accurate analysis using smaller sample volumes, fewer reagents, and greener operating conditions. This chapter examines how nanotechnology advances green analytical methods in alignment with the twelve principles of Green Analytical Chemistry. It discusses the contribution of nanomaterials to direct analysis, in-situ measurements, miniaturized and portable analytical devices, real-time monitoring, and waste reduction. Particular emphasis is placed on eco-friendly sample preparation techniques, including nano-enabled extraction and separation methods that significantly reduce solvent use, analysis time, and environmental burden. The chapter also highlights sustainable strategies for nanomaterial synthesis, including plant-based, microbial, biopolymer assisted, and microwave-assisted green routes. These approaches eliminate hazardous chemicals, operate under mild conditions, and yield biocompatible and functionally robust nanomaterials suitable for analytical applications. Furthermore, the role of green nanocatalysts, such as metal oxides, carbon-based materials, nanozymes, and photocatalysts, is discussed, demonstrating their ability to promote rapid, low-energy and environmentally friendly analytical reactions. Overall, this chapter highlights that integrating

C. Kothari (✉) · G. Sanghavi · K. Bhalodi
Department of Pharmaceutical Analysis, Institute of Pharmacy, Nirma University, Ahmedabad, Gujarat, India
e-mail: charmy.kothari@nirmauni.ac.in; charmyshah@gmail.com

G. Sanghavi
e-mail: 24mpa007@nirmauni.ac.in

K. Bhalodi
e-mail: krishnabhalodi@outlook.com

K. Bhalodi
Faculty of Pharmacy, Marwadi University, Rajkot, Gujarat, India

V. Bhusari and M. Ghante (eds.), *Green Analytical Chemistry in India*,
https://doi.org/10.1007/978-3-032-24564-9_6

nanotechnology with Green Analytical Chemistry provides a robust and practical framework for developing sustainable, efficient, and responsible analytical science for environmental, industrial, and biomedical applications.

Keywords Green Analytical Chemistry · Nanotechnology · Nanomaterials · Green synthesis · Nanocatalysts

1 Introduction

Green Analytical Chemistry (GAC) has emerged as an essential paradigm in modern analytical science, emphasizing environmental responsibility, sustainability, and operational efficiency. With increasing global awareness of environmental degradation, the need to reduce hazardous chemicals, minimize energy consumption, and adopt sustainable methodologies has become more urgent. With growing awareness of environmental impact, analytical chemistry is transitioning away from toxic solvents, large sample volumes, and energy intensive instruments toward greener alternatives. In this transformative landscape, nanotechnology plays a pivotal role by providing functional materials capable of enhancing sensitivity, reducing resource consumption, and enabling miniaturization [1, 2].

Nanomaterials owing to their unique physicochemical characteristics such as high surface area, enhanced reactivity, tunable particle size, and distinctive optical, magnetic, or catalytic properties have become prominent in the design of green analytical workflows. Their integration into extraction, separation, sensing, and catalysis platforms directly supports the principles of GAC. By enabling solvent free separations, rapid detection, selective enrichment of analytes, and low-energy catalytic transformations, nanotechnology not only strengthens analytical performance but also reduces the environmental footprint of laboratory operations.

The purpose of this chapter is to provide an in-depth examination of how nanomaterials contribute to greener analytical approaches, how sustainable synthesis routes enhance the safety and biocompatibility of nanomaterials, and how nanocatalysts are shaping chemical analysis across environmental and biomedical domains. Through critical discussion and evidence from recent studies, this chapter highlights advancements, challenges, and future perspectives in the intersection of nanotechnology and GAC [3, 4].

2 The 12 Principles of Green Analytical Chemistry

Direct Analytical Techniques

Direct analytical technique promotes analyzing samples in their original state, without relying on extensive preparation steps such as extraction, filtration, concentration, or derivatization. Direct analytical approaches reduce analysis time, minimize chemical usage, and prevent analyte loss or contamination. Nanomaterials greatly enhance the feasibility of direct analysis. Due to their superior surface reactivity, catalytic behavior, and signal amplification abilities, nano-enabled sensors can detect analytes directly in complex matrices such as biological fluids, wastewater, or food extracts. Techniques such as surface enhanced Raman spectroscopy (SERS), nano-based electrochemical sensors, and plasmonic detection systems exemplify this advancement.

Minimal Sample Size

Reducing sample volume lowers reagent consumption, decreases waste generation, and enhances safety particularly when dealing with hazardous biological or chemical samples. Modern nanosensors and nano-enabled analytical platforms operate efficiently with microlitre or nanolitre level samples. This is especially valuable in biomedical applications, where only minimal volumes of blood, serum, or tissue extracts may be available. Microfluidic devices integrated with nanomaterials further support this principle by enabling highly sensitive analysis in extremely small volumes.

In-Situ Measurements

In-situ analysis involves examining samples at the point of collection, thereby avoiding transportation, refrigeration, and long-term storage that may degrade analytes or introduce contamination. Nano-enabled portable sensors, field deployable spectroscopic devices, and paper-based nanobiosensors have transformed in-situ monitoring into a practical and efficient alternative. Such systems allow immediate assessment of environmental water samples, industrial effluents, soil contaminants, or biomedical markers at the sampling site, reducing energy consumption and improving data accuracy.

Minimal Waste Generation

Traditional analytical procedures often produce large volumes of hazardous waste, particularly when using organic solvents for extraction and chromatography. GAC emphasizes the reduction of waste at every analytical stage. Nanomaterials significantly contribute to this principle by replacing solvent intensive steps with sorbent based extractions. Magnetic nanoparticles, nanoadsorbents, graphene-based materials, and MOFs require only minute quantities of solvents or none at all to isolate analytes efficiently. Their high adsorption capacity allows multiple extraction cycles with minimal waste production.

Use of Safer Reagents

A core aim of GAC is to replace toxic, volatile, or environmentally harmful chemicals with safer alternatives. Water, ethanol, glycerol, ionic liquids, supercritical fluids, and biodegradable reagents are now preferred over hazardous solvents. Nanotechnology supports this shift by enabling reactions and sensing processes in aqueous media. For example, water dispersible carbon dots, biogenic metal nanoparticles, and polymeric nanosystems allow high-performance analysis without relying on harmful chemicals.

Energy Efficiency

Energy-efficient analytical methods reduce electricity consumption and carbon emissions. GAC encourages ambient temperature reactions, fast analytical cycles, and techniques that minimize instrument power usage.

Nanomaterials often serve as catalysts, accelerating reaction rates and lowering the required temperature for analytical processes. Microwave-assisted synthesis, sonochemical extraction, and nanocatalyzed reactions are prime examples, offering faster results at significantly reduced energy costs.

Automation and Miniaturization

Automation and miniaturization streamline workflows, improve reproducibility, and reduce reagent and solvent consumption. Lab-on-chip platforms, microfluidic devices, and portable analytical instruments exemplify this principle.

Nanomaterials enhance these platforms by improving sensitivity and selectivity in miniature detection systems. Their ability to operate efficiently within microchannels allows ultra-low reagent usage, rapid processing, and reduced overall waste.

Multi-analyte Capability

Analyzing multiple analytes in a single run reduces the need for repeated experiments, saving solvents, energy, reagents, and time. Nanotechnology facilitates multiplexed detection through engineered surfaces, functionalized nanoparticles, and advanced nanocomposites. Nano-enabled electrochemical and optical sensors can simultaneously measure various pollutants, biomarkers, or contaminants, thereby significantly improving analytical sustainability.

Use of Degradable or Reusable Reagents

GAC promotes the use of materials that can degrade naturally or be reused multiple times without generating hazardous residues. Biopolymers such as chitosan, cellulose, starch, and alginate based nanoparticles are biodegradable and environmentally benign.

Magnetic nanoparticles are particularly advantageous because they allow easy recovery using magnetic separation and can be reused across multiple extraction or catalytic cycles, thereby reducing both material consumption and waste [5].

Minimizing Sample Preservation Steps

Conventional sample preservation often involves chemical stabilizers, refrigeration, or freezing processes that require hazardous agents or high energy input. GAC encourages eliminating or minimizing these steps whenever possible. In-situ and real-time analysis made possible by portable nano devices reduces the need for preservatives. Direct and rapid analysis ensures that samples remain chemically intact without artificial stabilizers.

Real-Time Monitoring

Real-time analytical monitoring allows immediate detection of changes in chemical composition, making it highly valuable in environmental surveillance, industrial process control, and biomedical diagnostics. Nanosensors owing to their high sensitivity, catalytic enhancement, and rapid response time enable continuous and accurate measurements of gases, toxins, biomolecules, and pollutants. This reduces laboratory workload, eliminates repeated sampling, and minimizes chemical waste [6].

Eco-friendly Instrumentation

GAC advocates for analytical instruments that are energy efficient, compact and have minimal environmental impact. Portable, handheld, and battery operated devices incorporating nanotechnology reduce reliance on large, energy intensive laboratory instruments. Nano optical sensors, portable electrochemical devices, microfluidic chips, and paper-based nanoplatforms embody eco-friendly instrumentation by consuming less power, generating minimal waste, and offering sustainable analytical alternatives [7].

3 Role of Nanomaterials in Enhancing Green Analytical Methods

Nanotechnology plays an essential role in achieving GAC principles. Nanomaterials such as metal nanoparticles, carbon nanomaterials, quantum dots, and metal oxides have unique properties that allow analysis to be performed with higher efficiency and much lower environmental impact [8–10]. Because they are 1–100 nm in size and have a large surface to volume ratio, these nanomaterials enable analytical methods that are highly sensitive, selective, miniaturized, and compatible with aqueous or low toxicity systems fully aligning with the principles of GAC [11].

Key Ways in Which Nanomaterials Enhance Green Analytical Methods

Nanomaterials Improve Sensitivity While Using Fewer Chemicals

One of the most significant advantages of nanomaterials is their ability to dramatically enhance sensitivity even when used in very small quantities. Due to their large surface area and active binding sites, nanomaterials readily interact with analytes, resulting in strong measurable signals at trace or ultra-trace levels.

This high efficiency supports several GAC principles:

- Smaller sample volumes are required.
- Minimal quantities of reagents are needed.
- Hazardous chemicals can be eliminated or substituted with safer alternatives.
- Less waste is produced due to reduced material consumption.

For example, gold nanoparticles exhibit clear and strong color changes when they interact with biomolecules or pollutants, making them ideal for simple, low reagent colorimetric detection. In a similar way, quantum dots and carbon dots produce bright fluorescence and can operate efficiently in water-based systems. This eliminates or reduces the need for toxic organic dyes and supports safer, greener analytical methods [12, 13].

Nanomaterials Enhance Selectivity and Reduce Sample Preparation Steps

Complex samples like blood, soil, food, or wastewater often require extensive preparation steps before analysis. These steps typically involve using toxic solvents, heating, and multiple purification processes. Their surfaces can be easily functionalized with chelating agents (for metal ions), functional polymers, antibodies and aptamers, ionic, or other biomolecule groups. These added groups allow the nanomaterials to recognize and bind specific analytes with great precision.

These functional groups enable nanomaterials to selectively capture target analytes, thereby:

- reducing the number of preparatory steps;
- lowering solvent and reagent consumption;
- minimizing analyst exposure to toxic chemicals;
- improving analytical accuracy and reducing matrix interference.

For example, when magnetic nanoparticles functionalized with citric acid or amine groups, they can specifically capture heavy metals such as Pb^{2+} and Cd^{2+} from different environmental samples. Their selective binding makes the separation process much cleaner and more efficient. After extraction, the nanoparticles can be quickly removed using a simple magnet, avoiding filtration or centrifugation, and reducing both time and chemical use [14].

Nanomaterials for Eco-friendly Sample Preparation

Nanomaterials play a crucial role in making analytical workflows more environmentally friendly by improving sample preparation, which is generally the most solvent intensive and environmentally burdensome step. Nanotechnology has revolutionized this stage through techniques such as magnetic solid-phase extraction (MSPE), nano-dispersive solid-phase microextraction (nano-dSPME), and nano-adsorbent-based purification.

This nanomaterial assisted methods require minimal solvent and energy, reduce extraction time due to rapid nanoparticle–analyte interactions, and operate without heating or complex instrumentation. As a result, they generate fewer by products and produce less chemical waste.

Magnetic nanoparticles, carbon nanomaterials, and metal–organic frameworks (MOFs) have been widely adopted as green sorbents owing to their high surface area, reusability, and selective adsorption capabilities. These systems often allow simple magnetic separation instead of centrifugation or filtration, saving both time and energy. Overall, nanomaterial-based extraction and purification techniques support the core principles of Green Analytical Chemistry (GAC) by enhancing analytical efficiency while significantly lowering the environmental footprint [15].

Nanomaterials Support Miniaturization and Portable Analytical Devices

Miniaturizing analytical instruments is a key aspect of GAC because smaller systems naturally consume less solvent, reagents, and energy while generating minimal waste. This makes the analytical process not only more sustainable but also more economical and accessible.

Nanomaterials serve as essential building blocks in modern miniaturized analytical technologies. Their high surface area, strong catalytic activity, and excellent conductivity make them ideal for developing lab-on-a-chip microfluidic systems, paper-based analytical devices (PADs), wearable and portable nanosensors, and point-of-care diagnostic tools [16, 17].

These nano-enabled platforms function efficiently with microlitre to nanolitre sample volumes, require very low power input, and deliver rapid, sensitive, and selective results. Their portability allows in-field and real-time analysis, reducing the need for sample transportation, and laboratory infrastructure. Consequently, nano-enabled miniaturized devices make analytical testing more accessible, greener, and cost effective, aligning seamlessly with the goals of sustainable analytical science [18].

Nanomaterials Serve as Green Catalysts

Catalysis is central to many analytical reactions, and traditional catalysts often require high temperatures or corrosive chemicals. Nanocatalysts work differently; they can accelerate reactions efficiently at room temperature and even in water.

Common nanocatalysts used in these miniaturized systems include gold nanoparticles, silver nanoparticles, iron oxide nanoparticles, titanium dioxide (TiO_2), and cerium oxide (CeO_2). Each of these nanomaterials offers unique catalytic and surface properties that enhance sensitivity, speed, and selectivity in micro-scale analytical devices. Their ability to operate efficiently with minimal reagents makes them particularly valuable for sustainable, portable, and low power analytical technologies.

These nanocatalysts offer several important green benefits. Their fast reaction rates mean that less energy is required, helping to make analytical processes more efficient. They can often replace traditional toxic chemical catalysts, reducing hazards in the laboratory. Because they function well under mild conditions such as room temperature and neutral pH, they are suitable for analyzing environmentally sensitive or biological samples. An excellent example is TiO_2 nanoparticles, which act as powerful photocatalysts capable of degrading pollutants under sunlight. Such photocatalytic reactions eliminate the need for harsh oxidizing agents and are widely used in green sample cleanup and environmental analysis [19, 20].

4 Sustainable Synthesis of Nanomaterials for Analytical Applications

Sustainable or green synthesis of nanomaterials has emerged as a critical priority in modern nanotechnology. Conventional synthesis routes often rely on hazardous reducing agents, organic solvents, elevated temperatures, and energy intensive procedures that conflict with environmental and occupational safety. In contrast, GAC promotes the development of nanomaterials through methods that are inherently safer, non-toxic, energy efficient, and aligned with the principles of sustainability.

Green synthesis approaches aim to minimize ecological impact by using eco-friendly precursors, mild reaction conditions, renewable biological sources, and reduced energy input.

Such methods not only decrease pollution and hazardous waste but also improve the biocompatibility and regulatory acceptability of the resulting nanomaterials. Importantly, sustainably synthesized nanomaterials are better suited for integration into environmentally responsible analytical workflows, ensuring consistency with the goals and principles of GAC.

By adopting plant based, microbial, agro waste, or energy-efficient synthesis strategies, the field moves toward a holistic green analytical ecosystem where both the materials and the methods used are safe, clean, and fully sustainable.

Phytosynthesis of Nanomaterials

Plant mediated synthesis is one of the most widely adopted and environmentally friendly routes for producing nanomaterials. Plants naturally contain a diverse array of biomolecules such as polyphenols, flavonoids, tannins, terpenoids, alkaloids, proteins, sugars, and organic acids that function as powerful reducing, stabilizing, and capping agents. These phytochemicals can efficiently reduce metal ions to their nanoparticulate form without requiring toxic chemical reductants, thereby offering a safe, renewable, and cost-effective synthesis strategy.

Phytosynthesis is considered a truly green and sustainable method because it largely employs water as the reaction medium, eliminating the need for hazardous organic solvents. The natural phytochemicals replace harmful reducing agents like sodium borohydride hydrazine, significantly reducing the toxic burden associated with conventional synthesis.

Furthermore, the by products generated through plant mediated routes are generally biodegradable and nonhazardous, ensuring minimal environmental impact. The ability to conduct the synthesis at ambient temperature and pressure further enhances its sustainability by reducing energy consumption and simplifying the overall process.

Numerous studies have demonstrated the application of plant derived nanoparticles in analytical chemistry. For instance, silver nanoparticles synthesized using tea leaf extract have been successfully applied in the colorimetric detection of nitrite in water. Similarly, turmeric extract–derived gold nanoparticles exhibit strong surface plasmon resonance properties, enabling distinct and visually observable color changes that make them ideal for rapid colorimetric sensing applications. Additionally, iron oxide nanoparticles produced using neem extract have shown excellent performance in environmental applications, particularly in the magnetic extraction and removal of heavy metals from contaminated samples.

Microbial Synthesis

Microorganisms including bacteria, fungi, yeasts, and algae are capable of synthesizing nanoparticles through their natural metabolic pathways. This biological route, known as microbial synthesis or "microbiosynthesis," occurs via two main mechanisms: intracellular and extracellular synthesis.

- In intracellular synthesis, metal ions penetrate the microbial cell wall and are enzymatically reduced within the cytoplasm, leading to the formation of nanoparticles inside the cells.
- In extracellular synthesis, microorganisms secrete enzymes, proteins, and metabolites into the surrounding medium, which then reduce metal ions into nanoparticles outside the cells.

Both mechanisms rely on biological reducing and stabilizing agents naturally produced by the microbes, eliminating the need for synthetic toxic chemicals and making this approach inherently clean, safe, and environmentally friendly.

Microbial synthesis is considered highly sustainable because it operates under mild reaction conditions, often at room temperature, neutral pH, and atmospheric pressure. These gentle conditions significantly reduce energy consumption, minimize environmental impact, and simplify scalability. Microbial cultures are inexpensive to grow and can be easily scaled up, making this method suitable for industrial level nanoparticle production.

In analytical chemistry, microbial derived nanomaterials have gained considerable interest due to their biocompatibility, stability, and functional surface properties. For example, bacteria mediated zinc oxide nanoparticles have demonstrated high sensitivity in electrochemical sensing platforms due to their excellent electron transfer capabilities and high surface area. Similarly, fungi such as Fusarium oxysporum have been widely used to biosynthesize silver nanoparticles, which exhibit strong antimicrobial and optical properties.

These fungal derived AgNPs have proven valuable in colorimetric assays, biosensing, and simple field detection methods, where strong plasmonic signals enable rapid visual identification.

Biopolymer-Assisted Green Synthesis

Biopolymer-assisted synthesis represents another important green strategy for producing nanomaterials, utilizing naturally derived polymers such as chitosan, starch, alginate, cellulose, gelatin, and pectin. These biopolymers function simultaneously as reducing agents, stabilizing (capping) agents, and structure directing molecules, enabling the formation of nanoparticles under mild, environmentally friendly conditions. Because biopolymers are inherently biodegradable, nontoxic, renewable, and biocompatible, they eliminate the need for hazardous chemical stabilizers commonly used in conventional nanoparticle synthesis.

A major advantage of using biopolymers lies in their rich surface chemistry. Functional groups such as amino, hydroxyl, carboxyl, and sulfate groups enable strong interactions with metal ions during synthesis and also enhance the final nanomaterial's ability to bind analytes.

This makes biopolymer-assisted nanoparticles highly effective in sensing, adsorption, catalysis, and extraction based analytical applications. Their compatibility with biological and environmental samples further contributes to cleaner, safer, and more sustainable analytical workflows, fully aligning with the principles of GAC.

Several practical examples highlight their significance. Chitosan coated magnetic nanoparticles, for instance, are widely employed in sample preparation and drug extraction due to their strong metal binding capability, stability, and ease of magnetic separation. These properties reduce solvent consumption and eliminate complex separation steps. Similarly, cellulose nanofibers and cellulose derived nanomaterials have gained attention for fabricating eco-friendly sensors, benefiting from their mechanical strength, flexibility, biodegradability, and ability to immobilize sensing molecules.

Microwave-Assisted Green Synthesis

Microwave-assisted synthesis has emerged as a highly efficient and environmentally friendly method for producing nanomaterials. Microwave irradiation heats the reaction mixture rapidly and uniformly by directly interacting with polar molecules and ions. This volumetric heating mechanism drastically reduces reaction times from several hours in conventional heating to just a few minutes making the process significantly more energy efficient.

This method offers several green advantages. The rapid and uniform heating minimizes thermal gradients, which not only reduces energy consumption but also limits the formation of undesirable by products. Additionally, microwave-assisted processes often yield nanoparticles with smaller, more uniform sizes, improved crystallinity, and enhanced functional properties. Another key benefit is the compatibility of this method with water-based and solvent free systems, which eliminates the need for hazardous organic solvents and enhances the environmental safety of the overall process.

Microwave-assisted nanomaterials have found extensive applications in analytical chemistry. For instance, microwave synthesized TiO_2 nanoparticles are widely used in photocatalytic degradation studies due to their controlled morphology and high photocatalytic efficiency.

Their rapid and uniform formation under microwave conditions makes them ideal for environmental monitoring and pollutant degradation assays. Similarly, microwave reduced graphene oxide (MRGO) has gained importance in the development of highly sensitive electrochemical sensors. The rapid reduction process enhances conductivity and restores conjugated structures, improving the electrochemical performance of graphene-based sensing platforms. These examples clearly demonstrate how microwave-assisted synthesis supports fast, clean, energy efficient, and sustainable production of nanomaterials that are well suited for modern green analytical applications.

5 Green Nanocatalysts for Chemical Analysis

Green nanocatalysts represent a transformative advancement in analytical chemistry by enabling environmentally responsible and highly efficient reaction processes. Traditional catalytic systems often depend on harsh conditions, such as elevated temperatures, strong acids or bases, toxic organic solvents, and large amounts of precious metals which increase energy consumption, operational costs, and hazardous waste generation. These drawbacks conflict with the principles of GAC, which emphasize safety, sustainability, and waste minimization.

Nanotechnology provides a sustainable alternative through the development of nanocatalysts, which exhibit exceptional catalytic performance due to their nanoscale dimensions and high surface to volume ratios. Their extensive active sites and unique physicochemical properties allow reactions to proceed rapidly, selectively, and with significantly lower reagent requirements. In many cases, nanocatalyst driven reactions can occur at room temperature and in aqueous media, eliminating the need for harmful organic solvents. This shift not only reduces the environmental footprint of analytical processes but also enhances overall safety and efficiency.

From a green chemistry standpoint, nanocatalysts offer several critical advantages:

- reduced chemical consumption due to high catalytic efficiency at low dosages;
- energy savings by enabling low temperature or ambient condition reactions;
- cleaner reactions with fewer by products and reduced waste streams;
- faster analytical workflows that improve laboratory throughput;
- enhanced sensitivity and selectivity, enabling trace detection of pollutants, drugs, and biomarkers.

Their strong interactions with target analytes further boost analytical performance, allowing miniaturized, portable, and high precision sensing systems to operate reliably under eco-friendly conditions. These attributes make nanocatalysts indispensable tools in advancing sustainable analytical technologies and achieving full alignment with the goals of GAC.

Metal Oxide Nanocatalysts

Metal oxide nanoparticles like TiO_2, ZnO, Fe_3O_4, MnO_2, and CeO_2 are considered some of the most eco-friendly nanocatalysts. They are stable, low cost, and can often carry out catalytic reactions without requiring extra chemicals. Because of their strong activity and natural compatibility with water based systems, they are widely used in green analytical applications. These metal oxide nanocatalysts offer several green benefits. They can catalyze reactions just under sunlight through photocatalysis, which saves energy and avoids the need for artificial heating. Many of them are easy to make using green synthesis methods and are generally nontoxic.

ZnO nanoparticles are known for boosting fluorescence and electrochemical signals, making detections more sensitive. Fe_3O_4 nanoparticles act like peroxidase enzymes and can be easily separated with a magnet after the reaction, which simplifies the workflow [21].

Carbon-Based Nanocatalysts

Carbon-based nanomaterials such as graphene, graphene oxide (GO), carbon nanotubes (CNTs), and carbon dots (CDs) have become indispensable in sustainable catalysis and analytical chemistry. These nanostructures possess large surface areas, superior stability, lightweight construction, and remarkable electrical conductivity. Their ability to accelerate reactions without toxic chemicals makes them excellent candidates for eco-friendly analytical applications.

These nanomaterials are particularly sustainable because they can be synthesized from low cost, renewable, and green precursors such as fruit peels, plant extracts, and other biomass sources. Many carbon nanomaterials are also chemically robust and can be reused multiple times, substantially reducing analytic waste.

Carbon-based nanomaterials find a wide range of applications in green analytical techniques.

- Carbon dots possess catalytic and fluorescent properties, making them highly suitable for detecting biomolecules via oxidation based fluorescent assays [9, 22, 23].
- Graphene oxide accelerates electron transfer processes, significantly improving the sensitivity and speed of electrochemical sensors [14, 19, 24].
- Carbon nanotubes offer exceptional conductivity and reactivity, supporting advanced gas sensors, and environmental monitoring devices [9, 25, 26].

Their combination of biocompatibility, recyclability, and high performance explains why carbon-based nanomaterials are at the forefront of green sensor development.

Nanozymes

Nanozymes are nanomaterials that mimic the catalytic functions of natural enzymes while offering superior stability, cost effectiveness, and durability. Unlike biological enzymes, nanozymes do not degrade easily, tolerate harsh physical and chemical conditions, and require no special storage. Their discovery has significantly advanced green catalysis by reducing reliance on fragile, expensive enzymes and enabling more sustainable analytical processes.

Nanozymes exhibit various enzyme mimicking activities depending on their composition.

- Fe_3O_4 nanoparticles display peroxidase like activity, making them highly effective in colorimetric assays for pollutant or biomolecule detection [27].
- CeO_2 nanoparticles behave like oxidase enzymes, catalyzing oxidation reactions without requiring external cofactors or additional reagents [28].
- CuO nanoparticles demonstrate laccase like activity and are effective in degrading organic pollutants [3, 29].

These properties make nanozymes powerful tools for Green Analytical Chemistry, enabling robust assays that are substrate specific, energy efficient, and environmentally friendly.

Photocatalysts for Solar-Driven Analytical Chemistry

Photocatalysis represents one of the most environmentally sustainable catalytic strategies because it uses light especially sunlight as a clean, renewable energy source.

Photocatalysts such as graphitic carbon nitride (g-C_3N_4) and Ag/TiO_2 hybrids absorb light and convert it into chemical energy, driving a variety of catalytic reactions without external heating or chemical oxidants [30].

This strategy offers several green benefits:

- uses renewable solar energy that is freely available;
- eliminates the need for artificial heating or electrical input;
- avoids toxic reagents or oxidants;
- generates fewer by products and produces minimal waste;
- reduces operational costs and environmental impact.

Photocatalysts have wide-ranging analytical applications. They can degrade matrix interferences before chromatographic analysis, improving accuracy. They also support sunlight-driven formation of chromogenic products, making them valuable in low-cost colorimetric sensors. Additionally, photocatalysis is widely used in detecting organic pollutants, where contaminants are converted into analyzable species under light irradiation. Together, these applications highlight the powerful synergy between nanotechnology and solar energy in advancing clean, energy efficient, and sustainable analytical methods.

6 Case Studies: Nanotechnology for Environmental and Biomedical Analysis

1. **Case Study: Advancing Nanomaterial Synthesis through Green Chemistry for Sustainable Innovation** [31]

The work of de Oliveira et al. [31] highlights shift by demonstrating how silver, iron, and carbon-based nanoparticles can be synthesized through green chemistry principles emphasizing efficient resource use, renewable feedstocks, and low waste processes. These sustainably produced nanomaterials are applied in analytical sensing, environmental remediation, and biomedical diagnostics, clearly illustrating the expanding role of GAC in contemporary nanoscience.

Objective of the Study

The primary objective of the study was to systematically evaluate emerging eco-friendly strategies for synthesizing nanomaterials. The authors aimed to:

- understand green reaction mechanisms involved in nanoparticle formation;
- identify sustainable alternatives to hazardous reducing and stabilizing agents;
- correlate green synthesis approaches with analytical, environmental, and biomedical applications;

This holistic investigation sought to bridge sustainable chemistry practices with functional performance in analytical science.

Methodological Overview

Silver Nanoparticles (AgNPs)

Silver nanoparticles possess well documented catalytic, antimicrobial, and optical sensing capabilities. Green synthesis of AgNPs often employs plant extracts including Raphanus sativus and Sambucus ebulus rich in polyphenols and flavonoids that function as natural reducing and capping agents. Microbial systems such as Bacillus brevis and fungi like Talaromyces purpureogenus also facilitate nanoparticle formation through enzyme mediated reduction pathways, producing monodisperse, stable AgNPs. These materials serve as low cost, low waste analytical probes suitable for pollutant detection and biosensing.

Iron Nanoparticles (FeNPs)

Iron nanoparticles exhibit strong magnetic and catalytic properties, making them valuable tools in biosensing, water purification, and pollutant degradation. In green synthesis, Fe^{3+} is reduced to Fe^0 or mixed Fe(II)/Fe(III) oxides using plant derived polyphenols, such as those found in green tea. The utilization of industrial iron residues as precursors demonstrates how circular economy principles can be integrated into nanomaterial development. The resulting FeNPs are widely applied as magnetic sorbents for environmental analysis and as catalysts in degradation pathways.

Carbon Nanoparticles (CNPs)

Sustainable carbon-based nanomaterials such as graphene, carbon dots, and carbon quantum dots are increasingly synthesized from renewable precursors like honey, cyanobacteria, agricultural biomass, and food wastes. Hydrothermal, solvothermal, and microwave-assisted techniques support their generation without hazardous solvents and with reduced energy inputs. These carbon nanomaterials display excellent fluorescence, biocompatibility, and stability, making them highly suitable for bioimaging, drug monitoring, and green optical sensing.

A comparative assessment of nanomaterials based on their source, green synthesis method, size range, analytical relevance, and applications is presented in Table 1.

Environmental and Biomedical Significance

Green synthesized FeNPs and AgNPs provide safer, more sustainable pathways for microbial control, pollutant degradation, and water purification. Their antimicrobial properties make AgNPs effective for medical formulations, while FeNPs offer eco-friendly solutions for magnetic remediation and catalytic treatment of contaminants. Carbon-based nanomaterials, particularly carbon quantum dots, contribute significantly to biomedical diagnostics through fluorescence imaging, drug monitoring, and photoluminescent sensing.

These multifunctional nanostructures demonstrate that environmental sustainability and high analytical performance can coexist, reinforcing the alignment between GAC principles and modern nanotechnology.

Challenges and Future Directions

Despite significant advancements, several challenges persist in the adoption of green synthesis for nanomaterials:

- Reproducibility remains difficult due to natural variability in plant or microbial precursors.
- Scalability needs optimization to meet industrial and commercial demand.
- Long-term stability and storage conditions for biogenic nanoparticles require further study.

Table 1 Comparative assessment of nanomaterials synthesized using greener approaches

Nanomaterial type	Green source/ agent	Green synthesis method	Attributes	Primary application	Analytical relevance
Silver nanoparticles (AgNPs)	Plant extracts, microbial systems	Bioreduction using biological extracts	Eco-friendly, no toxic chemicals, aqueous media	Catalysis, antimicrobials, sensors	Used in environmental and biosensing due to plasmonic properties and eco-benefit
Iron nanoparticles (FeNPs)	Plant derived phytochemicals	Green co-precipitation influenced by biomolecules	Magnetic properties; reduced waste	Water treatment, magnetic separations	Acts as green sorbent in contaminant extraction and detection
Carbon nanomaterials (e.g., carbon dots, CQDs)	Biomass, fungi, bacteria	Bottom up hydrothermal or biological processes	Biocompatible; renewable feedstock	Bioimaging, fluorescence probes	Useful in eco-friendly optical/ electrochemical sensing

- Standardized toxicological frameworks are needed to evaluate safety across different biological systems.

Future research should focus on AI-guided optimization, lifecycle analysis of nanomaterials, integration of agricultural, and industrial waste streams as feedstocks, and the development of universal guidelines for safe and scalable green nanomanufacturing.

Critical Reflections on GAC

Green synthesis transforms nanomaterial production into a resource efficient, low toxicity process.

- Renewable biological feedstocks effectively replace hazardous reagents without compromising performance.
- Incorporating green nanomaterials into analytical systems enhances environmental safety while supporting biomedical and ecological innovations.

These insights reinforce the view that sustainable nanotechnology not only benefits analytical chemistry but also supports broader environmental health objectives.

Conclusion

This case study illustrates how green chemistry principles can fundamentally reshape nanomaterial design and production. By integrating plant-based, microbial, and biomass derived systems, researchers can generate non-toxic, efficient nanostructures suitable for environmental monitoring, pollutant remediation, and biomedical diagnostics. Aligning technological innovation with environmental stewardship exemplifies the true essence of GAC, ensuring that advances in nanotechnology remain responsible, sustainable, and future ready.

2. **Case Study: Nanotechnology Advances for Environmental and Biomedical Applications** [32]

Biomedical and environmental sciences are brought together by the revolutionary multidisciplinary platform of nanotechnology. The promise of artificial nanostructures with adjustable physicochemical characteristics for targeted medication delivery, ecological monitoring, and sustainable biomedical diagnostics is highlighted in the paper of Gangadhar and Subburaj [32]. Nanotechnology applications provide both environmental responsibility and analytical precision by integrating GAC concepts, such as atom economy, cleaner solvents, and decreased toxicity.

Objective of the Study

The primary objective of the study was to consolidate recent developments in nanomaterial-based biomedical and environmental systems, with a particular emphasis on:

- physicochemical features governing nano–bio interactions;
- sustainable synthesis and fabrication techniques;
- biosafety considerations;

- green analytical strategies linking biomedical nanoscience with environmental monitoring.

The work provides a comprehensive view of how eco-friendly nanotechnologies can simultaneously support environmental protection and medical innovation.

Methodological Framework: Nanomaterial Types and their Roles

Nanostructures and Functional Mechanisms

The study categorizes nanomaterials into key classes including polymeric nanoparticles, dendrimers, nanocapsules, liposomes, carbon nanotubes (CNTs), quantum dots (QDs), and nanodiamonds (NDs). Each class exhibits unique physicochemical traits such as surface charge, zeta potential, crystallinity, and quantum confinement that dictate reactivity, stability, and biological interactions.

- Quantum dots demonstrate size dependent fluorescence, offering superior sensitivity for imaging and analytical detection.
- Carbon nanotubes possess exceptional tensile strength and electrical conductivity, enabling pollutant adsorption, electrochemical detection, and biosensing.
- Nanodiamonds provide large surface areas and high chemical stability, facilitating adsorption of organic pollutants and biomolecular interactions.

Comprehensive characterization using TEM, SEM, AFM, XRD, UV-V is spectroscopy, FT-IR, and zeta potential analyses allow researchers to correlate nanoscale structure with analytical performance.

Biomedical Applications

Nanomaterials contribute to several biomedical domains, including advanced imaging, biosensing, controlled drug delivery, and tissue engineering:

- Liposomes and polymeric nanoparticles enhance drug encapsulation, improve pharmacokinetics, and increase bioavailability.
- Gold-mesoporous silica hybrid nanoparticles (Au-MSNs) enable simultaneous imaging and therapy (theranostics).
- PLGA-based nanofibrous scaffolds promote osteogenic differentiation and bone regeneration through biomimetic surface features.
- Quantum dots and nanodiamonds offer high resolution bioimaging due to their superior fluorescence stability.

These systems reflect the potential of nanotechnology to provide minimally invasive, highly sensitive solutions for disease monitoring and treatment.

Environmental and Analytical Applications

Nanotechnology also delivers powerful tools for environmental protection and analytical monitoring:

- Iron oxide nanoparticles (Fe_3O_4 NPs) provide magnetic separation capabilities, enabling efficient removal of heavy metals and pollutants from water.

- Carbon nanotubes and nanodiamonds, owing to their large surface area and tunable surface chemistry, serve as high-performance adsorbents for organic pollutants and toxic metals.
- Functionalization with carboxyl, amino, or thiol groups enhances selectivity toward persistent organic pollutants, heavy metal ions, and endocrine disruptors.
- When synthesized through biocompatible, green methods, these nanomaterials can also detect biomarkers of oxidative stress caused by heavy metal exposure bridging environmental monitoring with biomedical diagnostics.

Thus, the same nanomaterials contribute simultaneously to environmental remediation and biological analysis, exemplifying cross disciplinary value.

Integration into Biomedical and Analytical Systems

The successful integration of nanomaterials into analytical and biological systems depends on size, morphology, surface charge, and chemical functionality:

- PEGylation, polymer coatings, and ligand attachment enhance colloidal stability, reduce aggregation, and increase biocompatibility.
- Superparamagnetic iron oxide nanoparticles (SPIONs) improve MRI contrast due to high saturation magnetization and controlled relaxation properties.
- Quantum dots can be tuned across the 400–800 nm emission range, providing excellent spectral flexibility for multiplexed imaging.

Multifunctional nanostructures allow simultaneous detection, diagnosis, and therapy meeting GAC's goal of maximizing efficiency while minimizing reagent use and material waste.

Biosafety and Green Analytical Considerations

Ensuring biosafety remains central to the development of nanotechnologies. Common toxicity concerns arise from:

- oxidative stress and reactive oxygen species (ROS) generation;
- nanoparticle aggregation;
- surface charge mediated cellular damage;
- accumulation in organs and tissues.

The authors emphasize protective strategies such as:

- PEGylation to reduce immunogenicity and improve circulation time;
- biopolymer capping using chitosan, alginate, or proteins;
- green synthesis routes to reduce residual toxic precursors;
- comprehensive eco toxicological screening to assess environmental impact.

Regulatory frameworks including FDA, EMA, OECD, and ISO nanomaterial standards guide safety evaluation, exposure limits, and lifecycle assessment.

Challenges and Future Directions

Although the field is advancing rapidly, several barriers remain:

- Long-term biocompatibility and chronic exposure effects remain insufficiently studied.
- Large-scale production of uniformly green synthesized nanomaterials is still challenging.
- Reproducibility varies due to biological variability in plant or microbial precursors.
- Surface stability and functional integrity must be optimized for long-term storage.

Future directions include:

- AI driven optimization of green synthesis parameters;
- nanorobotics for precision medicine and targeted therapy;
- quantum dot biosensors with improved biocompatibility;
- life cycle analysis for sustainable nanomaterial design;
- integration of agricultural and industrial waste streams as low-cost nanomaterial precursors.

These innovations will help align nanoscience with global sustainability goals.

Significance for Environmental and Biomedical Analysis

The study demonstrates that nanotechnology offers powerful dual function platforms in analytical chemistry supporting molecular diagnostics, pollutant detection, regenerative medicine, and environmental remediation. Eco-synthesized Fe_3O_4, carbon-based nanomaterials, and polymeric nanostructures exemplify how green nanotechnology enhances both environmental safety and biomedical performance.

By integrating green synthesis, advanced material characterization, and biosafe implementation, Gangadhar and Subburaj [32] highlight how nanotechnology can evolve into a precision tool for sustainable environmental monitoring and healthcare diagnostics. This synergy forms the foundation of next-generation Green Analytical Nanotechnology, uniting scientific innovation with environmental stewardship and ethical responsibility.

3. **Case Study: Nanotechnology in Bioanalysis Innovations for Environmental and Biomedical Applications** [33]

Nanotechnology has emerged as a transformative force in bioanalysis, offering unprecedented improvements in analytical precision, sustainability, and efficiency. The review by Chopra et al. [33] underscores how nanomaterials particularly magnetic nanoparticles, carbon-based nanomaterials, and hybrid polymer–inorganic systems enhance sample preparation, extraction, and detection in both environmental and biomedical matrices.

Their unique physicochemical characteristics, including large surface area, tunable surface charge, high adsorption capacity, and exceptional catalytic behavior, provide powerful alternatives to solvent intensive and time consuming traditional analytical approaches. These innovations align closely with the fundamental principles of GAC.

Objective of the Study

The study aimed to evaluate nanotechnology driven bioanalytical techniques that:

- improve method sensitivity and selectivity;
- reduce solvent consumption and hazardous waste;
- simplify sample preparation workflows;
- integrate green nanomaterials into chromatographic, spectroscopic, and electrochemical analytical platforms.

Special attention was given to magnetic nanoparticles (MNPs), carbon nanomaterials (CNTs, GO, CQDs), molecularly imprinted polymers (MIPs), and metal–organic frameworks (MOFs) all of which are increasingly applied in green bioanalytical practice.

Methodological Framework and Technical Aspects

Conventional vs Nanotechnology-Based Sample Preparation

Traditional extraction techniques such as protein precipitation (PPT), liquid extraction (LLE), and solid-phase extraction (SPE) require high solvent volumes, prolonged processing, multiple purification steps, and often generate hazardous waste.

Nanotechnology provides greener alternatives:

- Magnetic solid-phase extraction (MSPE) enables rapid isolation without centrifugation.
- Micro solid-phase extraction (μSPE) minimizes sample and solvent volumes while maximizing extraction efficiency.

These techniques significantly improve environmental sustainability and analytical throughput.

Magnetic Nanoparticles (MNPs)

Functionalized Fe_3O_4 and γ-Fe_2O_3 nanoparticles serve as versatile sorbents with high affinity for diverse analytes. Their magnetic responsiveness eliminates centrifugation and filtration, reducing time and energy consumption.

Analytical performance highlights include:

- recovery over 80% for arsenic, opioids, antidepressants, and environmental pollutants;
- LOD values as low as 1 ng/mL in HPLC and GC-MS workflows;
- reusability across multiple cycles, supporting sustainable analytical practice;
- solvent free or low solvent operation, adhering to GAC principles.

These attributes position MNPs as model materials for green sample preparation and bioanalysis.

Carbon-Based Nanomaterials (CBNs)

Carbon nanotubes (CNTs), graphene oxide (GO), and carbon quantum dots (CQDs) demonstrate exceptional adsorption capacity, electroconductivity, and photoluminescence.

Key analytical advantages include:

- detection of phenolic drugs, NSAIDs, and fluoroquinolones with recoveries exceeding 100%;
- ultralow LODs achieved using green synthesized CQDs derived from natural resources like milk, lemon juice, and plant extracts;
- high stability, biocompatibility, and suitability for portable and point-of-care devices.

CBNs, especially CQDs, are increasingly preferred due to their inherently green and scalable synthesis.

Molecularly Imprinted Polymers (MIPs)

MIPs offer molecular level selectivity, created by forming polymeric networks around a template analyte. Their highly specific binding sites remain intact after template removal.

Performance indicators:

- tamoxifen detection at 0.01 μM using MIP-CNT composites;
- magnetic MIPs enabling valsartan extraction with 101–102% recovery;
- chiral MIP sensors achieving femtomolar sensitivity for pseudoephedrine.

Their stability, low cost, and customizability make MIPs highly valuable for pharmaceutical and forensic bioanalysis.

Metal–Organic Frameworks (MOFs)

MOFs, such as MIL-101(Cr) and Cu–MOF-199, provide exceptional porosity (up to 10,000 m^2/g), tunable chemistry, and high sorption capacity.

Applications include:

- extraction of NSAIDs and anticancer drugs from plasma;
- achieving LODs of 0.03–0.1 ng/mL via dispersive solid-phase extraction (DSPE) and UPLC–MS;
- enhancing selectivity through functionalized pores tailored for target analytes MOFs combine high analytical performance with low-energy synthesis, making them ideal green sorbents, cause of their stability, selectivity, and low-energy synthesis.

Comparative overview of nanomaterials highlighting key features, analytical techniques, and applications as shown in Table 2.

Environmental and Biomedical Implications

Nanotechnology supported bioanalytical tools offer dual roles in environmental and medical domains:

Environmental Applications

- MNPs and MOFs are effective catalysts and adsorbents for removing heavy metals, organic pollutants, and pharmaceutical residues from water.
- CNTs and GO adsorb persistent organic pollutants (POPs) owing to π–π interactions and surface functionality.

Table 2 Comparative assessment of nanomaterials using greener analytical approaches

Nanomaterial	Key feature	Analytical techniques	Green attribute	Applications
Magnetic NPs (Fe_3O_4, GO-Fe_3O_4)	Magnetic separation, high selectivity	MSPE, HPLC, GC-MS	Solvent free magnetic extraction; reusable sorbent minimizing waste	Drugs, metals, pollutants
Carbon-based NPs (CNTs, CQDs)	High conductivity, potential for green synthesis	μSPE, HPLC-MS	Biogenic/green synthesis; micro scale extraction reducing solvent use	Drugs, biomarkers
Molecularly imprinted polymers (MIPs)	Molecular recognition, high selectivity	Electrochemical, HPLC	Reusable polymeric sorbent; low solvent consumption; high selectivity	Pharmaceuticals
Metal–Organic Frameworks (MOFs)	High porosity, tunable surface chemistry	DSPE, UPLC-MS	Recyclable sorbent; low solvent dispersive extraction; sustainable synthesis routes	NSAIDs, anticancer drugs

Biomedical Applications

- Nanomaterials assist in drug monitoring, therapeutic drug management, and detection of disease biomarkers.
- MOFs and CQDs support fluorescence-based biosensing with high sensitivity.
- Biocompatible nanomaterials reduce interference and matrix effects in clinical assays.

This dual functionality highlights the increasing relevance of nanotechnology in global health and environmental sustainability.

Challenges and Future Outlook

Even though nano-enabled bioanalysis has great potential, there are still issues with standardizing toxicity assessment, enhancing biostability, and guaranteeing repeatable green synthesis. Future directions include developing multi analyte nanosensors for real time, in field bioanalytical monitoring, integrating nanotechnology with microfluidic systems, and processing data using artificial intelligence.

This case study demonstrates how nanotechnology significantly enhances bioanalytical chemistry by creating eco-friendly, highly sensitive, and selective analytical platforms. Magnetic, carbon-based, polymeric, and hybrid nanomaterials streamline sample preparation, reduce solvent usage, and provide exceptional analytical sensitivity addressing key biomedical and environmental needs. The integration of GAC ensures that nano-enabled bioanalysis supports both technological innovation and environmental stewardship, paving the way for a more sustainable analytical future.

7 Conclusion

This chapter highlights how GAC and nanotechnology work together to create analytical methods that are more sustainable, efficient, and environmentally friendly. Nanomaterials such as metal nanoparticles, carbon-based materials, and metal oxides support GAC principles by reducing solvent use, lowering energy consumption, and minimizing waste, while also improving sensitivity, selectivity, and speed of analysis. Their unique properties make it possible to work with very small sample volumes, perform direct or in-situ analysis, and develop portable, miniaturized sensing devices.

A major focus of the chapter is the green synthesis of nanomaterials, including plant-based (phytosynthesis), microbial, biopolymer assisted, and microwave-assisted methods. These sustainable approaches eliminate hazardous chemicals and promote safer, eco-friendly production of nanoparticles. The chapter also explains how green nanocatalysts such as metal nanoparticles, metal oxides, carbon nanomaterials, nanozymes, and photocatalysts enable faster reactions under mild conditions, reducing the need for toxic reagents and high energy inputs.

Finally, real case studies demonstrate how green nanotechnology is applied in environmental monitoring, pollutant removal, water analysis, drug detection, biosensing, and biomedical diagnostics. Together, the chapter shows that integrating GAC principles with nanotechnology leads to cleaner, safer, and more advanced analytical systems that support both scientific innovation and environmental sustainability.

References

1. Miladinović SM. Green Analytical Chemistry: Integrating Sustainability into Undergraduate Education. Anal Bioanal Chem. 2025;417(4):665–73. Available from: https://doi.org/10.1007/s00216-024-05680-4
2. Meher AK, Zarouri A. Green Analytical Chemistry-Recent Innovations. Analytica. 2025;6(1):10
3. Verma A, Gautam SP, Bansal KK, Prabhakar N, Rosenholm JM. Green Nanotechnology: Advancement in Phytoformulation Research. Medicines. 2019;6(1):39
4. Khan F, Shariq M, Asif M, Siddiqui MA, Malan P, Ahmad F. Green Nanotechnology: Plant-Mediated Nanoparticle Synthesis and Application. Nanomaterials. 2022;12(4), 673
5. Akbarzadeh A, Samiei M, Davaran S. Magnetic Nanoparticles: Preparation, Physical Properties, and Applications in Biomedicine. Nanoscale Res Lett. 2012;7(1):144. Available from: http://www.nanoscalereslett.com/content/7/1/144
6. Dhahi TS, Yousif Dafhalla AK, Tayfour OE, Mubarakali A, Alqahtani AS, Tayfour Ahmed AE, et al. Advances in Nano Sensors for Monitoring and Optimal Performance Enhancement in Photovoltaic Cells. iScience. 2024;27(4):109347. Available from: https://doi.org/10.1016/j.isci.2024.109347

7. Gałuszka A, Migaszewski Z, Namieśnik J. The 12 Principles of Green Analytical Chemistry and the Significance Mnemonic of Green Analytical Practices. TrAC - Trends Anal Chem. 2013;50:78–84
8. Burlec AF, Corciova A, Boev M, Batir-Marin D, Mircea C, Cioanca O, et al. Current Overview of Metal Nanoparticles' Synthesis, Characterization, and Biomedical Applications, with a Focus on Silver and Gold Nanoparticles. Pharmaceuticals. 2023;16(10):1410
9. Speranza G. Carbon nanomaterials: Synthesis, Functionalization and Sensing Applications, Nanomaterials 2021;11(4):1–99
10. Cotta MA. Quantum Dots and Their Applications: What Lies Ahead? ACS Appl Nano Mater. 2020;3(6):4920–4
11. Rabiee N. Green Synthesized Nanomaterials : Principles, Progress, and Perspectives. Green Nanomater Sustain Innov Divers Appl. Chapter 1: 2025:1–27
12. Muniyappan N, Pandeeswaran M, Amalraj A. Green synthesis of Gold Nanoparticles using Curcuma Pseudomontana Isolated Curcumin: Its Characterization, Antimicrobial, Antioxidant and Anti- inflammatory Activities. Environ Chem Ecotoxicol. 2021;3:117–24. Available from: https://doi.org/10.1016/j.enceco.2021.01.002
13. De Mendoza P, Echavarren AM. Synthesis of Arenes and Heteroarenes by Hydroarylation Reactions Catalyzed by Electrophilic Metal Complexes. Pure Appl Chem. 2010;82(4):20–80
14. Abo LD, Areti HA, Jayakumar M, Rangaraju M, Subashini S. Nanobiomaterials-Enabled Sensors for Heavy Metal Detection and Remediation in Eastewater Systems: Advances in Synthesis, Characterization, and Environmental Applications. Results Eng. 2025;27:105694. Available from: https://doi.org/10.1016/j.rineng.2025.105694
15. Nadkarni A, Rana D, Desai N, Benival D, Joshi V, Salave S, et al. Advanced Characterization and Sample Preparation Strategies for Nanoformulations. J Nanotheranostics. 2024;5(3):104–27
16. Gurkan UA, Wood DK, Carranza D, Herbertson LH, Diamond SL, Du E, et al. Next Generation Microfluidics: Fulfilling the Promise of lab-on-a-Chip Technologies. Lab Chip. 2024;1867–74
17. Leong SX, Leong YX, Koh CSL, Tan EX, Nguyen LBT, Chen JRT, et al. Emerging Nanosensor Platforms and Machine Learning Strategies Toward Rapid, Point-of-Need Small-Molecule Metabolite Detection and Monitoring. Chem Sci. 2022;13(37):11009–29
18. Perrucci M, Ali I, Mansour FR, Ulusoy HI, Ulusoy S, Kabir A, et al. Chemical Analysis using Miniaturized and Portable 3D Printed Systems: Where Are We Now? J Chromatogr Open. 2025;8(July):100241. Available from: https://doi.org/10.1016/j.jcoa.2025.100241
19. Salaheldeen M, M. Abu-Dief A, El-Dabea T. Functionalization of Nanomaterials for Energy Storage and Hydrogen Production Applications. Materials (Basel). 2025;18(4): 768.
20. Shah SIA, Ahmad W, Anwar M, Shah R, Khan JA, Shah NS, et al. Synthesis, Properties, and Applications of Fe_3O_4 and Fe_3O4-based Nanocomposites: A Review. Appl Catal O Open. 2025;203(8):207049. Available from: https://doi.org/10.1016/j.apcato.2025.207049
21. Xu S, Lee TR. Nguyen, M. D., Tran, H.-V., Xu, S. (2021). Fe_3O_4 Nanoparticles: Structures, Synthesis, Magnetic Properties, Surface Functionalization, and Emerging Applications. Applied Sciences, 11(23), 11301. https://doi.org/10.3390/app112311301
22. Bendicho C, Lavilla I, Pena-Pereira F, de la Calle I, Romero V. Paper-based Analytical Devices for Colorimetric and Luminescent Detection of Mercury in Waters: An overview. Sensors. 2021;21(22):7571
23. Liu J, Li R, Yang B. Carbon Dots: A New Type of Carbon-Based Nanomaterial with Wide Applications. ACS Cent Sci. 2020;6(12):2179–95
24. Jiříčková A, Jankovský O, Sofer Z, Sedmidubský D. Synthesis and Applications of Graphene Oxide. Materials (Basel). 2022;15(3): 920
25. Masroor S, Kumar A, Goswami A, Krishan H, Hussain S, Khan ME. Sustainable Approaches towards Nanomaterial Synthesis: Strategies, Properties, and Process Integration. Sustain Process Connect. 2025;1:0011
26. Hughes KJ, Iyer KA, Bird RE, Ivanov J, Banerjee S, Georges G, et al. Review of Carbon Nanotube Research and Development: Materials and Emerging Applications. ACS Appl Nano Mater. 2024;7(16):18695–13

27. Geleto SA, Ariti AM, Gutema BT, Abda EM, Abiye AA, Abay SM, et al. Nanocellulose/ Fe_3O_4/Ag Nanozyme with Robust Peroxidase Activity for Enhanced Antibacterial and Wound Healing Applications. ACS Omega. 2023;8(51):48764–74
28. Kumar A, Singh R, Pathak A, Guin D, Tripathi CSP. CeO_2 Nanosheets with Prominent Peroxidase-Mimicking Activity for the Colorimetric Sensing of H_2O_2, Glucose, and Ascorbic Acid. Next Nanotechnol. 2025;8:100239. Available from: https://doi.org/10.1016/j.nxnano.2025.100239
29. Farzin MA, Naghib SM. Laccase-mimicking Cu@CDs-Enabled Bioinspired Immunosensor using CuO Nanoparticles for Voltametric Detection of Serum Thyroxine Level in Patients with Hyperthyroidism or Hypothyroidism. Sensors and Actuators Reports. 2025;9:100291. Available from: https://doi.org/10.1016/j.snr.2025.100291
30. Bhanderi D, Lakhani P, Modi CK. Graphitic Carbon Nitride (g-C3N4) as an Emerging Photocatalyst for Sustainable Environmental Applications: A Comprehensive Review. RSC Sustain. 2023;2(2):265–87
31. de Oliveira ML, Rodrigues LM, Silva da Veiga MAM, Souza LRR. Advancing Nanomaterial Synthesis: Harnessing Green Chemistry for Sustainable Innovation. Green Anal Chem. 2024;1:100148
32. Gangadhar L, Subburaj S. Nanotechnology Advances for Biomedical Applications. Front Nanotechnol. 2025;7:1639506
33. Chopra A, Kumari Y, Kumar S, Sharma R, Bhatia R. Nanotechnology in Bioanalysis: Current Trends and aAplications. J Chromatogr Open. 2025;8:100254. Available from: https://doi.org/10.1016/j.jcoa.2025.100254

Biotechnological Approaches in Green Analytical Chemistry

Preeti D. Kulkarni and Manjushree K. Pawar

Abstract Biotechnology is increasingly recognized as a cornerstone of sustainable innovation, offering transformative solutions across analytical science, environmental monitoring, food systems, and pharmaceutical development. As global priorities shift toward eco-friendly practices, biotechnology provides versatile tools that reduce hazardous chemical use, improve analytical efficiency, and support greener industrial processes. Advances in enzyme-based and microbial biosensing, nano-enabled detection systems, and biodegradable polymer platforms have led to analytical technologies that are highly sensitive, portable, and aligned with the principles of green chemistry. These developments minimize waste generation, lower energy demands, and enable rapid, on-site measurements for diverse scientific and environmental applications. In parallel, microbial, cellular, and emerging cell free biosensors contribute significantly to sustainable environmental monitoring. Their ability to sense bioavailable pollutants, generate real-time signals, and function in resource-limited or field settings makes them exceptionally valuable for assessing water quality, soil contamination, air pollutants, and general ecosystem health. Modern improvements in genetic engineering microfluidics, and digital integration have enhanced the sensitivity, stability, and practicality of these biosensing systems, expanding their global relevance for environmental protection and public health. Biotechnology also continues to reshape food and pharmaceutical sectors through innovative approaches such as fermentation-based production, engineered probiotics, cellular agriculture, recombinant therapeutics, monoclonal antibodies, and mRNA vaccine platforms. These technologies enhance product safety, nutritional value, therapeutic efficacy, and manufacturing sustainability. Additionally, bioremediation, waste valorization, and circular bioeconomy strategies highlight biotechnology's role in reducing environmental burden while generating valuable bioproducts. Across these diverse applications, considerations such as biosafety, regulatory compliance,

P. D. Kulkarni (✉)
LSHGCT's Gahlot Institute of Pharmacy, Navi Mumbai, Maharashtra, India
e-mail: preeti.qara@gmail.com

M. K. Pawar
Department of Pharmaceutics, LSHGCT's Gahlot Institute of Pharmacy, Navi Mumbai, Maharashtra, India

V. Bhusari and M. Ghante (eds.), *Green Analytical Chemistry in India*,
https://doi.org/10.1007/978-3-032-24564-9_7

lifecycle assessment, and responsible innovation remain essential to ensure sustainable implementation. Overall, biotechnology provides a powerful, integrated framework for advancing environmentally responsible analytical methods, enabling next generation food and pharmaceutical systems. Its continued evolution is expected to drive significant progress toward global sustainability, environmental resilience, and improved human health.

Keywords Green analytical chemistry · Biosensors · Biodegradable polymers · Environmental monitoring · Biotechnological applications

1 Introduction

The rapid expansion of biotechnology has fundamentally reshaped the way modern science approaches food production, environmental protection, pharmaceutical innovation, and analytical chemistry. With rising global concerns surrounding climate change, public health, and resource depletion, there is a pressing need for technologies that are not only efficient but also environmentally responsible and economically feasible. Biotechnology addresses these demands by harnessing biological systems enzymes, cells, microorganisms, and biodegradable materials to deliver cleaner, safer and more sustainable solutions.

In analytical chemistry, traditional methods often rely on hazardous solvents, energy intensive processes, and complex instrumentation. The shift toward Green Analytical Chemistry has accelerated the development of biosensors, nano-enabled platforms, and biodegradable polymer-based devices. These innovations reduce chemical waste, minimize environmental impact, and enable rapid, on-site analysis. Enzyme-based, optical, electrochemical, and microfluidic biosensors now form an essential toolkit for laboratories aiming to implement greener workflows.

Parallel to these advances, microbial and cellular biosensors have gained prominence in environmental science due to their unique ability to sense bioavailable pollutants, respond to chemical stressors, and function autonomously in natural settings. Microbial fuel cell biosensors, genetically engineered bioreporters, and freeze-dried cell free systems offer portable, cost-effective options for real-time green monitoring. These platforms support sustainable water quality assessment, soil health evaluation, and detection of heavy metals, pesticides, organic contaminants, and biological oxygen demand.

Beyond sensing and analysis, biotechnology underpins major transformations in the food and pharmaceutical sectors. Food biotechnology enhances nutritional value, improves safety, and opens pathways for sustainable production through precision fermentation, probiotics, enzymes, and cellular bioeconomy development drives breakthroughs in drug discovery, recombinant protein production, monoclonal antibodies, advanced vaccines, and cell and gene-based therapies.

This chapter synthesizes these diverse yet interconnected domains, presenting a comprehensive overview of how biotechnology supports sustainability, innovation, and global well-being. By integrating green analytical tools, advanced biosensing systems, and modern biotechnological applications, the chapter underscores biotechnology's role as a cornerstone of future ready, environmentally aligned scientific and industrial advancements.

2 Enzyme-Based Biosensors for Sustainable Detection

Principles and Mechanisms

Fundamental Principles:

Enzyme-based biosensors are special tools that use natural ability of enzymes which are natural proteins that speed up reactions to detect and measure different substances as they act like a "key" that fits with a "lock" (the target substance). A transducer also called as sensor transforms this biochemical reaction into a signal which is measurable like electrical, optical or colorimetric responses. As these enzymes are specific, they do not require a lot of sample manipulation or chemicals or steps and this marks their importance in green chemistry by less pollution and waste and safer for the environment thus supporting the goals of green chemistry and sustainability [1–5].

Mechanisms of Detection:

The enzyme-based biosensors work under the mechanism of catalytic biosensing. When the enzyme comes in contact with analyte, it undergoes following three interactions:

Catalytic Conversion

The enzyme metabolizes the analyte into another product which is directly proportional to the amount of analyte. For example, in glucose biosensors the enzyme glucose oxidase converts glucose into gluconic acid and hydrogen peroxide which is then detected by using electrical methods [1, 2, 6].

Enzyme Inhibition or Activation

Some analytes can inhibit or activate the enzyme and the level of inhibition or activation shows how much analyte is present. Example is measurement of inhibition of acetylcholinesterase enzyme in pesticide detection by some analytes [1, 2, 8].

Change in Enzyme Properties

The analyte may change the enzymes' physical or chemical characteristics like its shape, charge or structure. These changes influence the signal output by the biosensor [1, 2, 5].

Signal Transduction Modalities

The measurable signals which are created by the transducer after the enzyme interacts with the analyte are detected by following methods:

Electrochemical These biosensors are sensitive, portable and selective which makes them very significant and common. They detect changes in voltage, current, resistance or capacitance caused from reactions by enzymes on the surface of electrode. Types: Amperometric, Potentiometric, Conductometric biosensors [1, 2, 5–7].

Electrochemical biosensors are used for detecting glucose, pesticides, and heavy metals in environmental and clinical samples [3, 9–11]. Recent advances include (1) Integration of nanoparticles—carbon nanotubes, graphene, metal nanoparticles for enhanced sensitivity and electron transfer [8, 9]. (2) Screen-printed electrodes and 3D-printed platforms for low-cost, mass production and portable diagnostics [5].

Optical Use optical (light)-based properties like absorbance, fluorescence or luminescence screen to detect enzyme-analyte reactions. The enzymatic optical biosensors can be used without labels that directly detects the natural change like absorbance or refractive index without adding any extra dye or marker. This process does not require any extra chemicals, shows results instantly as the reaction happens and one biosensor can test many analytes in the same experiment [1, 2, 9].

Colorimetric The distinct color changes are produced due to enzymatic reactions that are visible by naked eye or detected by using spectrophotometry [5, 10, 11].

Other Types and Configurations

Piezoelectric and Mass-Sensitive Biosensors

Piezoelectric biosensors, such as quartz crystal microbalance (QCM) platforms, detect mass changes resulting from enzyme-analyte interaction. These are suited for real-time protein and small molecule detection; recent advances achieve molecular resolution for cancer biomarkers [8].

Self-powered Biosensors

Self-powered enzymatic biosensors utilize biofuel cell systems, generating their own operational current from enzymatic breakdown of substrates like glucose. These eliminate external power needs and are being incorporated into wearable and implantable medical devices [9].

Microfluidic and Hybrid Designs

Microfluidic biosensors integrate enzymes within disposable, polymer-based chips, enabling rapid multiplex testing with minimal sample and reagent volumes [5].

Hybrid Biosensors

Co-immobilize multiple enzymes or combine enzymes with redox mediators/ conductive polymers, improving sensitivity and multi-analyte detection [3, 4].

Latest Innovations

Flexible, stretchable biosensors: Incorporating gels and nanocomposites enables biocompatible, skin adhesive devices for noninvasive monitoring [9].

CRISPR-based enzymatic biosensors: Enable highly specific detection with molecular precision for infectious diseases and cancer [8].

AI-enhanced biosensor analytics: Artificial intelligence aids in real-time data processing and efficacy evaluation for complex diagnostics [8].

Key Performance Enhancers

Nanomaterial scaffolds: Increase enzyme loading and electron transfer, boosting detection limits and stability [8, 9].

Advanced immobilization: Covalent binding, entrapment, and crosslinking improve enzyme stability and lifespan [1–4].

Tunable device structures: 3D configurations, microfluidics, and integrated electronics minimize diffusion barriers and enable decentralized testing [5, 9].

Biotechnological Approaches Supporting Green Analytical Chemistry

Use of Sustainable Materials and Design

Modern biosensors are now being made with eco-friendly materials like paper-based substrates which need minimal reagent [10, 11]. To make the sensors last longer and work better, enzymes can be fixed onto these supports in different ways like trapping them in nanomaterials or chemically attaching them to "green" surfaces. This improves stability, reusability and allows sensors to be made smaller and more portable [1, 2, 5].

Integration with Nanotechnology

Nanomaterials such as gold nanoparticles, carbon nanotubes, or metal oxides are very useful in biosensors. They help enzymes work more effectively by improving electron transfer and sensitivity. This means that biosensors can detect very small amounts of substances, work faster, and can be designed for miniaturized, point-of-care devices like simple test strips or portable kits [1, 2, 5].

Whole Cell and Genetic Engineering Strategies

Some biosensors use genetically modified microorganisms or cells that can sense pollutants directly. By adding or altering certain genes, these cells can detoxify harmful substances, improve detection of complex pollutants, and increase the sensor's specificity and functionality. This makes them useful for monitoring environmental contaminants [3].

Enzyme Recycling and Sensor Reusability

To make biosensors more sustainable, scientists are developing ways to reuse enzymes by protecting or reactivating them after each use. This reduces waste, lowers costs, and saves resources—making biosensors greener and more efficient [1, 2].

Examples and Applications

Glucose Biosensor (Amperometric) Utilizes glucose oxidase for diabetes monitoring; high selectivity, rapid response, low sample volumes [1, 2, 6, 7].

Urea Biosensor (Potentiometric) Urease-based sensors for urea in clinical and environmental samples, often incorporating nanoparticles to boost sensitivity [1, 2].

Environmental Monitoring Engineered microbial fuel cell-based biosensors quantify water quality parameters and heavy metals, using indigenous or modified bacteria interfaced with electrical circuits—converting biochemical energy directly into readable signals [3, 4].

Role in Green Analytical Chemistry

Enzyme-based biosensors exemplify Green Analytical Chemistry by:

- Reducing hazardous chemical usage and sample volume.
- Promoting environmentally benign materials and renewable biological reagents.
- Enabling rapid, on-site, and multiplexed analysis with minimal waste.
- Facilitating device integration for continuous and real-time monitoring in environmental applications [5–7, 10, 11].

Green Advantages

Reduced use of hazardous chemicals and solvents.

Minimal energy consumption due to mild operational conditions.

Examples: Portable glucose sensors for diabetes monitoring; acetylcholinesterase-based biosensors used to quickly test for pesticide residues in food like vegetables or in water. These are portable and give results in the field with almost no chemical waste (Table 1).

Limitations and Future Prospects

Enzyme stability, storage conditions, and cost can be hurdles, but advances in enzyme immobilization, genetic engineering, and nanotechnologies are addressing these challenges.

Table 1 Comparison of traditional versus enzyme-based detection methods in pesticide analysis

Method	Reagents used	Waste generated	Sensitivity	Green chemistry compliance
Conventional GC-MS	Organic solvents	High	High	Low
Enzyme-based biosensor	Aqueous buffer, enzyme	Minimal (biodegradable)	Moderate to high	High

3 Role of Biodegradable Polymers in Analytical Techniques

Biodegradable polymers are materials that can break down naturally over time, making them eco-friendly. In analytical chemistry, these polymers are used for preparing samples, making separation media, sensors, microfluidic devices, and disposable electrodes. Because their chemical structure can be easily modified, they can form films or gels, and they interact well with biological materials. These properties make them useful for creating special materials that can selectively absorb or detect certain chemicals, support enzyme activity, or convert chemical information into readable signals. Using biodegradable polymers helps achieve the goals of green chemistry such as reducing hazardous waste and using renewable materials. They can also be processed into analytical tools through methods like electrospinning, casting, or 3D printing, offering low-cost and sustainable alternatives. This chapter discusses the relationship between polymer structure and analytical performance, highlights recent advances, and explored future prospects for their use in analytical science [12].

Modern analytical chemistry is shifting toward environmentally friendly and affordable solutions. There is a growing demand for small, portable, and efficient devices that can perform accurate measurements with minimal waste. Biodegradable and bio-based polymers are ideal for this because they not only lower environmental impact but also provide flexible chemical structures that can be tailored for recognizing and detecting specific molecules. In recent years, numerous studies have shown that these polymers are increasingly being used in different areas such as sample preparation like in solid-phase extraction and microextraction, sensor fabrication in both optical and electrochemical, separations, and 3D printing of disposable analytical tools. This chapter brings together information from current research and gives useful insights for scientists working on eco-friendly analytical technologies [13].

Biodegradable Polymer Classes and Their Key Properties

Polysaccharides: This group includes natural polymers such as chitosan, cellulose and its derivatives, starch, and alginate. These materials are plentiful in nature and contain many hydroxyl or amine groups, which make them easy to modify chemically. They are hydrophilic, form films or gels easily, and are compatible with enzymes making them useful for immobilization processes. Chitosan, in particular has reactive amine groups that allow it to bind with other molecules and adjust to different pH levels, making it effective for selective adsorption applications [14].

Aliphatic polyesters: Examples are polylactic acid (PLA), polycaprolactone (PCL), and polyhydroxyalkanoates (PHAs). These polymers are thermoplastic, meaning they can be easily molded when heated, and they have adjustable mechanical strength. Their biodegradable nature makes them suitable for temporary analytical tools. PLA, when combined with carbon-based fillers, is frequently used to produce disposable electrochemical electrodes and filaments for 3D printing [15].

Protein-Based and Other Biopolymers: This class includes natural proteins such as silk fibroin, gelatin, and collagen. They are especially useful in applications that require biological compatibility or specific molecular interactions, such as in biosensors [5]. Key physical and chemical characteristics that make these materials suitable for analytical purposes include their surface functional groups (for chemical bonding or molecular recognition), porous structures (helpful in extraction or separation processes), and electrical properties (important when they are modified to conduct electricity, as in electrochemical analysis) [17].

Role in Sample Preparation and Extraction

Biodegradable polymers are increasingly being used as sorbent materials in various sample preparation methods such as solid-phase extraction (SPE), dispersive SPE, solid-phase microextraction (SPME), and liquid-phase microextraction.

SPE and Dispersive SPE Materials like chitosan and cellulose can be modified or crosslinked with special molecules (ligands) such as chelators or ion-exchange groups to create selective sorbents. These customized polymers can efficiently extract and isolate metals, pesticides, and pharmaceutical compounds from complex mixtures. For example, magnetic chitosan composites formed by combining chitosan with iron oxide allow fast and easy extraction through magnetic separation. Studies have shown that these materials significantly improve the recovery of trace organic compounds, especially when chitosan-based nanomaterials are used [18].

Microextraction (DIAME/MDLME Variants) In modern analytical chemistry, microextraction techniques such as DIAME (dispersive in-air microextraction) and MDLME (magnetic dispersive liquid microextraction) are increasingly adopting biopolymers as eco-friendly solid sorbents or as part of composite extraction materials. These natural polymers serve as sustainable alternatives to traditional sorbents made from non-renewable, fossil-based polymers. Recent studies and reviews highlight that using biodegradable materials in microextraction reduces environmental impact and supports Green Analytical Chemistry principles. Such materials can be engineered to possess desirable surface characteristics, porosity, and selectively, making them efficient for extracting a variety of analytes from complex samples while minimizing chemical waste [19].

Molecular Imprinting on Biopolymers Molecular imprinting is a technique used to design materials with specific recognition sites for target molecules. When applied to biodegradable polymers like cellulose or chitosan, molecularly imprinted polymer (MIP) layers can be formed that mimic the shape and chemical environment of desired analytes. This allows for selective binding, high recognition efficiency, and reusability in analytical applications. For example, imprinted chitosan and cellulose-based composites have shown excellent potential for detecting trace levels of organic and inorganic substances. These biopolymer-based MIPs combine biodegradability, biocompatibility, and high selectivity, making them greener alternatives to traditional synthetic polymer sorbents [20].

Analytical Performance and Considerations:

When carefully optimized, biodegradable polymer sorbents can perform nearly as well as conventional synthetic materials in terms of detection limits, selectivity, and analyte recovery. However, some challenges remain such as ensuring consistent quality across production batches and maintaining long-term stability. To ensure reliable analytical results, factors like crosslinker type, pore structure, chemical stability, and background interference must be evaluated and validated. Despite these challenges, biodegradable sorbents continue to attract attention for their balance between performance and environmental safety, marking a promising shift toward more sustainable analytical technologies [20].

Sensors and Transducers

Biodegradable polymers play a vital role in developing modern sensing devices. They are used not only in biochemical recognition where they help detect biological molecules but also in device fabrication, providing environmentally friendly and cost-effective materials. Their biocompatibility, film-forming ability, and tunable surface chemistry make them suitable for various types of sensors, including electrochemical, optical, and biosensors.

Electrochemical sensors: It is one of the most important applications of biodegradable polymers which detect chemical or biological substances through electrical signals. In recent years, 3D-printed electrodes made from polylactic acid (PLA) composites have become popular because they are not expensive, customizable, and environmentally sustainable. PLA filaments blended with conducive materials like carbon black or graphene form electrodes that can be printed on demand and used directly in experiments. To enhance their sensitivity and performance, these electrodes are often treated using chemical, thermal, or electrochemical activation. Such treatments improve the electrode's ability to transfer electrons, which is essential for accurate detection of analytes such as pesticides, pharmaceutical residues, and environmental pollutants. Researchers have also developed various surface modification like by coating electrodes with nanoparticles or enzymes to further increase selectivity and stability. These biodegradable electrodes have been successfully applied in fields such as environmental monitoring, food safety testing, and biomedical diagnostics. Another significant use of biodegradable materials is as immobilization layers. Polymers like chitosan and cellulose are frequently used to anchor biomolecules such as enzymes or antibodies onto electrode surfaces. Chitosan films, in particular, offer strong adhesion, chemical reactivity through amine groups, and good compatibility with biological molecules. This allows the development of sensors that are not only sensitive but also biocompatible and environmentally safe [14, 21].

Optical sensors and colorimetric strips: Biodegradable polymers are also valuable in optical and colorimetric sensing systems, which detect changes in light or color to identify specific substances. Materials such as cellulose paper or modified cellulose membranes are ideal for creating low-cost, disposable sensing platforms. These materials are often used in microfluidic devices, lateral flow assays, and smartphone-readable diagnostic strips. Their porous nature allows fluids to flow easily, facilitating rapid reactions and detection. Additionally, since cellulose and paper are biodegradable, these test strips can be safely discarded after use, minimizing environmental waste. Such platforms are commonly employed in healthcare diagnostics, environmental testing, and food quality control, providing a sustainable alternative to traditional plastic-based sensors [22].

Biosensors: Biosensors are analytical devices that combine a biological recognition element such as an enzyme, antibody, or DNA strand with a physical transducer that converts biological interaction into a measurable signal. Biodegradable polymers provide an excellent support matrix for these biological components because of their biocompatibility and chemical flexibility. Materials such as gelatin, chitosan, and silk fibroin can be used to form films, gels, or hydrogel matrices that immobilize enzymes or antibodies without damaging their activity. This helps maintain the biological molecule's natural structure and ensures consistent and stable sensor responses. For example, chitosan hydrogels are widely used in enzyme-based electrochemical biosensors due to their water retention ability, ionic conductivity, and mild processing conditions. Such sensors are often used in medical diagnostics, glucose monitoring, and environmental analysis. The use of biopolymers also enhance the reusability and sustainability of biosensors. Unlike conventional materials derived from fossil fuels, biodegradable matrices can degrade naturally, making them suitable for single-use or disposable biosensor systems [14].

Overall, the integration of biodegradable polymers into sensor technology supports the goals of Green Analytical Chemistry, reducing toxic waste, promoting renewable materials, and enabling cost-effective device production. Their combination of mechanical strength, flexibility, and biocompatibility makes them ideal for next generation analytical devices. As research advances, these materials are expected to play an even greater role in developing smart, sustainable, and miniaturized sensing systems for diverse scientific and industrial applications.

Chromatography, Membranes and Separation Media

Stationary Phases and Packing Materials: In recent years, researchers have increasingly explored the use of biodegradable polymers, such as modified cellulose and silica cellulose composites, as sustainable alternatives to traditional chromatographic packing materials. These eco-friendly materials can serve as stationary phases in techniques like high-performance liquid chromatography (HPLC) and solid-phase extraction. Cellulose, when chemically modified, offers multiple reactive sites that can be customized for selective adsorption of analytes. In addition, hybrid materials combining cellulose with silica or other inorganic supports provide a balance of mechanical strength, chemical resistance, and biodegradability. These materials also demonstrate good solvent stability and can withstand the high pressures typically used in chromatographic systems. However, large-scale industrial adoption is still limited by certain challenges, such as optimizing the mechanical durability and maintaining long-term structural stability under repeated use. Despite this, the trend toward Green Analytical Chemistry continues to drive interest in replacing petroleum-based polymers with biodegradable alternatives in separation science [23].

Membranes: Biodegradable polymer membranes have gained significant attention for their roles in filtration, pre-concentration, and selective separation processes.

Cellulose and alginate membranes are widely used due to their hydrophilic nature, flexibility, and ability to form stable films. These membranes are useful in enzyme reactors, flow-through analytical systems, and sample cleanup procedures. Researchers have also functionalized these membranes by attaching specific chemical or biological groups to their surfaces, allowing for targeted molecule capture and improved selectivity. Such modifications enhance the membrane's performance in complex analytical applications, including biosensing, drug purification, and environmental monitoring. Recent studies describe "intelligent" or stimulus responsive membranes made from cellulose derivatives that can respond to changes of pH, temperature, or ionic strength. These adaptive materials can alter their permeability or binding capacity-based on environmental conditions, making them promising for advanced separation technologies and controlled release systems. Overall, the combination of biodegradability, biocompatibility, and tunable surface chemistry makes cellulose and alginate-based membranes highly attractive for sustainable analytical and biotechnological applications. As fabrication methods improve and material properties are further optimized, these polymers are expected to replace conventional synthetic membranes in many laboratory and industrial processes [24].

Fabrication Techniques that Enable Analytical Devices

Developing analytical devices from biodegradable polymers requires effective fabrication methods that can shape, modify, and structure these materials for specific applications. Several modern fabrication techniques such as electrospinning, 3D printing, and casting or coating are widely used to create eco-friendly components for sensors, microfluidic systems, and sample preparation tools. These methods allow accurate control over the material's structure, porosity, and thickness, which directly influence the sensitivity, selectivity, and durability of analytical devices.

Electrospinning It is versatile technique used to produce very fine fibers or mats with a high surface area. During this process, a polymer solution is exposed to an electric field, which draws out ultrathin fibers that solidify into a mesh-like mat. Biodegradable polymers such as polylactic acid (PLA) and cellulose acetate are commonly used in electrospinning to create lightweight and porous materials. These electrospun mats have unique advantages—their large surface area and porous structure make them ideal for use as sorbents in sample extraction or as support layers in sensors and biosensors. They can effectively capture analytes from complex mixtures or provide a stable base for immobilizing enzymes and antibodies. Additionally, electrospinning allows easy blending of polymers with nanoparticles or biomolecules, improving their performance in analytical applications. Researchers are increasingly using electrospun biodegradable polymers for Green Analytical Chemistry, as they replace synthetic, non-degradable materials and reduce laboratory waste [17].

3D Printing (Fused Deposition Modeling, FDM) This is another promising fabrication method for producing analytical devices from biodegradable materials. In this process, a polymer filament often PLA or PLA mixed with carbon-based materials like graphite or graphene is melted and precisely deposited layer by layer to form a desired 3D-structure. This technique allows rapid prototyping of various components such as sensor housings, microfluidic channels, electrodes, and solid-phase extraction (SPE) cartridges. Because the process is computer controlled, complex and customized device designs can be fabricated quickly and at a low cost. 3D-printed biodegradable electrodes have shown excellent performance in detecting environmental pollutants, pharmaceutical residues, and biological compounds. Moreover, the ability to print devices on demand makes 3D printing an attractive approach for portable and disposable analytical systems. This technique supports the goals of sustainable manufacturing by minimizing material waste and using renewable polymer sources [21].

Casting and Coating These are simple yet effective fabrication methods used to produce thin films and membranes from biodegradable polymers. Materials such as cellulose and chitosan are dissolved in suitable solvents and then cast into flat molds or coated onto substrates, where they dry to form uniform films. These polymer films are widely applied in paper-based microfluidic devices, dipsticks, and diagnostic test strips. They provide excellent flexibility, transparency, and compatibility with biological and chemical reagents. Because of their smooth surface and adjustable thickness, cast or coated films are often used as functional layers in electrochemical and optical sensors. Also, coating biodegradable polymers onto conductive substrates can enhance the adhesion of biomolecules, improve signal stability, and increase device sensitivity. These features make casting and coating essential techniques for low cost, eco-friendly analytical platforms that can be easily fabricated in laboratory settings [22].

These fabrication techniques have revolutionized how biodegradable polymers are used in analytical chemistry. They allow researchers to create customized, disposable, and environmentally sustainable devices without relying on conventional plastics or non-renewable materials. As these technologies continue to advance, biodegradable polymers are expected to play an even greater role in next generation analytical devices, particularly in point-of-care diagnostics, environmental testing, and microfluidic-based analytical systems.

Green Analytical Chemistry and Sustainability Considerations

The use of biodegradable polymers in analytical science is closely aligned with the principles of Green Analytical Chemistry, which aims to reduce environmental impact while maintaining analytical efficiency. These materials help minimize reliance on non-renewable petrochemical plastics and offer easier disposal at the end of their use. Additionally, they often require fewer or less toxic solvents, particularly when combined with environmentally friendly sample preparation techniques such as microextraction and solvent free analytical methods. By using biodegradable polymers, laboratories can reduce chemical waste and design analytical systems that are more sustainable throughout their lifecycle. However, sustainability evaluation should go beyond the material's degradability alone. It is essential to consider life cycle analysis, biodegradation behavior under real environmental conditions, and the possible effects of additives or fillers used to improve performance. For instance, when a biodegradable base polymer is combined with non-degradable fillers such as carbon black or graphene, the final material may not fully break down, complicating claims of total biodegradability. Recent studies and reviews have highlighted the importance of balancing technical performance, cost-effectiveness, and environmental responsibility. While biodegradable polymers represent a major step toward sustainable analytical technologies, researchers must also assess trade-offs between durability, recyclability, and biodegradation rates to ensure genuine environmental benefits [12].

Practical Guidance and Methodological Considerations

When selecting or designing biodegradable polymers for analytical applications, several key factors must be considered to achieve both reliable performance and sustainability.

Selection Criteria: The choice of polymer should depend on the desired chemical functionality (example amine or hydroxyl groups), mechanical requirements (rigid versus flexible materials), and analytical purpose (hydrophobic versus hydrophilic interactions). The fabrication technique—whether 3D printing, casting, or electrospinning should also be chosen based on the type of device being developed and the nature of the analyte to be detected [15].

Modification Strategies: To enhance selectivity, conductivity, and porosity, biodegradable polymers can be modified using different strategies such as grafting, crosslinking, or nanoparticle incorporation. For instance, adding magnetic or carbon-based nanoparticles can improve both sensitivity and recovery efficiency. However, it is important to verify that no leachable monomers or unreacted crosslinkers remain in the final material, as these could interfere with analytical results or compromise biocompatibility. Researchers should always validate blank signals carefully to ensure measurement accuracy [18].

Stability and reproducibility: One challenge in using biodegradable polymers is their sensitivity to environmental conditions. These materials may hydrolyze, swell or lose integrity when exposed to moisture or high temperatures. Therefore, it is crucial to validate storage conditions, check for batch-to-batch variability, and monitor reusability over time. To ensure consistent performance, studies should report important analytical parameters such as sorbent capacity, breakthrough volume, kinetic behavior and lifetime of the material in the analytical setup. By following these practical guidelines, researchers can design biodegradable polymer systems that not only meet analytical performance standards but also contribute meaningfully to sustainability goals in laboratory and industrial environments [23].

Challenges and Limitations

Although biodegradable polymers have shown great promise in analytical chemistry, several challenges still limit their broader use.

Mechanical and chemical stability: Many natural polymers such as cellulose or chitosan are not as mechanically strong or chemically resistant as conventional materials like silica or crosslinked polystyrene. Under harsh solvent conditions or high pressures such as those used in high-performance liquid chromatography (HPLC)—these materials may deform, swell, or degrade. This restricts their use in certain high stress analytical applications where long-term structural integrity is essential [20].

Standardization and Reproducibility: One of the main concerns with biodegradable and natural polymers is their batch-to-batch variability. Factors such as natural source, extraction method, and degree of modification can lead to inconsistent physical or chemical properties. To ensure reliable analytical performance, it is necessary to develop standardized preparation protocols and quality controlled processing techniques for these materials [22].

End of life and composites: While many biodegradable polymers break down naturally, the inclusion of non-biodegradable additives or inorganic fillers (for example, carbon nanotubes or metal nanoparticles) can affect their overall degradability. When such composites are used, it becomes important to clearly define their end-of-life behavior and ensure transparent reporting of environmental impacts. Therefore, life-cycle assessments must carefully consider both the base polymer and any added components before labeling a material as fully biodegradable [16].

Future Directions

The future of biodegradable polymers in analytical chemistry looks promising, with several emerging research directions expected to expand their use and performance.

Smart and stimulus responsive materials: Next generation materials are being developed that can respond dynamically to environmental stimuli such as pH, temperature, or ionic strength. For instance, stimulus responsive cellulose or chitosan derivatives can be engineered to selectively capture or release analytes on demand. Such "smart materials" could enable 4D analytical devices capable of real-time adaptation and self-regulation [24].

Integration with digital and portable technologies: With the rapid growth of digital sensing, biodegradable sensors can be integrated with Internet of Things (IoT) platforms and portable electronic devices. Biopolymer-based sensor strips could easily connect to smartphone applications, allowing on-site monitoring of environmental pollutants, food safety indicators, or health biomarkers. This combination of biodegradability and digital readouts would make analytical monitoring both sustainable and accessible [25].

Standardization and regulatory adoption: For biodegradable polymers to become routine in analytical laboratories, standardized testing protocols, commercially reproducible sorbents, and validated performance data are essential. Establishing clear regulatory guidelines and reference materials will encourage the use of biopolymer-based devices in both academic and industrial settings [20].

Together, these directions highlight the growing potential of biodegradable materials to support greener, smarter, and more adaptable analytical platforms.

Biodegradable polymers represent a powerful and sustainable alternative to traditional synthetic materials in analytical chemistry. Their versatile chemical structures, ease of processing, and compatibility with modern fabrication techniques such as electrospinning, 3D printing, and casting—make them suitable for a wide range of applications, from sample preparation sorbents to low cost disposable sensors. Although some practical challenges remain, including issues of mechanical stability, standardization, and lifecycle management, continuous advancements in material science and green technology are steadily overcoming these barriers. As sustainability becomes a core focus of analytical research, biodegradable polymers are emerging as an essential and expanding toolkit for scientists aiming to design environmentally friendly, efficient, and affordable analytical systems.

4 Microbial and Cellular Biosensors for Green Monitoring

Microbial and cellular biosensors are innovative tools that use microorganisms like bacteria, yeast, or their components to detect harmful substances in the environment. These biosensors include various types, such as microbial fuel cell (MFC)-based sensors and advanced cell free or genetically engineered systems. They are low cost, sensitive, and can be easily used outside laboratories, making them highly suitable for environmental protection and sustainable monitoring. These biosensors can identify and measure pollutants such as heavy metals, pesticides, and toxic organic compounds. They are also useful in checking water quality by monitoring factors like oxygen levels and biological oxygen demand (BOD). Because of their efficiency, they can be used for continuous or on-site environmental observation. Three main aspects of microbial and cellular biosensors are as follows:

1. Basic concepts and designs: It discusses how these biosensors work, their sensing mechanisms, and the different types of devices used.
2. Environmental applications: It highlights their use in real-life environmental monitoring—for example, testing water, soil, and air quality, and assessing ecosystem health through case studies.
3. Methodologies and future aspects: It explores standardization, biosafety, environmental impact, and sustainability factors. It also looks at new directions such as integrating biosensors with the internet of Things (IoT), using synthetic biology for improvement, and developing cell-free alternatives for safer applications.

Environmental monitoring needs sensors that are affordable, sensitive to bioavailable fractions (not just total concentration), field-deployable, and low impact. Microbial/cellular biosensors excel at many of these: they transduce molecular interactions into measurable biological or electrical signals, often without expensive instrumentation; they can be engineered for selectivity; and many formats (paper-based, MFCs, whole-cell reporters) are amenable to low resource settings and reduced waste. Growing research including recent reviews and special issues—demonstrates rapid advances in sensitivity, circuit design, and portable platforms that make whole cell and related biosensors promising tools for sustainable environmental monitoring [26].

Platform Types and Sensing Principles (Mechanisms and Device Classes)

Whole-Cell (Living) Microbial Biosensors

Description and Mechanism Whole-cell biosensors (WCBs) use live microorganisms such as bacteria, yeast, or algae that are either naturally capable of genetically modified to give a measurable signal when exposed to specific substances or environmental stress. These signals can appear as changes in color, light, enzyme activity, or electrical output. In these systems, specific genes called reporter genes (lux, GFP, lacZ) are placed under the control of promoters that activate when the cell encounters certain chemicals or stress conditions like heavy metals or toxic organic compounds. Instead of directly measuring the chemical's amount, these biosensors show how much of it is biologically available and how it affects living cells [27].

Advantages

Low cost and easy production: WCBs are inexpensive and can be mass-produced easily.

Direct bioavailability detection: They indicate how much of a pollutant is biologically active rather than just present in the environment.

Versatile response: They can detect a wide range of pollutants, from heavy metals to pesticides and organic contaminants.

Environmentally relevant output: The biological response provides real-world data about toxicity or pollution effects.

Limitations

Variable sensitivity: These biosensors may not always detect very low levels of pollutants. Response time: Some may take several minutes to hours to show results. Environmental interference: The presence of other substances in samples can affect readings. Viability and safety concerns: Maintaining live cells requires biosafety precautions, and disposal needs to follow regulatory standards. Commercial challenges: Large-scale production faces technical barriers such as stability, storage, and reproducibility of cell-based systems [28].

Representative Designs and Improvements Modern WCBs are being enhanced through synthetic biology, which allows scientists to fine-tune genetic circuits, increase sensitivity, and reduce unwanted background signals. For example, genetically engineered bacteria have been developed to detect arsenic and mercury more efficiently. Recent innovations also integrate WCBs into paper-based platforms and microfluidic chips, making them portable and easy to use in the field. These formats simplify testing, improve cell stability, and make environmental monitoring faster and more practical [29].

Microbial Fuel Cell (MFC)-Based Biosensors

Principle: Microbial fuel cell (MFC) biosensors work on the ability of special microorganisms to generate electricity while breaking down organic matter. In this system, microbes oxidize a substrate such as organic compounds and release electrons, which travel to an electrode (the anode). This process produces an electrical current that changes in response to the amount of organic matter or the presence of toxic substances that interfere with microbial activity. Because the signal output is electrical, MFC-based biosensors can provide real time and continuous monitoring without requiring complex instruments. They are also suitable for remote environmental applications, and the small amount of electricity they generate can sometimes be used to power portable electronic devices or other low-energy systems [30].

Applications and strengths: MFC biosensors are highly useful for environmental assessment, especially for monitoring biochemical oxygen demand (BOD), organic load, and toxic pollutants such as heavy metals and pesticides. Variants like sediment MFCs and plant-based MFCs are adaptable to natural environments such as soil, sediments, and aquatic systems. Their ability to provide a continuous baseline signal makes them ideal for long-term environmental monitoring and for systems that need energy independence. Because they do not rely on external power sources, they are practical tools for autonomous sensing in remote or resource-limited areas [31].

Challenges: Despite their advantages, MFC biosensors face several technical and biological challenges. Maintaining biofilm stability (the layer of microbes that drives the sensor) is crucial for consistent performance. Issues such as fouling (clogging by organic or inorganic materials) and variations in microbial activity can affect signal accuracy. Also, achieving high sensitivity and selectivity –the ability to distinguish between different pollutants remains difficult. Translating small current changes into accurate quantitative data requires careful calibration and sometimes the use of additional sensors. Recent research continues to focus on improving MFC configurations, enhancing microbial performance, and developing robust systems that can function reliably in complex real-world environments [30].

Cell Free and Hybrid Biosensors

Cell free biosensors are innovative tools that function without the need for living cells. Instead, they use biological components such as enzymes, nucleic acids, or transcription and translation machinery extracted from cells. These systems can detect a wide variety of targets including DNA, RNA, small molecules, and heavy metals. They are prepared in freeze-dried or paper-based formats, which makes them portable, easy to handle, and suitable for use outside laboratory settings. Because no live organisms are involved, they are safe, easy to transport, and do not pose biosafety risks. Cell free biosensors offer high specificity and quick detection, often giving visible color changes or other clear readouts without the need for expensive instruments. These systems have developed rapidly in recent years and are seen as promising green monitoring tools that can meet regulatory safety standards while remaining effective in the field [32].

Trade-offs and limitations: They perform well in terms of safety and detection speed, but some may need cold storage or temperature control unless they are freeze-dried (lyophilized). The sensitivity and stability of these biosensors for certain chemicals or biological targets are still being improved through ongoing research. Despite these limitations, cell-free and hybrid biosensors are becoming increasingly popular due to their portability, eco-friendliness, and potential for large-scale environmental applications.

Sensing Transduction Modes and Device Architectures

Optical reporters: In this approach, biosensors detect and report signals through visible changes such as luminescence, fluorescence, or pigment variation. These visual indicators make the results easy to observe either directly in a laboratory or even through smartphone-based imaging tools. Optical biosensors are especially suitable for endpoint assays, where a final color or light change indicates the presence of a specific analyte. They are simple, cost-effective, and ideal for field or classroom testing without requiring complex instruments [29].

Electrochemical transduction: Electrochemical biosensors work by converting biological reactions into electrical signals. This type includes microbial fuel cell (MFC) sensors, which measure changes in current or voltage, and bioelectrochemical sensors, where microbial or enzymatic reactions produce measurable electrical outputs. These sensors provide continuous and real-time data, making them useful for long-term environmental monitoring and automated logging systems. Because the output is electrical, they can be integrated with digital platforms for remote data collection, allowing for efficient monitoring of pollutants or biological oxygen demand (BOD) in various environments [30].

Colorimetric and Paper-Based Microfluidic Devices

These devices often use freeze-dried (lyophilized) cells or cell-free biological reagents embedded in paper strips or microfluidic chips. When exposed to a target substance, they produce a visible color change that can be easily interpreted. Such biosensors are lightweight, portable, and environmentally friendly, making them excellent for on-site testing. They require only minimal reagents and generate very little waste, offering a sustainable and affordable solution for large-scale environmental monitoring, especially in resource-limited settings [33].

Applications in Green Environmental Monitoring (Case Studies and Performance)

Water Quality Monitoring (Surface Water, Wastewater, Potable Supplies)

Detection of Bioavailable Heavy Metals: Whole-cell biosensors (WCBs) have been genetically modified to detect metal ions such as mercury (Hg), arsenic (As), and calcium (Cd) using metal responsive regulatory genes like ArsR and MerR. These biosensors provide readings of the bioavailable fraction of metals—the portion that can actually interact with and affect living organisms rather than the total concentration. For example, microbial mercury sensors have been successfully used near artisanal mining areas, where contamination is common. Such tools are valuable because they help identify health relevant exposure even when total metal levels appear similar. However, sensitivity can sometimes be limited, and in such cases, the accuracy can be improved using sample pre-concentration or signal amplification strategies [29].

Monitoring of Pesticides and Organic Pollutants: Microbial and enzyme-based biosensors have been designed to detect toxic organic compounds, including pesticides and herbicides. Engineered bacteria that contain specific degradation pathways can convert these compounds into detectable products, allowing rapid and selective measurement. Recent advancements have demonstrated portable field-use biosensors capable of detecting organophosphates and herbicides, even outside laboratory settings. These biosensors often produce measurable color or light changes that can be read using smartphones or paper-based devices, making them highly useful for on-site environmental surveillance [34].

Biochemical Oxygen Demand (BOD) and General Toxicity Testing: The earliest microbial biosensors were developed to measure biochemical oxygen demand (BOD)—an essential indicator of water pollution. Early models used bacteria such as Pseudomonas to detect organic load in wastewater. Modern versions of these biosensors, including microbial fuel cell (MFC)-based and whole-cell designs, now offer rapid and cost-effective alternatives to traditional, time consuming BOD tests. MFC-based biosensors are particularly promising for continuous wastewater monitoring and industrial effluent control. They can operate for long periods with minimal maintenance while providing real-time data on water quality and pollution trends. Their low cost, simple setup, and compatibility with automated systems make them attractive tools for sustainable environmental management [35].

Soil and Sediment Monitoring

Sediment MFCs and benthic sensors: Sediment microbial fuel cells (SMFCs) are advanced biosensors that serve dual purposes—they generate small amounts of energy while also sensing environmental conditions such as organic load and redox changes in sediments. These systems use electroactive bacteria living in the sediment water interface to produce electrical signals that correspond to pollution or microbial activity levels. Recent developments have led to the creation of portable SMFC prototypes capable of outdoor operation. These devices can function continuously in wetlands, riverbeds, and estuarine environments, even powering small electronic sensors for field monitoring. Their ability to work in natural ecosystems without an external power source makes them sustainable tools for long-term environmental observation [31].

Bioavailability assessment in soils: In soil environments, whole-cell biosensors are used to measure the bioavailable fractions of heavy metals and other contaminants, which represent the portion actually accessible to loving organisms. This approach provides a more realistic evaluation of ecological risks compared to conventional chemical testing, which only measures total concentrations. To improve performance and stability, biosensors are often embedded in protective materials such as alginate beads or soil compatible cartridges. These encapsulation techniques protect the microbial cells and allow the sensors to be used directly in soil samples or in situ probes without loss of sensitivity. Such configurations make biosensors especially useful for on-site soil health monitoring, contamination assessment, and agricultural sustainability studies [36].

Air and Indoor Environmental Monitoring

Detection of VOCs and Airborne toxins: Microbial biosensors are increasingly being explored for detecting volatile organic compounds (VOCs) and other airborne toxic substances commonly found in industrial or indoor environments. These biosensors often use engineered microorganisms, such as yeast strains, that produce visible color or luminescence changes when exposed to specific VOCs. The main challenges in this area involve maintaining cell stability and activity under fluctuating environmental conditions, such as humidity and temperature, which can affect sensor performance. To overcome this, modern designs employ hybrid systems that integrate biological sensing elements with sorption membranes or cell free reagents, allowing VOCs to be captured and analyzed efficiently on moist surfaces or paper-based substrates [37].

Case Studies (Representative Recent Demonstrations)

Whole-cell mercury sensor for mining sites: Genetically engineered microbial sensors have been applied in artisanal mining regions to detect bioavailable ionic mercury in soil and water. These biosensors are designed to produce measurable optical or electrical responses that correspond to mercury concentrations. Such systems are used to map contamination hotspots, assist in environmental cleanup decisions, and provide early warnings in areas of high exposure. Their performance depends on sensitivity to the bioavailable form of mercury rather than total metal concentration, which makes them valuable for ecological risk assessment and real-time monitoring in complex field conditions [29].

MFC-based BOD sensors for wastewater: Microbial fuel cell (MFC) biosensors have been used to monitor biochemical oxygen demand (BOD) in wastewater treatment plants and industrial effluents. The electrical current generated by microbes in these systems correlates with the organic load of the water sample. These biosensors can provide continuous and real-time data, serving as an early warning system for pollution events such as septic tank failures or process disruptions in treatment facilities. In some designs, the electricity generated is even used to partially power telemetry or data transmission units, enhancing their suitability for automated field use [30].

Paper-based cell free biosensors for pesticides and metals: recent advances in cell free biosensing technology have led to the development of paper-based strips for detecting pesticides and heavy metals. These biosensors use lyophilized (freeze-dried) reagents that produce a visible color change upon exposure to contaminants. The results can be easily read using a smartphone camera, enabling rapid on-site testing without laboratory equipment. Their low cost, portability, and excellent biosafety make them ideal for environmental monitoring in low resource regions and for community-based pollution assessment programs [32].

Practical, Regulatory and Sustainability Considerations (Design Guidelines, Biosafety, Lifecycle)

Analytical Performance and Validation For biosensors to be used in real-world monitoring, their performance must be carefully validated. Key parameters include limit of detection (LOD), precision, accuracy, selectivity, response time, and stability (shelf-life and operational duration). In the case of continuous devices like microbial fuel cells (MFCs), additional factors such as biofilm development, signal drift, and fouling are also evaluated. Studies emphasize the need for rigorous field testing under actual environmental conditions and comparison with standard analytical chemistry methods to ensure reliability.

Biosafety and Regulatory Framework Live-cell biosensors must comply with biosafety and genetic material regulations to prevent risks such as environmental release or gene transfer. Risk mitigation strategies include using non-replicating or killed cells, encapsulation, or auxotrophic strains that depend on specific nutrients to survive. Developers should follow local GMO and biosafety regulations, as approval standards vary by country. Maintaining biosafety remains a key step for commercial deployment [38].

Sustainability and Lifecycle Aspects Green biosensor design focuses on low-impact materials and energy-efficient operation. Examples include biodegradable housings instead of plastics and MFC systems that generate their own power. Proper end of life management is essential to avoid hazardous waste from components such as metal electrodes. While biosensors reduce chemical and energy use compared to traditional methods, full lifecycle assessment (LCA) is needed to measure their net environmental benefits [35].

Standardization and commercialization: Major challenges in biosensor commercialization include reproducibility, shelf stability, and regulatory validation. Progress can be supported by modular designs (e.g., replaceable cartridges) and partnerships with water and environmental agencies for field trials. Recent perspectives also recommend establishing community wide standards for testing, performance reporting, and regulatory consistency to accelerate broader adoption [38].

Future Directions and Research Opportunities

Synthetic Biology and Circuit Engineering Future biosensors will rely heavily on synthetic biology tools to enhance sensitivity and minimize background noise. Advanced circuit designs using signal amplification and logic-based sensing will allow the detection of multiple analytes at once and enable time controlled reporting. Recent work on optimized ArsR circuits and engineered reporter genes shows great potential for improving sensor precision and adaptability [39, 40].

Cell Free Platforms as a Regulatory Friendly Systems The development of cell free and freeze-dried paper-based biosensors offers a safer and more flexible alternative to living cell systems. These platforms reduce biosafety concerns while allowing rapid, distributed testing in the field. Continued progress in multiplexing and stability will make these sensors more practical for routine environmental and health monitoring [40].

Integration with IoT and Remote Telemetry Combining microbial fuel cells (MFCs) and electrochemical biosensors with Internet of Things (IoT) technologies enables continuous, real-time data collection from remote sites. Such integration supports long-term environmental surveillance, particularly in hard-to-reach areas, by using low power telemetry systems for wireless monitoring and reporting [31, 40].

Standardization and Lifecycle Assessment (LCA) To fully realize the environmental benefits of biosensors, standardized testing protocols and inter-laboratory comparisons are essential. Establishing clear LCA frameworks will help measure the true sustainability of biosensor production, usage, and disposal. These steps are critical for ensuring reliable performance, regulatory acceptance, and large-scale deployment of biosensing technologies [35, 40].

5 Applications of Biotechnology in Food, Environment, and Pharmaceuticals

Biotechnology plays a vital role in advancing modern science and industry, influencing food production, environmental protection, and pharmaceutical innovation. In the food sector, biotechnology covers a wide range of processes, from traditional fermentation and enzyme-based technologies to advanced methods like genetic engineering, cellular agriculture, and microbial consortium design. These innovations help improve crop yield, nutritional quality, and sustainability, ensuring more efficient and resilient food systems. In the environmental field, biotechnology uses microorganisms, enzymes, biosensors, and genetically modified systems to manage pollution, promote waste valorization, support bioenergy production, and enable real-time environmental monitoring. Such approaches contribute to eco-friendly solutions for resource recovery and sustainable development. In pharmaceutical applications, biotechnology has revolutionized the creation of biopharmaceuticals, vaccines, and therapeutic molecules through recombinant DNA technology, monocloncal antibody production, gene and cell therapies, and mRNA-based platforms. Emerging technologies such as CRISPR and other gene editing tools, along with synthetic biology and cell free expression systems, have accelerated the pace of drug discovery and vaccine development. Innovations in bioprocess optimization and analytical technologies support efficient large-scale production while maintaining safety and quality standards. This chapter provides an overview of recent advancements in biotechnology across these three domains, highlighting representative case studies, key implementation including regulation, biosafety, and lifecycle assessment, and future research directions that aim to enhance global sustainability and human health [41].

Biotechnology in Food

Traditional and Modern Fermentation Fermentation is one of the earliest and most established applications of biotechnology. It plays a vital role in food preservation, flavor enhancement, nutritional improvement, and the production of functional ingredients. Modern developments in strain selection, metagenomics of starter cultures, and advanced fermentation methods such as fed-batch and continuous processes have led to higher yields, consistent product quality, and improved safety. Genetically engineered yeasts and bacteria are widely used to produce vitamins, bioactive compounds, and flavor molecules on an industrial scale. Recent research has emphasized metagenomics analysis and microbial consortia engineering for optimizing fermentation and controlling microbial activity more precisely. Example: Modified *Saccharomyces* strains have been developed to enhance aroma production or to reduce anti-nutritional compounds in plant-based foods, improving both sensory quality and health value [41].

Enzymes and Biocatalysis in Food Processing Enzymes such as proteases, amylases, lactases, and lipases are increasingly used to replace chemical processing methods due to their specificity, efficiency, and lower environmental impact. Advances in industrial enzyme engineering through directed evolution and rational design have produced enzymes that are thermostable, highly active, and substrate specific. These enzymes are commonly used in industries like dairy, baking, brewing, and juice clarification, offering improved product quality while supporting the demand for clean label and sustainable processing. The global enzyme market continues to expand due to the push for energy efficient and eco-friendly technologies [42].

Microbial-Based Nutrition and Probiotics Probiotics represent another important application of biotechnology in the food sector. Engineered or naturally selected microbial strains are designed for enhanced survival, targeted health benefits, and specific bioactivities, such as bile salt metabolism or vitamin biosynthesis. Advanced selection techniques, including omics-based profiling, have improved both the efficacy and safety of probiotic products, helping them gain regulatory acceptance. Beyond live microbes, new categories like postbiotics and paraprobiotics which are derived from fermentation but do not contain live organisms are gaining attention for consumers seeking the functional benefits of microbes without live-cell formulations [41].

Cellular Agriculture and Novel Foods Cellular agriculture, which includes the production of cultured meat and fermentation derived dairy proteins, offers an innovative route to sustainable protein generation. Similarly, single cell proteins from microalgae and fungi provide nutrient rich alternatives with a much lower land and water footprint compared to conventional livestock farming. Advances in bioprocess scale-up, serum-free media design, and biomaterial scaffolds have improved texture, yield, and product safety. Regulatory frameworks and techno-economic assessments are ongoing to ensure that cultured meat and precision fermentation products meet food safety and labeling standards. Together, these technologies represent a promising step toward ethical, resource efficient, and climate friendly food production [42].

Food Safety: Detection and Control Modern biotechnology has transformed the way foodborne risks are detected and managed. Tools such as PCR-based diagnostics, CRISPR-enabled detection systems, and biosensors enable rapid identification of pathogens and toxins in food samples. Additionally, phage therapy, bacteriocins, and other biotechnological interventions are being explored as natural alternatives to chemical preservatives. Technologies like DNA barcoding and next generation sequencing (NGS) allow precise traceability of ingredients, improving transparency across the food supply chain. These advances reduce response times during contamination events and strengthen global food safety and quality assurance systems [43].

Sustainability and Valorization of Food Waste Biotechnology also plays a key role in reducing food waste by converting agricultural and food by-products into valuable materials. Through microbial and enzymatic bioprocessing, waste biomass can be transformed into bioplastics, enzymes, organic acids, and single cell proteins, contributing to a circular bioeconomy. The concept of the biorefinery integrates these processes, using pre-treatment, enzymatic hydrolysis, and integrated conversion systems to maximize the recovery of energy and valuable compounds from waste. Such sustainable valorization strategies support both environmental conservation and economic growth, aligning with global goals for resource efficiency and waste reduction [44].

Biotechnology in the Environment

Bioremediation and Biodegradation Bioremediation involves the use of native or genetically engineered microorganisms an enzymes to breakdown or detoxify pollutants such as petroleum hydrocarbons, chlorinated solvents, pesticides, and heavy metals. Through biosorption and biotransformation, these biological agents convert harmful contaminants into less toxic or harmless forms. Recent advancements in genomics and metabolic engineering have improved the selection and design of microbial strains capable of acting on a wider range of pollutants, including complex contaminant mixtures. Modern strategies include both in situ methods, such as bioaugmentation and biostimulation, and ex situ processes involving biopiles, bioreactors, and constructed wetlands [6]. Example applications include the treatment of hydrocarbon-contaminated soils through bioaugmentation and the use of enzyme-based systems for breaking down dyes and chemicals in textile industry effluents [45].

Wastewater Treatment and Bioenergy In wastewater treatment, microbial consortia and engineered strains are utilized to degrade organic matter, reducing pollution while generating bioenergy. Technologies such as microbial fuel cells (MFCs) employ electroactive microbes to convert organic waste into electricity, offering a sustainable alternative to conventional treatment systems. MFCs can also function as self-powered biosensors, providing real-time measurements of biochemical oxygen demand (BOD) and water toxicity. Recent studies have focused on improving electrode materials, biofilm performance, and reactor design to enhance energy recovery and ensure continuous environmental monitoring [46].

Biosensors for Environmental Monitoring Modern biosensors, including enzyme, antibody, nucleic acid, and whole-cell-based systems, are powerful tools for detecting environmental pollutants such as heavy metals, pesticides, and pathogens. These biosensors are designed for field use, providing sensitive and rapid detection without the need for sophisticated laboratory equipment. Recent advancements in paper-based platforms, cell free biosensors, and smartphone integrated readout systems have made monitoring more portable and cost-effective. Reviews highlight the potential of biosensors in supporting sustainable development goas by enabling point of need environmental monitoring, although challenges remain in ensuring long-term stability, reproducibility, and standardization [43].

Synthetic Biology and CRISPR for Environmental Applications The integration of synthetic biology an CRISPR gene editing technologies is expanding the capabilities of environmental biotechnology. Through genetic modification, microorganisms can be engineered to degrade pollutants more efficiently or to contain biosensing genetic circuits that signal the presence of contaminants. These engineered systems enhance both biodegradation performance and biosafety by preventing uncontrolled spread of modified microbes. However, while CRISPR-enabled tools show great promise, ethical and regulatory considerations remain important to prevent potential environmental risks. Current research continues to refine microbial consortia and modular genetic networks for targeted and safe environmental applications [47].

Waste Valorization and Circular Bioeconomy Biotechnology plays a central role in the conversion of waste materials from industrial and municipal sources into value added bioproducts such as bioplastics (PHAs), biofuels, and chemical feedstocks within biorefinery frameworks. This approach supports a circular bioeconomy, emphasizing the reuse of resources while minimizing waste generation. Recent techno-economic and life-cycle assessments demonstrate significant potential for reducing greenhouse gas emissions and improving resource recovery. Studies on agro-waste valorization have showcased successful examples of large-scale implementation, although further work is needed to address scaling, cost-efficiency, and policy integration challenges [48].

Biotechnology in Pharmaceuticals

Recombinant Proteins, Monoclonal Antibodies and Biologics Modern biotechnology relies heavily on recombinant expression systems such as *E.coli*, yeast, and mammalian cell lines to produce therapeutic proteins, enzymes, hormones, monoclonal antibodies, and recombinant vaccines. These biologically derived products form the backbone of today's pharmaceutical industry. Advances in continuous bioprocessing, process intensification, and the use of platform cell lines (for example, Chinese Hamster Ovary—CHO cell) have significantly shortened production timelines and lowered manufacturing costs. As a result, there has been a rapid increase in the number of approved biologics and a growing development pipeline across therapeutic areas. Industry analyses consistently highlight the expansion of biologics as one of the most dynamic sectors in pharmaceutical innovation [49]·

Vaccines: From Recombinant to mRNA Biotechnology has also revolutionized vaccine development, shifting from traditional recombinant protein or viral vector vaccines to mRNA-based platforms. These next generation vaccines allow faster design and production, as demonstrated during the COVID-19 pandemic, where mRNA vaccines provided rapid, scalable solutions for global immunization. Beyond infectious diseases, the scope of mRNA therapeutics is expanding into protein replacement therapies and personalized cancer treatments. Ongoing research continues to refine delivery systems, stability, and manufacturing processes for these platforms. Recent reviews emphasize the need for standardized production frameworks to ensure safety, efficacy, and regulatory compliance in mRNA-based pharmaceuticals [50].

Cell and Gene Therapies Advances in biotechnology have transformed modern medicine through cellular and gene-based therapies, including CAR-T, stem cell, and viral vector mediated gene therapies. These treatments offer new hope for diseases once considered incurable by directly targeting genetic or cellular causes. However, challenges remain in regulatory approval, manufacturing scalability, and long-term safety evaluation. The high cost of production and complex logistics also present barriers to widespread adoption. Despite these hurdles, current reviews highlight significant progress in clinical success rates, regulatory adaptation, and cost reduction strategies for next generation cell and gene therapy platforms [49].

Microbial Fermentation for Small Molecule APIs and Intermediates Microbial and enzymatic fermentation technologies are increasingly used for the production of active pharmaceutical ingredients (APIs) and their chiral intermediates, such as statins and antibiotics. These biotechnological approaches offer greater stereoselectivity, higher yields, and environmentally friendly alternatives to traditional petrochemical syntheses. Advances in synthetic biology and metabolic pathway engineering have enabled optimization of microbial strains for higher productivity and reduced byproduct formation. Case studies demonstrate successful replacement of chemical synthesis routes with greener, biocatalytic processes, underscoring biotechnology's contribution to sustainable pharmaceutical manufacturing.

Drug Discovery and Screening Modern drug discovery leverages high throughout screening technologies using engineered cell lines, organoids, and microphysiological (organ-on-chip) systems to better predict how drugs will behave in the human body. These platforms improve translational predictivity, helping researchers identify effective candidates earlier in the development process. Cutting-edge tools such as CRISPR-based genetic screens and single cell omics have accelerated target identification and mechanism of action studies, providing deeper insights into disease biology. Ongoing research on cell models and organoids continues to reduce late stage clinical failures, making the overall drug development process more efficient and cost-effective [49].

Cross-cutting Themes: Regulation, Biosafety, Sustainability, Scale-Up

Regulatory and Biosafety Considerations Biotechnological innovations operate within a wide range of regulatory frameworks, covering areas such as food safety, genetically modified organism (GMO) governance, environmental release, and biologics approval. Emerging technologies including engineered microbes, gene drives, and cell-based foods present new regulatory challenges requiring adaptive policies and transparent oversight. To ensure biosafety, multiple containment strategies are implemented, such as physical barriers, genetic safeguards like kill switches, auxotrophic strains, and cell free systems that prevent the release of living organisms into the environment. Researchers and industries are encouraged to engage early with regulatory agencies and to maintain clear, evidence-based risk assessments throughout the development process [47].

Lifecycle Assessment and True Sustainability While biotechnology offers solutions to reduce resource consumption and carbon emissions, its sustainability must be evaluated through comprehensive lifecycle assessments (LCAs). These assessments help identify trade-offs, such as energy demands for culture media or waste management in bioprocessing. Recent LCA studies emphasize the importance of defining system boundaries, feedstock sourcing, and end of life management when measuring environmental impact. By integrating LCA frameworks, biotechnology can better quantify its net sustainability benefits and support the transition toward a circular and resource efficient bioeconomy [44].

Scale-Up, Reproducibility and Cost Successfully moving biotechnological innovations from laboratory research to industrial production requires addressing several key challenges. These include optimizing bioreactor design, managing downstream purification costs particularly for biologics, and ensuring consistent raw material quality and strain reproducibility. Comprehensive techno-economic analyses and industrial case studies help identify major cost drivers such as culture media expenses, purification methods, and regulatory compliance requirements. By understanding these factors, industries can design more efficient production systems and reduce the economic barriers to large-scale commercialization [51].

Ethics, Social Acceptance and Governance The successful adoption of biotechnology depends not only on technical performance but also on public trust and ethical governance. Public acceptance of innovations such as genetically modified crops or cultured meat is influenced by perceived benefits, transparent risk communication, and confidence in regulatory oversight. To ensure responsible innovation, researchers and policymakers are encouraged to adopt multidisciplinary frameworks that integrate social science, ethics, and stakeholder participation. This inclusive approach promotes dialogue between scientists, regulators, and the public, fostering a balanced understanding of the risks and benefits associated with emerging biotechnologies.

Future Directions and Research Opportunities

Cell Free Synthetic Biology Future research in cell free systems focuses on developing portable diagnostics and modular biosynthesis platforms that function without live cells, thereby minimizing biosafety concerns. These systems enable rapid, on-site detection of contaminants and scalable production of biomolecules under controlled conditions [52].

Engineered Microbial Consortia and Community Engineering Designing and optimizing microbial consortia for targeted biotransformations presents significant opportunities in waste treatment, soil restoration, and bioresource recovery. Such systems can be tailored for complex environmental or industrial processes, promoting sustainable and circular bioeconomy practices [45].

Data-Driven Strain and Process Design Integrating artificial intelligence (AI) and machine learning (ML) with multi-omics data is reshaping how strains and bioprocesses are designed. These computational tools accelerate the design build test learn cycle, improving prediction accuracy and reducing development timelines for industrial and therapeutic biotechnology applications.

Decentralized Biomanufacturing The concept of modular and small footprint biomanufacturing aims to make vaccine and biologic production more accessible in low resource regions. By using portable, automated systems, decentralized production can enhance global health resilience and reduce dependence on large centralized facilities [49].

6 Conclusions

Biotechnology represents a transformative, cross-disciplinary toolkit driving innovation across food systems, environmental management, and pharmaceutical development. Advances in synthetic biology, genome editing, cell free technologies, and bioprocess intensification are accelerating the translation of laboratory research into real world solutions. However, as the field progresses, it also introduces new challenges related to regulation, biosafety, ethical governance, and lifecycle sustainability. Addressing these issues requires collaborative, multidisciplinary research to ensure biotechnology continues to deliver safe, equitable, and sustainable benefits for society and the planet. Biotechnology has become a central driver of sustainable innovation, offering solutions tailored to the global need for environmental protection, efficient resource utilization, and improved human health. Across analytical chemistry, environmental monitoring, food systems, and pharmaceuticals, biological tools and processes are reshaping traditional practices and enabling cleaner, greener, and more adaptable technologies.

Advances in biosensors ranging from enzyme-based platforms and biodegradable polymer systems to microbial fuel cells and cell free diagnostics exemplify the shift toward environmentally conscious analytical methods. These devices reduce chemical load, limit hazardous waste, and provide rapid, field ready alternatives to conventional laboratory techniques. Their integration with nanomaterials, microfluidics, AI-based analytics, and IoT systems further enhances performance, accessibility, and sustainability.

In the environmental sector, microbial and cellular biosensors offer real-time insights into pollutants, toxicity, and ecosystem health. Their ability to sense bioavailable contaminants and operate autonomously makes them indispensable for continuous monitoring and management of water, soil, and air quality. These tools support global goals for sustainable development by enabling early detection, informed decision making and community-based environmental stewardship.

Within the food and pharmaceutical industries, biotechnology enhances production efficiency, safety, and sustainability from precision fermentation, probiotics, and cultured foods to mRNA vaccines, recombinant therapeutics, and cell/gene therapies. Bioremediation, waste valorization, and circular bioeconomy frameworks further demonstrate biotechnology's potential to transform environmental and industrial systems in eco-friendly ways.

As these technologies advance, considerations such as biosafety, regulatory frameworks, lifecycle assessment, ethical governance, and commercial scalability become increasingly important. Addressing these dimensions ensures that biotechnological innovations remain safe, equitable, and environmentally responsible.

Overall, biotechnology stands as a powerful, cross-disciplinary engine for progress. Through the integration of Green Analytical Chemistry, microbial biosensing, and sustainable food, environmental, and pharmaceutical technologies, biotechnology continues to expand its role in shaping a healthier, more resilient, and more sustainable future.

References

1. Melo RLF, Neto FS, Dari DN, Fernandes BCC, Freire TM, Fechine PBA, Soares JM, Dos Santos JCS. A comprehensive review on enzyme-based biosensors: Advanced analysis and emerging applications in nanomaterial-enzyme linkage. International Journal of Biological Macromolecules. 2024;264(2):130817.
2. Kumar A, Maiti P. RSC Advances: "Paper-based sustainable biosensors, Material advances, Royal society of chemistry, 2024; 5, 3563-3586.
3. Mor R. A review on development of enzymatic biosensors for industrial applications. World Journal of Advanced Research and Reviews. 2024;23(1):2582–2590.
4. Naresh V, Lee N. A Review on Biosensors and Recent Development of nanostructured materials-enabled biosensors. Sensors 2021;21(4):1109.
5. Wijayanti SD, Tsvik L, Haltrich D. Recent advances in electrochemical enzyme-based biosensors in food and beverage analysis. Foods2023;12:3355.
6. Huang CW, Lin C, Nguyen MK, Hussain A, Bui XT, Ngo HH. A review of biosensor for environmental monitoring: principle, application, and corresponding achievement of sustainable development goals. Bioengineered 2023;14(1):58–80
7. Gupta AK, Krasnoslobodtsev AV, Fueling the future: The emergence of self-powered enzymatic biofuel cell biosensors. Biosensors 2024;14:316.
8. Joshi KM, Salve S, Dhanwade D, Chavhan M, Jagtap S, Shinde M et al. Advancing protein biosensors: redefining detection through innovations in materials, mechanisms, and applications for precision medicine and global diagnostics. RSC Advances, Royal society of chemistry. 2025;15: 11523–11536.
9. Sharma M, Mahajan P, Alsubaie AS, Khanna V, Chahal S, Thakur A, et al. Next-generation nanomaterials-based biosensors: Real-time biosensing devices for detecting emerging environmental pollutants. Materials today sustainability 2025;29:1–32.
10. Xiao X, Ulstrup J. Towards continuous potentiometric enzymatic biosensors. Current opinion in electrochemistry2024;46: 1–7
11. Fan YF, Guo ZB, Ge GB. Enzyme-Based Biosensors and Their Applications. Biosensors 2023; 14:476.

12. Kim MS, Chang H, Zheng L, Yan Q, Pfleger BF, Klier J, Nelson K et al. A review of biodegradable plastics: chemistry, applications, properties, and future research needs. Chemical reviews. 2023;123(16):9915-9939. 12
13. Samir A, Ashour FH, Hakim AAA, Bassyouni M. Recent advances in biodegradable polymers for sustainable applications. Materials degradation 2022;6(68):1-28. 13
14. Bocchetta P, Othman A, Gupta M, Andriani G, Martin P, Kumar Y, Joly N et al. Chitosan in electrochemical biosensors: nanostructuring and methods of synthesis. European polymer journal 2024;213:113092. 14
15. Bikiaris ND, Koumentakou I, Samiotaki C, Meimaroglou D, Varytimidou D, Karatza A, Kalantzis Z et al. Recent advances in the investigation of poly(lactic acid) (PLA) Nanocomposites: Incorporation of various nanofillers and their properties and applications. Polymers. 2023;15(5):1196 15
16. Jha S, Akula B, Enyioma H, Novak M Amin V, Liang H, Biodegradable biobased polymers: A review of the state of the art, challenges, and future directions. Polymers 2024;16(16):2262
17. Qian F, Jia R, Cheng M, Chaudhary A, Melhi S, Mekkey SD, Zhu N et al. An overview of polylactic acid (PLA) nanocomposites for sensors. Advanced composites and hybrid materials 2024;7(75):5
18. Martello L, Rapti A, Bikiaris DN, Lambropoulou DA. Synthesis and evaluation of a chitosan nanomaterial as efficient sorbent for determination of fungicide residues in waters and wine by liquid chromatography high resolution mass spectrometry. Analytical methods, royal society of chemistry, 2024;16:873-883. 18
19. Abbasi S, Haeri SA. Biodegradable materials and their applications in sample preparation techniques—a review. Microchemical journal 2021;171:106831 19
20. Badawy ME, El-Nouby MAM, Kimani PK, Lim LW, Rabea E. A review of the modern principles and applications of solid phase extraction techniques in chromatographic analysis. Analytical sciences 2022;38(12):1457–1487. 20
21. Rodrigues LDS, Rocha GSD, Fornasier F, Matos RF, Nicolini JV, Ferraz HC. 3D-Printed Electrodes Based on Polylactic Acid and Carbonaceous Materials for Electrochemical Sensors and Biosensors: Fabrication and Surface Activation via Chemical, Electrochemical, and Laser/Plasma Methods. Electroanalysis 2025;37:1–25.
22. Etale A, Onyianta AJ, Turner SR, Eichhorn SJ. Cellulose: A Review of Water Interactions, Applications in Composites, and Water Treatment. Chemical reviews 2023;123(5):2016–2048.
23. Baidurah S. Methods of analyses for biodegradable polymers: A review Polymers 2022;14(22):4928.
24. Chang S, Weng Z, Zhang C Jiang S, Duan G. Cellulose-Based Intelligent Responsive Materials: A Review. Polymers 2023;15:3905.
25. Meher AK, Zarouri A. Green analytical chemistry-recent innovations. Analytica 2025;6(1):10.
26. Capin J, Chabert E, Zuniga A, Bonnet J. Microbial biosensors for diagnostics, surveillance and epidemiology: Today's achievements and tomorrow's prospects. Microbial biotechnology 2024;17(11):1–7.
27. Mendoza ASG, Acosta MFM, Sanchez JAM, Vazquez LEC. Principles and challenges of whole cell microbial biosensors in the food industry. Journal of food science 2024;89:5255–5269.
28. Huang Z, Gustave W, Bai S, Li Y, Li B, Elcin E, Jiang B et al. Challenges and opportunities in commercializing whole cell bioreporters in environmental application. Environmental research 2024;262:119801.
29. Zevallos-Aliaga D, Graeve SD, Obando Chavez P, Vaccari NA, Gao Y, Peeters T, Guerra DG. Highly Sensitive Whole-Cell Mercury Biosensors for environmental monitoring. Biosensors 2024; 14(5):246.
30. Liu C, Cheng L, Jia H. Application of microbial fuel cell-based biosensor in environmental monitoring — critical review. Electroanalysis 2024;36(10):e202400100.
31. Y hubenova, Bardarov I, Hubenova E, Slavcheva E. Sediment microbial fuel ells capable of powering outdoor environmental monitoring sensors. Sensing and biosensing research 2024;46:100695

32. Green TP, Talley JP, Bundy BC. Recent advances in developing cell-free protein synthesis biosensors for medical diagnostics and environmental monitoring. 2025;15(8): 499.
33. Noviana E, Ozer T, Carrell CS, Link JS, McMahon C, Janh I, Henry CS. Microfluidic paper based analytical devices: from design to applications. Chemical reviews 2021;121(13):11835–11885.
34. Mendes F, Machado BO, Castro BB, Sousa MJ, Chaves SR. Harnessing the power of biosensors for environmental monitoring of pesticides in water. Applied microbiology and biotechnology 2025;109(92):1–16.
35. Williams A, Aguilar MR, Arachchillage KGGP, Chandra S, Rangan S, Gupta SG, Vivancos JMA. Biosensors for Public Health and Environmental Monitoring: The Case for Sustainable Biosensing. ACS sustainable chemistry and engineering journal 2024;12(28):10296–10312.
36. Liu M, Yang W, Zhu W, Yu D Innovative applications and research advances of bacterial biosensors in medicine. Frontiers in microbiology 2025;16:1–14.
37. Thiruvengadam M, Chi HY, Choi HJ, Jung BS, Lee SB, Park Y, Jeon D et al. sustainable and smart nano-biosensors: Integrated solutions for healthcare, environmental monitoring, agriculture, and food safety. Industrial crops and products 2025;233:121337.
38. Huang Z, Gustave W, Bai S, Li Y, Li B, Elcin E, Jiang B et al. Challenges and opportunities in commercializing whole cell bioreporters in environmental application. Environmental research 2024;262;119801.
39. Guo W, Liu MQ, Yang XQ Guo YY, Hui CY. Optimized genetic circuitry and reporters for sensitive whole cell arsenic biosensors: advancing environmental monitoring. Applied and environmental microbiology. 2025;91(8):
40. Chakma R, Hossain MK, Paramasivam P, Bousbih R, Amami M, Toki GFI, Haldhar R, Karmaker AK. Recent applications, challenges and future prospects for microbial fuel cells: A review. Global challenges 2025;9(5): 2500004
41. Ranjha MMAN, Shafique B, Khalid W, Nadeem HR, Mueen-ud-din G, Khalid MZ. Applications of Biotechnology in Food and Agriculture. Proceedings of the National Academy of Sciences, India Section 2022;92(1):11–15.
42. Liu X, Wu D, Shao Y, Wu Y. New food sources and production systems: a comparison of international regulations and China's advancements in novel foods with synthetic biology. Food science and human wellness 2024;13(5):2519–2542.
43. Huang CW, Lin C, Nguyen MK, Hussain A, Bui XT, Ngo HH. A review of biosensor for environmental monitoring: principle, application, and corresponding achievement of sustainable development goals. Bioengineered 2023;14(1):58–80.
44. Ezeorba TPC, Okeke ES, Mayel MH, Nwuche CO, Ezike TC. Recent advances in biotechnological valorization of agro-food wastes (AFW): Optimizing integrated approaches for sustainable biorefinery and circular bioeconomy. 2024;26:101823.
45. Kuppan N, Padman M, Mahadeva M, Srinivasan S, Devarajan R. A comprehensive review of sustainable bioremediation techniques: Eco friendly solutions for waste and pollution management. Waste management bulletin 2024;2:154–171
46. Mendes F, Machado BO, Castro BB, Sousa MJ, Chaves SR. Harnessing the power of biosensors for environmental monitoring of pesticides in water. Applied microbiology and biotechnology 2025;109(92):1–16.
47. Prado JV, Ortiz CC, Perez ABM, Torres CF, Yanez MV, Deng L, Ostria CB, et al. Advances in environmental biotechnology with CRISPR/Cas9:bibliometric review and cutting-edge applications. Discover applied sciences 2025;7(167):1–33.
48. Ezeorba TPC, Okeke ES, Mayel MH, Nwuche CO, Ezike TC. Recent advances in biotechnological valorization of agro-food wastes (AFW): Optimizing integrated approaches for sustainable biorefinery and circular bioeconomy. 2024;26:101823.
49. Chancellor D, Barrett D, Jatkoe LN, Millington S, Eckhardt F. The state of cell and gene therapy in 2023. Molecular therapy 2023;31(12):3376–3388.
50. Żak MM. Clinical development of therapeutic mRNA applications. Molecular therapy 2025;33(6):2583–2609.

51. Odouard IC, Kang SY, Mo J, Anderson G, Alexander GC. Research and development investments for biologics independently developed by US biotechnology startups. Health affairs scholar 2025;3(7):1–8.
52. Y Xiao, Du Z, Li Y, Cao L, Zhu B, Kitaguchi T, Huang C. A review on the application of biosensors for monitoring emerging contaminants in the water environment Sensors 2025;25(16):4945.

Computational Approaches in Green Analytical Chemistry

Rajesh B. Patil and Sadhana S. Raut

Abstract Analytical chemistry, traditionally rooted in empirical experimentation, is undergoing a transformative shift toward computational and sustainable innovation. The increasing global emphasis on environmental responsibility has catalyzed the evolution of Green Analytical Chemistry (GAC)—a framework prioritizing waste minimization, energy efficiency, and reduced use of hazardous substances. This chapter explores how computational tools such as chemometrics, chromatographic simulators, molecular modeling, and artificial intelligence (AI) are revolutionizing analytical method development by enabling predictive, data-driven, and resource-efficient workflows. Strategies like Design of Experiments (DoE), multivariate data analysis (MVA), and response surface modeling are discussed in the context of Quality by Design (QbD) principles to demonstrate systematic, risk-based optimization. Further, the integration of machine learning and molecular modeling is shown to enhance predictive accuracy, solvent selection, and green metrics evaluation. Collectively, these advancements redefine the analytical landscape by coupling scientific rigor with environmental stewardship, marking a decisive move from traditional trial-and-error methodologies to intelligent, sustainable, and simulation-guided practices.

Keywords Green Analytical Chemistry (GAC) · Computational chemistry · Chemometrics · Design of Experiments (DoE) · Quality by Design (QbD)

1 Introduction

Analytical chemistry has long been the cornerstone of scientific investigation, providing reliable tools for the qualitative and quantitative determination of substances in complex matrices [1]. Over the past two decades, however, the field

R. B. Patil · S. S. Raut (✉)
Department of Pharmaceutical Chemistry, Sinhgad Technical Education Society's, Sinhgad College of Pharmacy, Pune, India
e-mail: sadana.105@gmail.com

V. Bhusari and M. Ghante (eds.), *Green Analytical Chemistry in India*,
https://doi.org/10.1007/978-3-032-24564-9_8

has witnessed a paradigm shift, driven not only by the demand for greater sensitivity, selectivity, and throughput but also by the growing awareness of environmental sustainability [2]. Traditional method development, particularly in chromatography, spectroscopy, and sample preparation, is often resource-intensive, relying on a "trial-and-error" approach that consumes large volumes of hazardous solvents, generates chemical waste, and requires significant experimental time [3]. In light of these challenges, the integration of modeling, simulation, and data-driven strategies has emerged as an indispensable approach to designing robust, efficient, and greener analytical methods [4].

The increasing demand for eco-friendly practices in analytical chemistry has led to the emergence of Green Analytical Chemistry (GAC), which emphasizes minimal use of hazardous substances, waste reduction, and energy efficiency [3]. Computational tools provide robust support to GAC by enhancing method development, optimizing existing analytical techniques, and reducing the need for extensive experimentation [2]. With the growth of digital technology, India has seen significant adoption of such computational approaches across academia and industry [5]. The concept of GAC has revolutionized the field of analytical science by integrating environmental considerations into routine laboratory practices. It encourages the development of methods that minimize the use of toxic reagents, reduce energy consumption, and generate minimal waste, aligning with the principles of sustainable development [4]. In this context, computational approaches have emerged as powerful enablers for GAC, allowing scientists to simulate, predict, and optimize analytical methods before they are practically implemented.

The application of chemometric modeling, chromatographic simulators, molecular modeling, and artificial intelligence (AI) offers a transformative pathway for analytical method development. These tools allow researchers to virtually explore experimental conditions, predict method outcomes, and optimize analytical protocols without excessive laboratory experimentation. Such strategies are fully aligned with the principles of Green Analytical Chemistry (GAC), which emphasize waste minimization, safer solvent use, energy efficiency, and reduced environmental footprint while ensuring method robustness and regulatory compliance. In particular, the concept of Quality by Design (QbD), as recommended by ICH guidelines (Q8–Q14), encourages systematic risk-based method development, where modeling and simulation play a central role in predicting method performance and ensuring lifecycle robustness [6].

One of the most significant enablers in this space is chemometric modeling, which combines statistical, mathematical, and computational methods to analyze chemical data and guide experimental design. Approaches such as Design of Experiments (DoE) allow systematic variation of multiple factors (pH, gradient time, temperature, solvent composition) to understand their impact on analytical performance. Designs like full factorial design (FFD), fractional factorial design (FrFD), central composite design (CCD), and Box–Behnken design (BBD) provide a structured methodology for screening, optimization, and robustness testing. These experimental design approaches, when integrated with multivariate analysis tools, significantly reduce the experimental burden in HPLC/UHPLC method development, stability testing, and robustness mapping. Moreover, the ability of DoE to generate predictive response surface models enables analysts to identify optimal conditions with minimal experimental effort, exemplifying both efficiency and sustainability [7].

Beyond DoE, multivariate data analysis (MVA) has become indispensable in handling the vast datasets generated from modern analytical platforms such as NIR, FTIR, UV–Vis, LC–MS, and NMR. Methods like Principal Component Analysis (PCA), Partial Least Squares (PLS/PLSR), and PLS-DA not only aid in calibration and quantitative modeling but also assist in impurity profiling, authenticity testing, metabolomics, and stability studies. By extracting hidden patterns, identifying outliers, and reducing data dimensionality, MVA ensures that analysts can obtain meaningful insights from complex spectral or chromatographic fingerprints. When combined with regression methods such as Multiple Linear Regression (MLR) or Principal Component Regression (PCR), chemometrics extends beyond descriptive analysis to predictive modeling, thereby enabling proactive method development [8].

Another vital dimension is Response Surface Methodology (RSM), which has been widely adopted in QbD-driven method development. By fitting mathematical models to experimental data and visualizing them as 2D contour plots or 3D response surfaces, RSM provides intuitive maps of factor–response relationships. Applications in HPLC mobile phase optimization, extraction parameter tuning, and robustness mapping (similar to DryLab®) have shown how simulation can eliminate redundant trials, saving both solvents and experimental time. The growing integration of machine learning (ML) and advanced chemometric approaches further enhances predictive accuracy. Methods such as artificial neural networks (ANNs), Support Vector Machines (SVMs), genetic algorithms (GAs), and random forests (RF) are increasingly being used to predict retention times, optimize chromatographic conditions, and classify spectral data. Such models excel in capturing nonlinear relationships that traditional regression may miss, especially in big data contexts generated from hyphenated techniques (LC–MS, GC–MS, LC–NMR) [9].

The field has also embraced chromatographic simulators such as DryLab®, ChromSword®, ACD/ChromGenius™, ACD/LC Simulator, and Virtual Column®, which enable virtual experimentation with chromatographic variables. By simulating changes in temperature, gradient time, pH, or solvent ratios, these tools generate predictive resolution maps, allowing analysts to identify robust operating conditions before performing a single physical experiment. Notably, DryLab® has gained regulatory acceptance (FDA, EMA) as part of QbD-based submissions, underlining the credibility and utility of such tools [10]. By reducing laboratory experimentation, chromatographic simulators not only accelerate method development but also directly contribute to greener practices by minimizing solvent use and waste generation.

Complementing chemometrics and simulation, molecular modeling techniques such as molecular docking, molecular dynamics (MD), quantum chemical calculations, and quantitative structure–retention relationship (QSRR) modeling provide mechanistic insights into analyte–stationary phase interactions, solvent compatibility, and detection parameters. These approaches are especially valuable in chiral separations, biosensor development, and stationary phase selection, where experimental screening is often laborious and costly. For instance, molecular docking can predict enantioselective interactions on chiral stationary phases, while MD simulations reveal dynamic solute–solvent–surface interactions that govern retention and selectivity. Furthermore, quantum chemical methods (DFT, ab initio) can predict UV/Vis absorption maxima, IR frequencies, or NMR shifts, guiding the selection of detection wavelengths and validation markers. By integrating QSPR/QSAR modeling, molecular approaches also support virtual solvent screening, enabling early elimination of hazardous or non-sustainable solvent candidates [11].

The integration of artificial intelligence in green metrics prediction marks the most recent and impactful development. AI-driven green score predictors leverage large datasets of solvents, reagents, and analytical parameters to forecast sustainability metrics such as E-factor, PMI (Process Mass Intensity), atom economy, biodegradability, and toxicity indices. Models based on regression, tree-based ensembles (Random Forest, Gradient Boosting), neural networks, and hybrid QSPR–ML approaches are increasingly being deployed to evaluate the greenness of analytical procedures. Tools like AGREE, GAPI, and DOZN 2.0 provide benchmark metrics, which, when combined with ML algorithms, offer automated decision support for greener method design. Importantly, such frameworks not only assess the current greenness of analytical workflows but also suggest modifications for improvement, thereby serving as proactive sustainability guides [12].

In this context, solvent substitution tools such as Hansen Solubility Parameters (HSPiP), COSMO-RS, UNIFAC, and EPI Suite™ complement QSPR/QSAR approaches by virtually predicting solvent–solute compatibility, solvation energies, and environmental fate properties. Their ability to propose eco-friendly alternatives to conventional solvents (e.g., ethanol–water replacing acetonitrile, or deep eutectic solvents replacing dichloromethane) demonstrates the power of computational substitution in analytical method design. These approaches have been further enriched by the incorporation of multi-criteria decision analysis (MCDA) and Pareto optimization, ensuring that solvent choices balance analytical performance with environmental sustainability [13].

Overall, the convergence of chemometrics, simulation, molecular modeling, and AI represents a paradigm shift in analytical chemistry, moving from empirical trial-and-error to data-driven, predictive, and sustainable method development. Such approaches not only accelerate method design and reduce experimental burden but also embed the principles of Green Analytical Chemistry into the very foundation of analytical workflows. As global laboratories, including those in India, increasingly adopt these methodologies, the future of analytical chemistry is poised to become not only more efficient and robust but also environmentally responsible, meeting the dual imperatives of scientific progress and sustainability.

2 Modeling and Simulation for Method Development

2.1 Chemometric Modeling

2.1.1 Experimental Design Approaches (DoE)

Chemometric modeling has emerged as an indispensable tool in modern analytical method development, providing a systematic and statistically sound approach to experimental planning, data interpretation, and process optimization. One of the most widely adopted chemometric strategies is Design of Experiments (DoE), which allows for the structured exploration of multiple experimental variables simultaneously. Unlike traditional one-factor-at-a-time methods, DoE provides deeper insights into factor interactions and main effects, thus enabling efficient method optimization and robustness evaluation [7]. This not only reduces the number of experiments required but also ensures reliable and reproducible method development in compliance with Quality by Design (QbD) principles [14].

Among DoE approaches, the Full Factorial Design (FFD) is fundamental, as it explores all possible combinations of factors and their interactions. Although comprehensive, it becomes resource-intensive when the number of factors is high, and is therefore most suitable for preliminary optimization involving few variables [15]. Fractional Factorial Design (FrFD) addresses this limitation by evaluating a subset of experimental combinations, making it useful in early screening phases where the goal is to identify critical factors with fewer experiments [16]. These approaches have been widely applied in chromatographic method development, particularly in optimizing conditions such as pH, gradient time, and solvent composition. For example, Hibbert [17] demonstrated that FrFD was successfully applied in liquid chromatography to identify significant parameters affecting retention time and resolution, thereby reducing solvent consumption and analysis time.

For more detailed optimization, Response Surface Methodology (RSM) designs are often employed. Central Composite Design (CCD) and Box–Behnken Design (BBD) are commonly used to model quadratic relationships between factors and responses. CCD provides flexibility and good predictive ability, especially when curvature effects are significant, while BBD is more economical, avoiding extreme experimental combinations that may be impractical [18]. A case study by Bezerra et al. [19] highlighted the use of RSM with CCD for optimizing HPLC mobile phase composition for pharmaceutical compounds, where the response surface plots provided visual insights into factor–response interactions, ultimately improving peak resolution and reducing run time. Similarly, Ferreira et al. [20] reported that BBD was used in the optimization of multicomponent analysis of drugs in complex formulations, demonstrating efficiency in robustness testing. Taguchi Design, on the other hand, focuses on achieving robustness by identifying conditions that minimize variability and enhance reproducibility. Roy [21] reported its successful use in ruggedness testing of spectroscopic methods, where external noise factors (e.g., instrument variation, environmental fluctuations) were accounted for, ensuring reliable method performance.

The application of these experimental design approaches in chromatography and spectroscopy highlights their versatility in pharmaceutical analysis. For instance, Orlandini et al. [22] applied DoE and chemometric modeling to optimize UHPLC methods for drug stability testing under QbD principles, ensuring regulatory compliance with ICH guidelines. Another case study by Moeder et al. [23] demonstrated that CCD-based RSM was used for the optimization of pesticide analysis in environmental samples, achieving enhanced sensitivity and lower limits of detection. Furthermore, the predictive power of these designs minimizes trial-and-error experimentation, aligning with the principles of Green Analytical Chemistry (GAC) by reducing solvent use, waste generation, and experimental costs [3]. As regulatory frameworks increasingly emphasize risk-based approaches such as QbD, chemometric modeling using DoE continues to play a central role in ensuring analytical methods are not only efficient and reproducible but also environmentally sustainable.

2.1.2 Multivariate Data Analysis (MVA)

Multivariate Data Analysis (MVA) techniques are increasingly used to process complex datasets in pharmaceutical sciences, particularly when dealing with spectra, chromatograms, or metabolomic fingerprints. Among these, Principal Component Analysis (PCA) is one of the most widely applied, providing dimensionality reduction, visualization of hidden patterns, and detection of outliers in analytical data [24]. A notable case study demonstrated the use of PCA in Near-Infrared (NIR) spectroscopy to evaluate batch-to-batch variation of paracetamol tablets, where PCA successfully differentiated between compliant and non-compliant samples, improving process control under Process Analytical Technology (PAT) frameworks [25]. Similarly, Partial Least Squares Regression (PLSR) has been extensively used in developing calibration models for quantitative analysis. For instance, PLS models were applied to Fourier Transform Infrared (FTIR) spectra for predicting the active pharmaceutical ingredient (API) content in tablets, showing strong predictive accuracy and reducing reliance on destructive reference methods [26].

Beyond calibration, MVA has also proven valuable in drug authentication and classification. Partial Least Squares Discriminant Analysis (PLS-DA) has been employed to distinguish authentic vs. adulterated herbal medicines, providing a rapid screening tool for regulatory compliance [27]. Another case study reported the application of Principal Component Regression (PCR) for impurity profiling in liquid chromatography–mass spectrometry (LC–MS) datasets, demonstrating its ability to handle correlated variables and provide robust regression models [9]. In the field of metabolomics, PCA and PLS were used to monitor the stability of biological samples during storage, with clear separation of fresh vs. degraded profiles, thus aiding stability studies [28]. These applications underscore the versatility of MVA in pharmaceutical analysis, enabling improved quality control, rapid screening, and enhanced interpretability of complex analytical data.

2.1.3 Response Surface Methodology (RSM)

Response Surface Methodology (RSM) is increasingly adopted in analytical method development because of its ability to mathematically model complex relationships between experimental inputs and measured responses [7]. A classic case study by Singh and Ahuja optimized HPLC conditions for simultaneous estimation of amlodipine and atorvastatin [29]. By varying mobile phase composition, pH, and flow rate systematically, the RSM model identified an optimal set of conditions that provided excellent resolution between peaks with minimal baseline noise. The 3D response surfaces and contour plots illustrated clear factor–response interactions, reducing the number of experimental trials needed compared to traditional one-factor-at-a-time approaches. Importantly, this optimization not only improved robustness but also minimized solvent consumption, thereby contributing to greener chromatographic practices.

Another practical application of RSM was demonstrated in ultrasound-assisted extraction of nutraceuticals from plant matrices. Wang and Weller reported how extraction yield of bioactives such as polyphenols was strongly influenced by temperature, solvent-to-solid ratio, and ultrasound time [30]. RSM modeling revealed significant curvature effects, and the quadratic model predicted optimal extraction conditions that were experimentally validated. Such applications highlight the efficiency of RSM in both analytical and preparative workflows. In a regulatory framework, particularly under ICH Q8 (QbD paradigm), RSM is also applied to establish a "design space." Molnár and colleagues showed that RSM-based robustness mapping in DryLab® for UHPLC significantly reduced method development time while identifying a wide operational range for gradient time and temperature [31]. Together, these examples establish RSM as a cornerstone for robust, efficient, and environmentally sustainable method development.

2.1.4 Machine Learning and Advanced Chemometrics

Machine Learning (ML) has moved analytical chemistry into the predictive era by handling nonlinear, high-dimensional datasets. A case study by Rani et al. demonstrated how Artificial Neural Networks (ANNs) were employed to optimize RP-HPLC conditions for pharmaceutical compounds [32]. Unlike traditional RSM, which assumes polynomial relationships, ANNs were able to capture nonlinear factor–response interactions, resulting in superior prediction accuracy for retention times and resolution. This marked an important shift toward AI-assisted method development.

Support Vector Machines (SVMs) have also been applied to address classification challenges. Wu et al. used SVM models on spectral fingerprints of herbal medicines, successfully differentiating between authentic and adulterated samples [33]. This has significant implications for quality assurance, especially in markets where counterfeit drugs are prevalent. Likewise, Genetic Algorithms (GAs) have been integrated with chromatographic simulations, as shown by Gratzfeld-Hüsgen and colleagues, to rapidly search a large experimental space and determine near-optimal gradient programs [34]. These methods drastically reduce trial-and-error and improve the efficiency of method development.

Meanwhile, Random Forests (RFs) have proven their utility in analyzing large spectral datasets. Fernández Pierna and Dardenne applied RFs to NIR spectroscopy for classification of pharmaceutical solid dosage forms, distinguishing between different polymorphs and assessing content uniformity with high accuracy [35]. This demonstrates how ensemble learning methods are enabling advanced chemometrics in PAT (Process Analytical Technology) and big-data analytics from LC–MS and NMR platforms. Collectively, ML and advanced chemometric methods are enhancing predictive accuracy, ensuring robustness, and accelerating decision-making in analytical chemistry.

2.1.5 Multivariate Curve Resolution (MCR)

Multivariate Curve Resolution (MCR) techniques, particularly MCR-ALS (Alternating Least Squares), are invaluable for deconvolution overlapping peaks and spectra in complex datasets. Tauler's pioneering application in HPLC–DAD showed how MCR-ALS could resolve co-eluting drug degradation products during stability studies without requiring additional chromatographic separation [36]. The method allowed quantification of hidden analytes that would otherwise be masked in conventional analysis, improving both efficiency and interpretability of stability-indicating assays.

In a related study, de Juan et al. applied MCR to LC–MS datasets for impurity profiling of active pharmaceutical ingredients (APIs) [37]. MCR successfully extracted pure component mass spectra from overlapping signals, enabling the characterization of unknown impurities without complete chromatographic resolution. This represents a significant advance in impurity profiling and regulatory compliance. Furthermore, Veselkov and colleagues employed MCR in metabolomics, applying it to complex NMR datasets of urine samples [38]. Their recursive alignment and deconvolution approach separated overlapping resonances, revealing metabolic signatures associated with disease states. These case studies emphasize how MCR extends beyond routine analysis to enable biomarker discovery, impurity characterization, and stability studies—all without the need for time-consuming or resource-heavy separation techniques. By maximizing information extraction from existing data, MCR aligns strongly with principles of Green Analytical Chemistry.

The summary of all above techniques under chemometric modeling is given in Table 1 with its GAC relevance, advantages and disadvantages.

Table 1 Comparative overview of chemometric and computational approaches in analytical method development and their relevance to Quality by Design (QbD) and Green Analytical Chemistry (GAC) principles

Technique/approach	Primary purpose	Key features/methodology	Advantages	Limitations	Typical applications	Green Analytical Chemistry (GAC) relevance
Full Factorial Design (FFD)	Comprehensive factor study	Evaluates all factor combinations	Complete interaction mapping	Resource-intensive for large variable sets	Preliminary optimization in chromatography	Enhances understanding, reduces rework
Fractional Factorial Design (FrFD)	Screening of critical factors	Evaluates subset of FFD combinations	Fewer experiments, saves time	May overlook higher-order interactions	Early-phase screening	Reduces experimental runs, solvent, and energy
Central Composite Design (CCD)	Quadratic modeling, optimization	Combines factorial and axial points	Flexible, accurate prediction	Moderate experiment count	HPLC and UHPLC optimization	Reduces solvent and runtime
Box–Behnken Design (BBD)	Response surface modeling	Avoids extreme experimental points	Cost-effective, robust	Limited curvature estimation	Robustness testing	Minimizes waste, energy consumption
Taguchi Design	Robustness assessment	Orthogonal array-based	Tolerant to external noise	Limited interaction study	Ruggedness testing of spectroscopic methods	Reduces error-related reanalysis
Principal Component Analysis (PCA)	Data simplification and pattern recognition	Eigen decomposition	Detects hidden trends, outliers	No predictive capacity	NIR and PAT monitoring	Enables non-destructive testing
Partial Least Squares Regression (PLSR)	Quantitative calibration	Covariance-based regression	Handles multicollinearity	Requires calibration dataset	FTIR quantification of APIs	Reduces destructive reference tests
Partial Least Squares Discriminant Analysis (PLS-DA)	Classification	Discriminant modeling using PLS	High classification power	Sensitive to sample imbalance	Herbal drug authentication	Rapid quality screening, low waste

(continued)

Table 1 (continued)

Technique/approach	Primary purpose	Key features/ methodology	Advantages	Limitations	Typical applications	Green Analytical Chemistry (GAC) relevance
Principal Component Regression (PCR)	Regression of correlated data	Combines PCA and regression	Stable against collinearity	May lose interpretability	Impurity profiling (LC–MS)	Improves data reuse efficiency
Response Surface Methodology (RSM)	Optimization of multivariate systems	Quadratic polynomial modeling	Visual response surfaces, curvature detection	Requires prior screening	HPLC and extraction optimization	Fewer experiments, solvent saving
Artificial Neural Networks (ANNs)	Nonlinear prediction	Learns complex factor–response patterns	Handles nonlinearities	Needs large training data	HPLC optimization	Predictive modeling reduces trials
Support Vector Machines (SVMs)	Classification	Margin-based learning	High accuracy	Sensitive to kernel choice	Herbal fingerprint classification	Rapid non-destructive screening
Genetic Algorithms (GAs)	Optimization	Evolutionary parameter search	Efficient exploration	Randomness may affect reproducibility	Chromatographic simulation	Saves resources via virtual experimentation
Random Forests (RFs)	Classification, regression	Ensemble tree-based	Robust, interpretable	Needs many trees	NIR polymorph classification	Enables data-based analysis over lab work
Multivariate Curve Resolution (MCR-ALS)	Signal deconvolution	Alternating least squares	Resolves overlapping peaks	Needs good initial estimates	HPLC–DAD, LC–MS, NMR analysis	Maximizes information from existing data

(continued)

Table 1 (continued)

Technique/approach	Primary purpose	Key features/ methodology	Advantages	Limitations	Typical applications	Green Analytical Chemistry (GAC) relevance
Integration with QbD	Systematic method design	Risk-based approach	Ensures robustness, regulatory compliance	Needs statistical expertise	Stability and impurity studies	Promotes efficient, compliant workflows
Integration with GAC	Sustainability enhancement	Eco-efficiency assessment	Minimizes waste, solvent, energy	Requires metric evaluation tools	Green method validation	Core of sustainable analytical development

2.2 *Chromatographic Simulators*

Chromatographic simulators represent an innovative paradigm in analytical method development by enabling virtual experimentation before physical trials. These tools integrate advanced mathematical models, retention databases, and artificial intelligence to predict chromatographic behavior under varying conditions [39]. By simulating variables such as gradient time, pH, solvent composition, and temperature, simulators help identify robust analytical conditions, reducing solvent use, waste generation, and development time. Their adoption aligns with Quality by Design (QbD) principles and Green Analytical Chemistry (GAC), as they minimize trial-and-error approaches while supporting lifecycle management. Regulatory agencies such as the FDA and EMA increasingly recognize the use of such tools in submissions, making them valuable assets in modern pharmaceutical research and quality control [40].

2.2.1 DryLab® (Molnár-Institute)

DryLab® is the most widely used and validated chromatographic simulator for method development and robustness testing. It models HPLC and UHPLC separations under varying parameters such as gradient slope, pH, temperature, and solvent composition [31]. A notable case study involved the optimization of a stability-indicating UHPLC method for monoclonal antibodies, where DryLab® predicted resolution maps and reduced the number of experimental runs from 120 to just 12, saving time and resources [41]. Another example demonstrated its utility in impurity profiling of active pharmaceutical ingredients (APIs). By simulating chromatographic conditions, DryLab® identified critical separation windows for degradation products, reducing risk of co-elution [42].

Importantly, DryLab® is widely accepted by regulatory bodies. Molnár and colleagues reported its integration into QbD-based submissions to the FDA and EMA, where resolution maps served as a "design space" for method lifecycle management [43]. Additionally, its application in gradient elution methods for peptide analysis demonstrated superior reproducibility, showing how DryLab® supports biologics and small-molecule workflows alike. Overall, DryLab® exemplifies how simulation can accelerate robust method development while meeting global regulatory standards.

2.2.2 ChromSword®

ChromSword® specializes in automated HPLC method development, supporting both isocratic and gradient elution. Its strength lies in integration with LC instrumentation such as Agilent and Shimadzu systems, allowing real-time method optimization [44]. A case study applied ChromSword® for the separation of steroidal compounds, where its predictive algorithms optimized gradient slope and solvent ratios to achieve baseline separation in fewer trials compared to traditional approaches [45]. Another application in pharmaceutical impurity analysis demonstrated how ChromSword® reduced screening experiments by predicting selectivity differences across columns and solvents [46].

ChromSword®'s automation modules make it particularly valuable for labs with high-throughput requirements. In food chemistry, it was used to optimize the separation of pesticide residues, where its simulation platform predicted elution profiles for over 20 analytes in complex matrices [47]. These applications highlight ChromSword® as a versatile tool bridging automation, simulation, and experimental validation.

2.2.3 ACD/ChromGenius™ (ACD/Labs)

ACD/ChromGenius™ uses a knowledge-based approach for retention prediction, relying on physicochemical properties and structural descriptors of analytes [48]. This makes it particularly powerful in early-phase pharmaceutical research, where experimental data may be limited. A case study showed its application in predicting retention times of newly synthesized drug candidates, enabling faster column and mobile phase selection during preclinical development [49]. Another report highlighted its role in predicting chromatographic behavior of structurally related impurities, improving early impurity profiling strategies [50].

In addition, ChromGenius™ supports QbD initiatives by integrating with method development workflows. For example, it has been used to establish retention trends across a series of structurally diverse compounds, helping analysts design robustness studies [51]. By reducing dependency on extensive trial runs, ChromGenius™ provides a green and cost-effective pathway for chromatographic scouting and early-phase analytical development.

2.2.4 ACD/LC Simulator (ACD/Labs)

The ACD/LC Simulator extends predictive modeling by simulating chromatographic conditions, such as mobile phase composition, flow rate, and gradient programs [52]. One application was in stability studies of antibiotics, where the LC Simulator predicted elution behavior of degradation products and minimized the number of confirmatory experiments [53]. Another study demonstrated its use in impurity profiling of complex mixtures, particularly in the separation of alkaloids, where the simulator predicted optimal mobile phase gradients [54].

By offering scenario testing, ACD/LC Simulator reduces the need for multiple physical runs and supports robust method development. Its predictive capability has also been leveraged for bioanalytical assays, where analysts optimized LC–MS conditions virtually, minimizing matrix effects [55]. Collectively, the LC Simulator enhances decision-making in stability and impurity analysis while conserving resources.

2.2.5 ChromX (Emerging Tools)

ChromX is a relatively new simulation platform that integrates AI-driven algorithms for multi-parameter chromatographic optimization [56]. Unlike traditional simulators, ChromX uses machine learning models trained on large chromatographic datasets, allowing faster predictions of retention and resolution under diverse experimental conditions. A recent study applied ChromX to peptide separations, where it successfully predicted optimal gradient programs and solvent ratios without extensive manual tuning [57].

Another application in metabolomics showed how ChromX could optimize chromatographic conditions for complex biological extracts, reducing run-to-run variability while improving reproducibility [58]. These early case studies suggest ChromX has significant potential in big-data-driven workflows, particularly in biopharmaceuticals and omics research.

2.2.6 Virtual Column® (ACE—Advanced Chromatography Technologies)

Virtual Column® serves as a decision-support tool by using column selectivity databases to recommend the most suitable stationary phase for a given separation [59]. It does not simulate chromatograms directly but provides guidance on column chemistry based on analyte properties. A case study highlighted its role in selecting columns for the separation of basic pharmaceutical drugs, where it predicted stationary phases that minimized peak tailing [60].

Another example demonstrated its utility in method scouting, where Virtual Column® helped researchers identify the best column chemistry for separating isomeric impurities, saving significant method development time [61]. By guiding column selection, Virtual Column® complements simulators like DryLab® and ChromSword®, ensuring that analysts begin their method development with the right stationary phase.

2.3 Molecular Modeling

Molecular modeling has become an indispensable tool in analytical method development, providing atomic- and molecular-level insights into interactions that govern chromatographic and spectroscopic behavior [62]. By applying computational quantum chemistry, molecular docking, and statistical simulations, analysts can rationalize and predict analyte interactions with stationary phases, solvents, and detection reagents. These approaches not only accelerate method development but also reduce experimental costs and align with Quality by Design (QbD) and Green Analytical

Chemistry principles [63]. The integration of molecular modeling with chemoinformatics and artificial intelligence has further extended its potential, enabling predictive, data-driven workflows in pharmaceutical analysis and bioanalytical research [64].

2.3.1 Molecular Docking

Molecular docking predicts binding interactions between analyte molecules and chromatographic stationary phases or detection reagents. In chiral chromatography, docking has been used to study enantioselectivity by simulating differential interactions of enantiomers with chiral selectors. For example, docking of propranolol enantiomers on β-cyclodextrin-based stationary phases accurately predicted elution order, supporting experimental chiral HPLC results [65]. Similarly, docking has been applied to predict selectivity of drug metabolites on derivatization reagents in LC–MS workflows, guiding reagent choice for sensitive detection [66].

In solid-phase extraction (SPE), docking simulations between flavonoids and polymeric sorbents revealed binding site preferences, which guided the selection of more selective SPE cartridges [67]. In biosensor applications, molecular docking between analytes and immobilized receptors has been used to design sensor surfaces with enhanced specificity [68]. Such studies highlight docking as a predictive complement to experimental optimization, especially in complex mixtures where multiple interactions compete.

2.3.2 Molecular Dynamics (MD) Simulations

MD simulations allow real-time evaluation of solute–solvent–stationary phase interactions. This has been applied to rationalize peak tailing in reversed-phase HPLC: simulations of caffeine interactions with C18 phases showed persistent hydrogen bonding with residual silanol groups, explaining experimental asymmetry [69]. In another case, MD simulations of peptides in hydrophilic interaction chromatography (HILIC) revealed how acetonitrile content modulates solvation layers, influencing retention behavior [70].

Mobile phase optimization also benefits from MD. For instance, simulations demonstrated how buffer pH altered ionization states of basic drugs, which corresponded to observed shifts in retention times [71]. In gas chromatography, MD simulations of volatile compounds with ionic liquid stationary phases predicted selectivity trends, later validated experimentally [72]. By explaining adsorption, interaction strength, and analyte stability, MD serves as a powerful complement for chromatographic troubleshooting and design.

2.3.3 Quantum Chemical Calculations (DFT, Ab Initio, Semi-empirical)

Quantum chemical methods provide detailed electronic structure information useful in spectroscopic method development. Density Functional Theory (DFT) calculations have been used to predict UV/Vis absorption maxima of flavonoids, which closely matched experimental values and guided wavelength selection for HPLC–UV methods [73]. Similarly, ab initio methods have been applied to calculate IR frequencies of pharmaceuticals, improving identification of polymorphic forms [74].

In NMR spectroscopy, calculated chemical shifts of alkaloids using semi-empirical methods correlated with observed spectra, supporting structural confirmation and impurity identification [75]. Another application involves predicting pKa values of drug candidates through quantum chemistry, which informed mobile phase pH selection for LC separations [76]. These examples illustrate how electronic-level predictions can significantly streamline method validation and analyte characterization.

2.3.4 Quantitative Structure–Retention Relationship (QSRR) and QSAR

QSRR approaches link molecular descriptors with chromatographic retention, allowing prediction of retention times. In one study, hydrophobicity and polarity descriptors were used to predict LC retention of 200 pharmaceutical compounds, achieving excellent correlation with experimental data [77]. QSAR models have similarly been applied in predicting separation selectivity of steroid derivatives, providing guidance before physical experiments [78].

Another case study applied QSRR to pesticides in LC–MS/MS, where descriptor-based models predicted gradient elution times, reducing the need for extensive scouting [79]. Retention time databases built using QSRR are increasingly used in metabolomics for annotating unknowns, particularly in high-resolution LC–MS studies [80]. By guiding solvent strength, gradient slope, and stationary phase choice, QSRR serves as a predictive framework for chromatography-driven workflows.

2.3.5 Monte Carlo and Statistical Simulations

Monte Carlo simulations introduce stochastic variability into molecular modeling, providing probabilistic insights into adsorption and distribution processes. In chromatography, Monte Carlo methods have been applied to optimize adsorption isotherms of amino acids on ion-exchange resins, enabling better prediction of overload effects [81]. In SPE, Monte Carlo simulations predicted partition coefficients of antibiotics between aqueous and solid phases, improving recovery efficiencies [82].

Another application involved robustness testing of HPLC methods, where Monte Carlo simulations generated thousands of virtual runs by varying pH, temperature, and gradient slopes within tolerance ranges [83]. This probability-based approach identified method conditions most resilient to variability, complementing traditional robustness studies. Such simulations provide a statistically rigorous foundation for ensuring reproducibility in analytical workflows.

2.3.6 Molecular Electrostatic Potential (MEP) and Surface Mapping

MEP analysis visualizes charge distribution, hydrogen-bonding regions, and hydrophobic surfaces, aiding interpretation of chromatographic retention. For example, MEP mapping of β-blockers revealed charge localization around amine groups, correlating with their retention shifts in ion-pair chromatography [84]. In another study, MEP maps of enantiomeric amino acids explained their differential interactions with chiral selectors, supporting experimental chiral separation results [85].

Surface mapping has also been applied to design derivatization reagents. For instance, analysis of carbonyl-rich pharmaceuticals identified favorable hydrogen-bonding regions, guiding derivatization strategies for improved GC detection [86]. In ion-pair LC, MEP helped predict interaction strengths of analytes with various ion-pairing reagents, leading to improved selectivity [87]. These applications show how electrostatic visualization tools bridge molecular properties with chromatographic performance.

2.3.7 Hybrid Approaches (Chemoinformatics + Molecular Modeling)

Hybrid approaches integrate molecular descriptors, docking, MD, and machine learning for predictive analytical chemistry [88]. For example, hybrid QSRR–machine learning models trained on docking and MD-derived descriptors predicted retention times of peptides with higher accuracy than conventional models [89]. In another application, chemoinformatics combined with DFT-derived descriptors was used to optimize gradient conditions for small-molecule separations [90].

Emerging AI platforms now combine docking, MD, and deep learning to predict GC and LC outcomes in silico. A case study in metabolomics applied hybrid modeling to predict retention of thousands of metabolites, dramatically reducing experimental burden [91]. Such integrative approaches signal the future of chromatographic method development, where in silico predictions guide experimental validation in a highly efficient cycle.

3 Artificial Intelligence in Predicting Green Metrics

Artificial Intelligence (AI) has emerged as a transformative tool in analytical chemistry, particularly in assessing and predicting green metrics such as solvent waste, atom economy, process mass intensity (PMI), and environmental impact factors. Unlike conventional assessment methods, which are retrospective and often qualitative, AI-driven models can analyze large datasets, uncover hidden correlations, and provide predictive insights during the early stages of method development. This allows chemists to integrate sustainability by design into analytical workflows. Recent advances in machine learning, deep learning, and hybrid computational approaches have made it possible to simulate the greenness of analytical methods before experimental validation, reducing cost, time, and environmental burden. Furthermore, AI-powered predictors are increasingly integrated with cheminformatics tools, regulatory databases, and green chemistry software, making them indispensable for the future of sustainable analytical sciences.

3.1 Machine Learning Models

3.1.1 Regression-Based Models

Regression-based approaches, such as Linear Regression (LR), Multiple Linear Regression (MLR), and Partial Least Squares Regression (PLSR), have long been used to model chemical processes. In green chemistry, these models predict quantitative green metrics like E-factor, atom economy, and solvent waste by correlating experimental parameters (solvent polarity, reaction time, temperature, catalyst type) with sustainability outcomes. A classic case study demonstrated the use of MLR to predict solvent waste in HPLC methods by correlating solvent composition and run time with waste volume, providing a quick decision-support tool for selecting greener methods [12].

PLSR has gained traction in complex chemical datasets where predictor variables are highly correlated. For instance, Sarkar et al. applied PLSR to link process parameters (e.g., pH, heating time, solvent polarity) with biodegradability indices in pharmaceutical effluents, enabling early identification of greener solvent alternatives [92]. Similarly, PLSR has been used in green extraction methods for natural products, helping select conditions that maximize extraction yield while minimizing energy use and solvent consumption [93].

These regression approaches are often integrated with chemometric tools such as SIMCA and MATLAB PLS Toolbox, which further enhance interpretability. While regression models offer transparency and ease of use, they struggle with nonlinear systems, leading researchers to combine them with more sophisticated AI models [94].

3.1.2 Tree-Based Models

Decision Trees (DTs) provide simple, interpretable rule-based models to classify analytical methods as green or non-green. In pharmaceutical development, DTs have been used to evaluate compliance with the Green Analytical Procedure Index (GAPI), where rules are based on solvent toxicity, energy input, and waste generation [95]. For instance, a DT model was applied to predict whether new HPLC methods for antihypertensive drugs would pass GAPI evaluation before actual validation, saving experimental resources.

Random Forests (RFs) improve robustness by aggregating multiple decision trees. RFs have been applied in solvent substitution studies, ranking solvents by their predicted environmental impact scores derived from REACH and EPA toxicity datasets [96]. Case studies in green sample preparation showed RF models could successfully identify eco-friendly solid-phase extraction solvents while maintaining analytical efficiency [97].

Gradient Boosting Machines (GBM, XGBoost, LightGBM, CatBoost) are particularly powerful for complex, nonlinear chemical systems. For example, GBM models predicted PMI and solvent toxicity for over 1,200 industrial reactions from the ACS Green Chemistry Institute database, outperforming regression and RF models [98]. Another study used XGBoost to rank solvents for biodegradability in chromatography, providing predictive insights that reduced the need for experimental solvent screening [99].

3.1.3 Neural Networks (NNs/Deep Learning)

Artificial Neural Networks (ANNs) capture complex nonlinear dependencies in chemical datasets. In green chemistry, ANNs have been applied to predict sustainability scores from reaction descriptors, including molecular weight, logP, hydrogen bond donors/acceptors, and polarity [100]. A study by Kumar et al. used ANN models to optimize green synthetic routes of active pharmaceutical ingredients (APIs), reducing waste and energy consumption [101].

Deep Neural Networks (DNNs) and their variants—Convolutional Neural Networks (CNNs) and Recurrent Neural Networks (RNNs)—have opened novel applications. CNNs can analyze chemical images, chromatograms, and spectral data to directly predict greenness indices without traditional descriptors [102]. For instance, a CNN-based approach classified NMR spectra of solvents as green or hazardous, achieving over 90% accuracy. RNNs, particularly LSTMs, have been applied to time-series data of chromatography systems, predicting solvent consumption and energy profiles over continuous runs [103].

Although deep learning models are often considered "black boxes," recent integration of explainable AI (XAI) approaches, such as SHAP values, has improved interpretability by identifying which chemical features most influence green predictions [104].

3.1.4 Support Vector Machines (SVMs/SVR)

Support Vector Machines (SVMs) are highly effective for small-to-medium datasets, common in analytical chemistry. In green metrics, SVM classifiers have been used to predict whether analytical methods comply with AGREE scores, outperforming regression models in categorical predictions [105].

Support Vector Regression (SVR) extends this to continuous predictions. For example, SVR models predicted E-factors for pharmaceutical reactions, achieving strong generalization on small datasets where neural networks overfitted [106]. Another application involved SVR predicting solvent volatility and biodegradability indices, guiding chromatographic solvent selection [107].

SVM-based approaches remain attractive for laboratories lacking access to big data, where they provide accurate and computationally efficient solutions for predicting greenness.

3.1.5 Hybrid/Advanced Approaches

Hybrid approaches combine cheminformatics descriptors, molecular modeling, and AI. The Quantitative Structure–Green Relationship (QSGR) integrates QSAR/QSPR with ML to predict eco-toxicity and biodegradability of solvents and reagents [108]. For example, hybrid models were used to screen green solvents for ionic liquid replacements in chromatography, balancing separation efficiency with environmental impact [109].

Another frontier is multi-objective optimization. Using genetic algorithms integrated with ML, researchers optimized chromatographic methods by balancing resolution, run time, and solvent waste. A case study in green supercritical fluid chromatography showed that hybrid AI approaches reduced solvent use by 35% without compromising separation efficiency [110].

Transfer learning has also gained attention: pre-trained models on large chemical toxicity datasets (e.g., EPA ECOTOX) were fine-tuned to predict analytical greenness, offering higher predictive power in small experimental datasets [111]. Such hybrid strategies represent the future of AI-driven sustainable analytical chemistry.

3.2 Development of Green Score Predictors

The success of AI models in predicting green metrics depends heavily on data quality and availability. Several databases provide essential descriptors: PubChem, ChemSpider, and ChEMBL for molecular properties; EPA ECOTOX and ECHA REACH for toxicity; and cheminformatics tools like RDKit, Mordred, and PaDEL-Descriptor for molecular descriptors [112]. These resources feed into green metrics calculators such as AGREE, GAPI, NEMI, DOZN 2.0, and CHEM21 Toolkit, which generate quantitative or semi-quantitative sustainability labels [113].

Machine learning frameworks in Python (scikit-learn, XGBoost, TensorFlow, PyTorch) and R (caret, mlr3, randomForest) have been widely applied to train predictive models. Workflow-based platforms like KNIME and Orange make AI tools accessible to non-experts by providing visual pipelines for classification and regression tasks [114].

Visualization and decision-support tools, including AGREE wheels, Tableau dashboards, and R Shiny apps, enable real-time interpretation of green predictions. Furthermore, deployment tools such as Streamlit, Flask APIs, and cloud-based ML platforms (AWS SageMaker, Azure ML) are now being used to develop interactive green score predictor applications that can be directly integrated into laboratory workflows [115].

These advancements illustrate how AI is transforming sustainability assessment from a post-experimental evaluation to a proactive design strategy, significantly advancing the field of Green Analytical Chemistry. Figure 1 indicates steps involved in development of green score predictors.

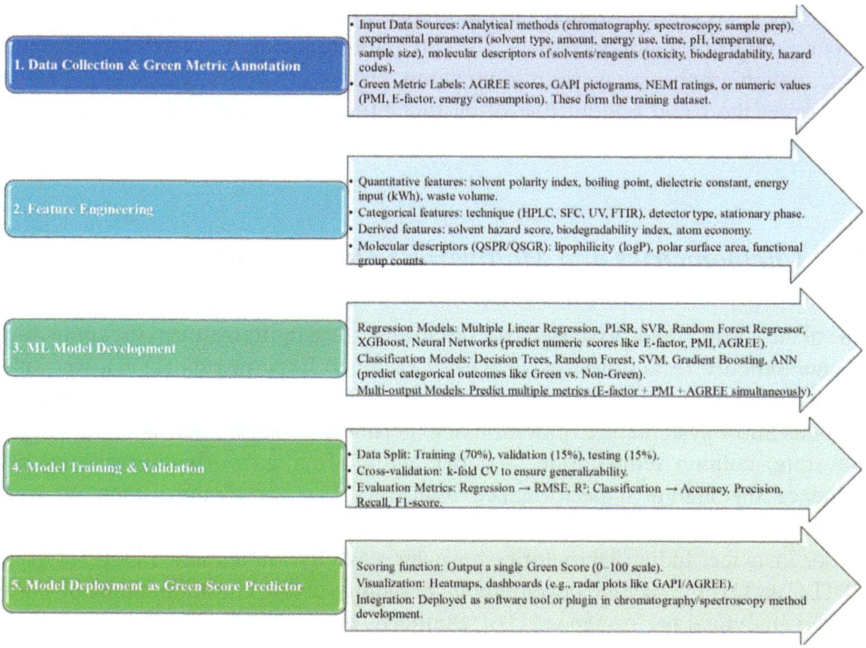

Fig. 1 Steps in developing green score predictors

4 Data-Driven Optimization of Analytical Processes

The integration of data-driven approaches in analytical chemistry has transformed the paradigm of method development, particularly within the context of Green Analytical Chemistry (GAC). Traditional optimization strategies often relied on empirical trial-and-error experimentation, which was both resource-intensive and environmentally unsustainable due to excessive use of solvents, reagents, and energy. With the advent of computational tools, artificial intelligence (AI), and machine learning (ML), analytical method optimization can now be achieved in a more efficient, predictive, and sustainable manner. These approaches not only reduce the ecological footprint of analytical methods but also enhance accuracy, reproducibility, and robustness.

Data-driven optimization encompasses the use of historical experimental data, molecular descriptors, process parameters, and sustainability indicators (such as solvent toxicity, energy consumption, and waste generation) to construct predictive models. These models assist in selecting the most suitable conditions for chromatography, spectroscopy, and sample preparation, while minimizing the need for resource-intensive laboratory trials. The core philosophy aligns with the principles of GAC, which emphasize waste minimization, energy efficiency, and safety, thereby addressing the triple bottom line of sustainability: environmental, economic, and social impact.

4.1 Chemometrics and Multivariate Optimization

One of the earliest data-driven approaches in analytical optimization is the application of chemometric techniques, particularly Design of Experiments (DoE), Response Surface Methodology (RSM), and Partial Least Squares Regression (PLSR). These methods allow systematic exploration of experimental factors such as solvent ratios, flow rate, column temperature, and pH, while simultaneously quantifying their influence on analytical performance indicators (e.g., resolution, retention time, sensitivity).

For instance, Indian laboratories have successfully implemented DoE for green HPTLC and HPLC method development, reducing solvent consumption while maintaining analytical performance [116]. Similarly, chemometric-assisted UV–Vis and FTIR spectroscopy optimization has been employed to reduce sample preparation steps, thereby minimizing hazardous chemical use [117]. PLSR models, in particular, help correlate process variables with green metrics, such as the Analytical Eco-Scale or GAPI, enabling simultaneous optimization of analytical efficiency and sustainability.

4.2 Machine Learning Models in Process Optimization

Machine learning offers powerful frameworks for predictive optimization of analytical methods. Regression-based models such as Multiple Linear Regression (MLR) and Support Vector Regression (SVR) have been used to predict solvent waste, energy consumption, and atom economy, allowing chemists to simulate outcomes prior to experimentation. More advanced models, including Random Forest (RF), Gradient Boosting (XGBoost, LightGBM), and Artificial Neural Networks (ANNs), are particularly effective in capturing complex, nonlinear interactions between process parameters and green metrics [118, 119].

A notable example comes from CSIR-CIMAP, Lucknow, where AI-driven metabolomic fingerprinting was applied to authenticate medicinal plants [120]. While the primary goal was authentication, the same data-driven ML pipeline—utilizing descriptors, classification models, and feature ranking—can be extended to green/non-green method classification in chromatography and spectroscopy. Similarly, a recent QbD-based HPTLC optimization study in India demonstrated that annotated datasets of solvents and conditions can serve as input features for training predictive models, highlighting the synergy between QbD and ML [121].

4.3 Quantum Descriptors and Green Metrics Integration

Beyond traditional experimental descriptors, modern computational chemistry provides quantum chemical descriptors such as dipole moment, molecular electrostatic potential, and HOMO–LUMO energy gaps. When integrated with ML models, these descriptors enable Quantitative Structure–Green Relationship (QSGR) studies, which link molecular structure with ecological parameters such as toxicity, biodegradability, and persistence [122].

For example, descriptors generated from tools like RDKit, Mordred, and PaDEL-Descriptor have been successfully combined with AI algorithms to predict solvent toxicity and biodegradability. By extending this framework to analytical chemistry, it becomes possible to evaluate solvent systems or mobile phases not only on performance metrics (resolution, efficiency) but also on environmental impact—a critical step in method greening. Such hybrid approaches are gaining traction in India and globally, with researchers increasingly exploring AI–cheminformatics integration for green chromatography optimization [123].

4.4 Real-Time Optimization and Digital Twins

The emergence of real-time monitoring systems and digital twins represents the next frontier in data-driven Green Analytical Chemistry. A digital twin is a virtual replica of a laboratory process that continuously updates based on real-time data, enabling adaptive optimization of analytical workflows. For instance, ML algorithms embedded within chromatography instruments can dynamically adjust mobile phase gradients or column temperature to balance separation efficiency and solvent usage [124].

Pilot studies in pharmaceutical industries in India have begun exploring such adaptive optimization for HPLC and LC–MS-based quality control, where predictive algorithms reduce run times and solvent waste without compromising compliance standards. Coupled with visualization dashboards (heatmaps, radar plots, AGREE wheels), these systems provide decision support to analysts and regulatory bodies alike, ensuring sustainable compliance.

Data-driven optimization offers a transformative pathway toward sustainable analytical chemistry. By leveraging chemometrics, ML, quantum descriptors, and real-time digital tools, laboratories can significantly reduce environmental impact while enhancing performance and robustness. The shift from empirical experimentation to predictive, model-based optimization not only accelerates method development but also aligns with the global mandate for green and sustainable science. In the Indian context, emerging applications from CSIR institutes and pharmaceutical industries showcase how AI-driven optimization can be embedded within everyday laboratory practices. As datasets grow and computational tools become more accessible, the vision of fully autonomous, green analytical method design is increasingly within reach.

5 Virtual Screening for Green Solvent Selection

The selection of solvents plays a pivotal role in analytical chemistry, both from the perspective of method performance and environmental sustainability. Solvents often account for the largest fraction of chemical waste generated in laboratories and industries, making their choice a major determinant of the environmental footprint of analytical methods [125]. Conventional solvents such as acetonitrile, dichloromethane, and chloroform are widely used because of their favorable physicochemical properties, but they pose significant risks due to toxicity, volatility, and persistence in the environment [125, 126]. In line with the principles of green chemistry, there is an increasing demand for computational strategies that can identify safer and more sustainable alternatives without compromising analytical efficiency [127].

Virtual screening has emerged as a powerful tool in this context, allowing researchers to evaluate large libraries of potential solvents through computational models rather than resource-intensive laboratory testing [127, 128]. Two major computational approaches dominate this field: Quantitative Structure–Activity Relationship (QSAR) and Quantitative Structure–Property Relationship (QSPR). While QSAR primarily links molecular structure with biological or toxicological activities [125, 126], QSPR focuses on predicting physicochemical and thermodynamic properties such as boiling point, viscosity, solubility, or partition coefficients [129]. The integration of QSAR and QSPR methods allows chemists to simultaneously evaluate both performance-related and safety-related parameters of solvents. For instance, QSPR can help identify solvents with optimal polarity and volatility for chromatographic separations, while QSAR can ensure these solvents also possess low aquatic toxicity and high biodegradability [128, 130].

By combining QSAR and QSPR in solvent screening, chemists achieve a more holistic approach to green solvent design and selection. This dual assessment enables the rational replacement of hazardous solvents with bio-derived alternatives, ionic liquids, or deep eutectic solvents that maintain analytical efficiency while reducing ecological and health impacts [126, 129]. Moreover, the predictive capabilities of these models significantly accelerate the solvent discovery pipeline, reduce experimental costs, and align with global regulatory frameworks such as REACH and OECD guidelines for chemical safety assessment [128, 129]. Thus, virtual screening using QSAR/QSPR represents a forward-looking strategy to reconcile analytical performance with environmental stewardship. The differentiation of QSAR and QSPR is given in Table 2 as follows.

5.1 Computational Solvent Screening

5.1.1 QSAR

Quantitative Structure–Activity Relationship (QSAR) modeling has emerged as a valuable computational tool in the selection of safe and sustainable solvents for analytical chemistry [131–133]. By correlating structural descriptors of candidate solvents with their biological or ecotoxicological activities, QSAR enables early prediction of potential hazards without the need for extensive animal or environmental testing [132, 134]. The general workflow of QSAR applied to solvent screening is illustrated in Fig. 2, which outlines the steps involved in descriptor calculation, model development, validation, and prediction of solvent safety profiles. This approach is particularly relevant in Green Analytical Chemistry, where solvents not only need to support chromatographic separation, spectroscopic clarity, or sample preparation efficiency but also minimize risks to human health and the environment [135].

For instance, QSAR models have been employed to screen alternative solvents in high-performance liquid chromatography (HPLC) and gas chromatography (GC) by predicting mutagenicity or aquatic toxicity, thereby replacing harmful solvents such as acetonitrile or dichloromethane with bio-based or ionic liquid alternatives [109, 136]. Similarly, in sample extraction techniques like liquid–liquid extraction (LLE) or solid-phase microextraction (SPME), QSAR-guided evaluation of solvents ensures that solvents with high extraction efficiency also possess low toxicological impact [4].

Table 2 Differentiation of QSAR and QSPR in virtual solvent selection

Aspect	QSAR	QSPR
Focus	QSAR establishes a relationship between molecular structure and biological or toxicological activity	QSPR establishes a relationship between molecular structure and physicochemical as well as environmental properties
Endpoints	QSAR predicts biological hazards such as cytotoxicity, mutagenicity, aquatic toxicity, and dose–response measures like LD_{50} and EC_{50}	QSPR predicts solvent properties including flash point, boiling point, vapor pressure, solubility, biodegradability, logP, viscosity, melting point, and life cycle indices
Applications in solvent selection	QSAR is applied to ensure that solvents are non-toxic to humans and ecosystems, thereby avoiding hazardous solvents	QSPR is applied to ensure that solvents possess desirable physicochemical and green characteristics, making them safe, stable, biodegradable, and non-volatile
Data type	QSAR relies on experimental activity values obtained from bioassays and eco-toxicity studies	QSPR relies on physicochemical measurements obtained experimentally or from chemical databases such as PubChem and ChemSpider
Main use in green chemistry	QSAR is mainly used to screen solvents for minimal biological and ecotoxic impact	QSPR is mainly used to screen solvents for safe physicochemical performance along with environmental compatibility
Examples	QSAR has been used to predict the aquatic toxicity of ionic liquids and to exclude solvents with mutagenic alerts	QSPR has been used to predict the boiling points of green bio-based solvents and to screen ionic liquids for low volatility and high biodegradability

Recent studies have shown that QSAR predictions for endpoints such as cytotoxicity, biodegradability, and fish LC50 values can effectively filter solvent candidates before experimental trials, significantly reducing both cost and environmental burden [137, 138]. Thus, QSAR acts as a preventive screening mechanism, ensuring that solvents used across diverse analytical platforms align with the principles of green chemistry while maintaining analytical performance.

Practical applications of QSAR-based solvent screening highlight its effectiveness. For example, Voutchkova et al. demonstrated the use of computational descriptors to identify green solvents as substitutes for traditional aprotic polar solvents in organic synthesis, showing that QSAR-driven predictions successfully reduced toxicity while maintaining reaction efficiency [138]. In another case, Byrne et al. applied QSAR and other in silico tools to evaluate bio-based solvents for chromatographic methods, where candidates such as 2-methyltetrahydrofuran and ethyl lactate were identified as environmentally preferable alternatives to petroleum-derived solvents [136]. These case studies illustrate how computational solvent screening bridges

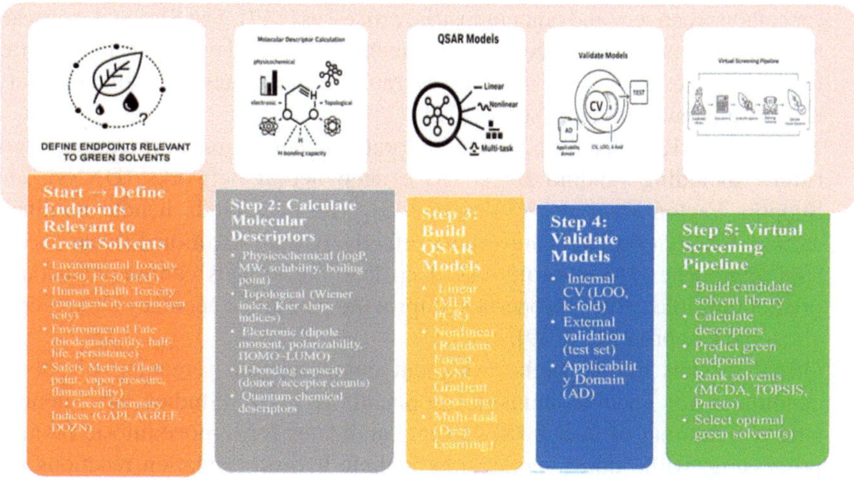

Fig. 2 Quantitative Structure–Activity Relationship (QSAR) workflow for screening and designing green solvents

theory and practice, making QSAR a powerful enabler of sustainable analytical chemistry.

Applications of QSAR in Virtual Solvent Selection:

(1) Quantitative Structure–Activity Relationship (QSAR) applications in virtual solvent selection represent a powerful computational approach that combines molecular descriptors with machine learning algorithms to predict solvent properties and optimize solvent choice for various applications [139].

(2) QSAR-based virtual solvent selection enables screening of large molecular libraries without experimental synthesis or testing [140]. The approach uses mathematical models to establish relationships between molecular structure and solvent properties, then applies these models to predict the behavior of novel solvents from large chemical databases [141]. These models can effectively reduce screening libraries from 10^5 to 10^7 chemical structures down to 10^1 to 10^3 candidates for experimental validation [142].

(3) In pharmaceutical development, QSAR methods facilitate rational solvent selection for crystallization processes, drug formulation, and solid form design [143]. Studies have developed QSAR models relating solvent structure to crystal morphology using 2D and 3D molecular descriptors, enabling prediction of crystal aspect ratios based on solvent molecular features [144]. These applications extend to polymorph screening, coformer selection, and counterion identification in pharmaceutical solid form development [145].

(4) QSAR approaches support sustainable solvent selection by incorporating environmental, health, and safety (EHS) parameters into predictive models [146]. Computer-aided solvent design methodologies use QSAR to predict various EHS criteria, enabling identification of green thermomorphic solvent systems

for catalyst recovery and other industrial processes [147]. The SUSSOL software applies artificial intelligence clustering techniques based on physical properties to identify greener solvent alternatives, ranking candidates according to safety, health, and environmental scores [12].

(5) Modern QSAR solvent selection incorporates advanced machine learning algorithms including random forest (RF), deep neural networks (DNNs), and Support Vector Machines (SVMs) [148]. These methods handle nonlinear structure–activity relationships more effectively than traditional linear regression approaches [133]. Multi-task deep learning models can simultaneously predict multiple solvent properties, improving predictive accuracy compared to single-task approaches [149].

(6) QSAR models enable high-throughput computational screening of solvent candidates for specific industrial applications [150]. Studies have demonstrated screening of over 8000 solvent candidates for lignocellulosic biomass processing, while other applications include hydroformylation reactions and separation processes [151]. The Conductor-like Screening Model for Real Solvents (COSMO-RS) method combined with QSAR approaches provides thermodynamic property predictions for solvent selection [152].

(7) In pharmaceutical co-crystal formation, QSAR models predict suitable solvent-coformer combinations based on molecular descriptors [153]. These computational screening approaches successfully identify appropriate solvents for co-crystal synthesis, reducing experimental screening requirements and improving discovery of new crystalline materials with enhanced solubility and bioavailability properties [154].

5.1.2 Quantitative Structure–Property Relationship (QSPR)

Quantitative Structure–Property Relationship (QSPR) modeling serves as a powerful approach in predicting the physicochemical and environmental properties of solvents, making it highly applicable in the development of green analytical methods. Unlike QSAR, which emphasizes biological activity, QSPR correlates molecular descriptors of solvents with measurable properties such as boiling point, flash point, viscosity, vapor pressure, solubility, and biodegradability [129, 155]. These predictions enable chemists to identify solvents that not only perform effectively in analytical procedures but also align with green chemistry principles. The overall steps involved in QSPR-guided solvent selection are illustrated in Fig. 3, which highlights the virtual screening workflow adopted to predict solvent properties and identify sustainable alternatives.

For example, in liquid chromatography, QSPR models have been used to screen bio-derived solvents and ionic liquids with favorable polarity, volatility, and solubility profiles, thereby providing safer alternatives to conventional organic solvents like methanol or acetonitrile [109, 136]. In spectroscopic techniques such as UV–Vis and NMR, QSPR-guided solvent selection ensures adequate transparency, low volatility, and chemical stability, which are critical for accuracy and reproducibility [156].

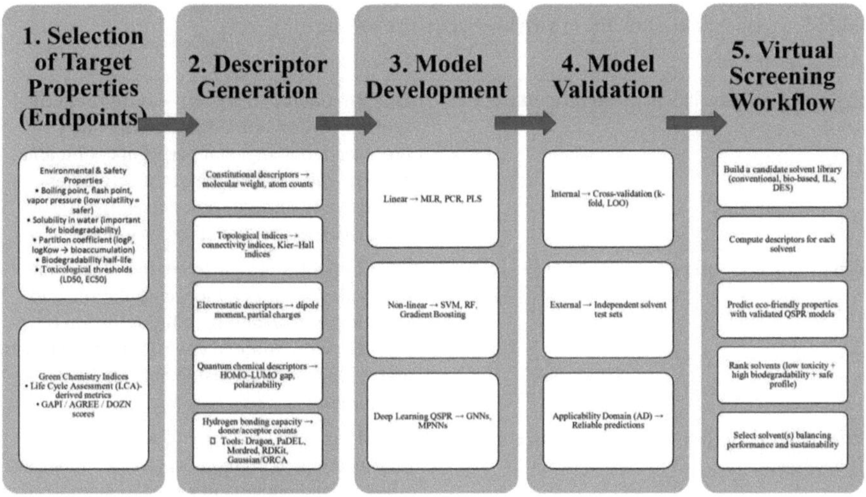

Fig. 3 QSPR virtual screening workflow

Similarly, in sample preparation methods like dispersive liquid–liquid microextraction (DLLME) and supercritical fluid extraction (SFE), QSPR models assist in predicting solvent miscibility and partition coefficients, allowing rapid identification of eco-friendly candidates with minimal experimental effort [4, 157].

Recent research has demonstrated the utility of QSPR in predicting life cycle indices and biodegradation rates of solvents, further strengthening its role in sustainability assessments [97, 158]. Thus, QSPR not only facilitates the efficient design of solvent systems for diverse analytical applications but also ensures that these solvents meet environmental safety and regulatory standards.

Applications of QSPR in Virtual Solvent Selection

Quantitative Structure–Property Relationship (QSPR) modeling has emerged as a powerful computational tool in virtual solvent selection for green chemistry, drug development, and industrial processes. Some important applications are enlisted in Table 3.

5.2 *Solvent Substitution Tools*

In recent years, computational solvent substitution tools have gained wide application in Green Analytical Chemistry, as they help predict solvent–analyte compatibility while minimizing environmental and health hazards. Hansen Solubility Parameters (HSPiP) is one of the most widely used tools, which evaluates solvent–solute interactions based on dispersion, polar, and hydrogen-bonding forces [13]. By comparing the "solubility sphere" of a solute with potential solvents, HSPiP enables chemists

Table 3 Applications of QSPR in virtual selection of solvent

Application	Description
Prediction of physicochemical properties [159, 160]	QSPR models can predict solvent properties such as boiling point, flash point, vapor pressure, viscosity, and partition coefficient (logP), which are essential for screening eco-friendly and safe solvents without exhaustive experimental measurements
Environmental safety assessment [161, 162]	QSPR enables virtual screening of solvents for biodegradability, bioaccumulation, and aquatic toxicity, aiding in the selection of solvents with minimal environmental hazards and alignment with sustainable chemistry principles
Pharmaceutical formulation and drug solubility [163, 164]	QSPR predicts solubility of active pharmaceutical ingredients (APIs) in various solvents and co-solvent systems, supporting solvent choice for crystallization, formulation, and drug delivery
Process optimization in chemical industries [165]	In large-scale manufacturing, QSPR allows rapid solvent ranking for extraction, separation, and reaction media, optimizing efficiency while reducing cost and environmental risks
Green chemistry and substitution of hazardous solvents [109, 166]	QSPR-based solvent design identifies safer alternatives to toxic solvents by predicting life-cycle indices and eco-toxicity parameters, enabling regulatory compliance and greener process development

to identify greener replacements with reduced toxicity—for example, substituting dichloromethane with ethyl lactate or bio-alcohols in extraction and chromatography [167].

Similarly, Conductor-like Screening Model for Real Solvents (COSMO-RS) applies quantum chemical descriptors and statistical thermodynamics to predict solvation energies, partitioning, and miscibility of solvents [152]. This method allows virtual screening of large solvent databases to propose eco-friendly alternatives, such as using water–ethanol mixtures instead of acetonitrile in HPLC or deep eutectic solvents (DES) instead of volatile organic solvents in sample preparation [168].

Other important computational tools include Universal Quasichemical Functional Group Activity Coefficient (UNIFAC) models, which estimate activity coefficients of solvents based on group contribution methods [169], and SPARC or EPI Suite™ from the US EPA, which predict physicochemical and environmental fate properties of solvents [170]. In addition, cheminformatics-based QSPR/QSAR models are increasingly applied to design and pre-select safer solvents by correlating molecular descriptors with toxicity, biodegradability, or greenness scores [171]. Together, these tools reduce experimental trial-and-error and accelerate the transition toward safer solvent use in analytical techniques such as chromatography, spectroscopy, and extraction.

6 Conclusion

The conclusion of the book chapter focuses on the integration of computational and data-driven strategies within analytical chemistry, emphasizing their pivotal role in enabling greener, more efficient, and robust method development. By harnessing tools like chemometric modeling, chromatographic simulators, molecular modeling, and artificial intelligence, the field is transitioning away from traditional, resource-intensive trial-and-error practices toward predictive, sustainable workflows. This paradigm shift not only accelerates the method design process and reduces environmental impact but also aligns closely with the foundational principles of Green Analytical Chemistry. Globally and increasingly within Indian scientific and industrial settings such approaches ensure that analytical chemistry meets the dual imperatives of advancing scientific progress while upholding ecological responsibility.

References

1. Kellner R, Mermet JM, Otto M, Widmer HM, editors. *Analytical Chemistry.* 2nd ed. Weinheim: Wiley-VCH; 2004.
2. Armenta S, Garrigues S, De la Guardia M. Green Analytical Chemistry. *TrAC Trends Anal Chem.* 2008; 27 (6): 497–511.
3. Gałuszka A, Migaszewski Z, Namieśnik J. The 12 principles of green analytical chemistry and the significance mnemonic of green analytical practices. *TrAC Trends Anal Chem.* 2013; 50: 78–84.
4. Tobiszewski M, Mechlińska A, Namieśnik J. Green Analytical Chemistry—theory and practice. *Chem Soc Rev.* 2010; 39 (8): 2869–78.
5. Rao RN. Computational tools and chemometrics in modern analytical chemistry. *Curr Sci.* 2018; 114 (3): 452–60.
6. International Conference on Harmonisation (ICH). ICH Harmonised Tripartite Guideline: Pharmaceutical Development Q8(R2); Quality Risk Management Q9; Pharmaceutical Quality System Q10; Development and Manufacture of Drug Substances Q11; Analytical Procedure Development Q14. Geneva: ICH; 2005–2023.
7. Montgomery DC. Design and analysis of experiments. 9th ed. Hoboken: Wiley; 2017.
8. Esbensen KH, Swarbrick B. Multivariate Data Analysis—An introduction to multivariate data analysis and chemometrics. 6th ed. Oslo: CAMO Software; 2018.
9. Brereton RG. Chemometrics for pattern recognition. Chichester: John Wiley & Sons; 2009.
10. Molnár I. Computerized optimization in chromatography by linear regression. *J Chromatogr A.* 2002; 965 (1–2): 175–94.
11. Cramer RD. Computational models and molecular design. *J Comput Aided Mol Des.* 2012; 26 (1): 35–43.
12. Pena-Pereira F, Wojnowski W, Tobiszewski M. AGREE—Analytical GREEnness metric approach and software. *Anal Chem.* 2020; 92 (14): 10076–82.

13. Hansen CM. Hansen solubility parameters: a user's handbook. 2nd ed. Boca Raton: CRC Press; 2007.
14. International Conference on Harmonisation (ICH). ICH Harmonised Tripartite Guideline: Pharmaceutical Development Q8(R2). Geneva: ICH; 2009.
15. Antony J. *Design of experiments for engineers and scientists*. 2nd ed. Oxford: Elsevier; 2014.
16. Hibbert DB. Experimental design in chromatography: A tutorial review. *J Chromatogr B*. 2012; 910: 2–13.
17. Hibbert DB. Design of experiments in pharmaceutical analysis: A review. *J Pharm Biomed Anal*. 2016; 130: 65–79.
18. Myers RH, Montgomery DC, Anderson-Cook CM. *Response surface methodology: process and product optimization using designed experiments*. 4th ed. Hoboken: Wiley; 2016.
19. Bezerra MA, Santelli RE, Oliveira EP, Villar LS, Escaleira LA. Response surface methodology (RSM) as a tool for optimization in analytical chemistry. *Talanta*. 2008; 76 (5): 965–77.
20. Ferreira SL, Bruns RE, Ferreira HS, Matos GD, David JM, Brandão GC, et al. Box–Behnken design: An alternative for the optimization of analytical methods. *Anal Chim Acta*. 2007; 597 (2): 179–86.
21. Roy RK. *A primer on the Taguchi method*. 2nd ed. Dearborn: Society of Manufacturing Engineers; 2010.
22. Orlandini S, Pasquini B, Del Bubba M. Analytical quality by design (AQbD) in pharmaceutical development. *TrAC Trends Anal Chem*. 2013; 42: 206–14.
23. Moeder M, Schrader S, Winkler M, Popp P. Application of experimental design for the optimization of an analytical method for pesticides. *J Chromatogr A*. 2000; 873 (1): 95–106.
24. Wold S, Esbensen K, Geladi P. Principal component analysis. *Chemometr Intell Lab Syst*. 1987; 2 (1-3): 37–52.
25. Blanco M, Villarroya I. NIR spectroscopy: a rapid-response analytical tool. *TrAC Trends Anal Chem*. 2002; 21 (4): 240–50.
26. Haaland DM, Thomas EV. Partial least-squares methods for spectral analyses. 1. Relation to other quantitative calibration methods and the extraction of qualitative information. *Anal Chem*. 1988; 60 (11): 1193–202.
27. Balabin RM, Smirnov SV. Melamine detection by mid- and near-infrared spectroscopy with chemometrics. *Food Chem*. 2011; 126 (1): 412–7.
28. Trygg J, Holmes E, Lundstedt T. Chemometrics in metabolomics. *J Proteome Res*. 2007; 6 (2): 469–79.
29. Singh B, Ahuja N. Response surface optimization of HPLC conditions for analysis of amlodipine and atorvastatin. *J Pharm Biomed Anal*. 2007; 43 (2): 722–8.
30. Wang L, Weller CL. Recent advances in extraction of nutraceuticals from plants. *TrAC Trends Anal Chem*. 2006; 25 (9): 777–91.
31. Molnár I, Horváth C, Rieger HJ. DryLab® applications for UHPLC method development. *J Chromatogr A*. 2013; 1293: 1–13.
32. Rani S, Sharma P, Kaur P. Artificial neural networks for optimization in HPLC method development. *Anal Chim Acta*. 2014; 822: 29–36.
33. Wu HL, Wang T, Yu RQ. SVM-based classification of herbal medicine spectra. *Anal Chim Acta*. 2008; 614 (2): 196–204.
34. Gratzfeld-Hüsgen A, Naik S, Mistry B. Application of genetic algorithms for chromatographic optimization. *J Chromatogr A*. 2001; 937 (1-2): 33–43.
35. Fernández Pierna JA, Dardenne P. Random forests in NIR spectroscopy for pharmaceutical quality control. *Anal Chim Acta*. 2008; 705 (1-2): 151–9.
36. Tauler R. Multivariate curve resolution applied to liquid chromatography-DAD. *J Chromatogr A*. 1995; 703 (1-2): 229–48.
37. De Juan A, Jaumot J, Tauler R. Multivariate Curve Resolution (MCR) in LC-MS data analysis. *Anal Chim Acta*. 2014; 818: 1–14.
38. Veselkov KA, Lindon JC, Ebbels TM, Crockford D, Volynkin VV, Holmes E, et al. Recursive segment-wise peak alignment of NMR spectral data for MCR-based metabolomics. *Anal Chem*. 2009; 81 (1): 56–66.

39. Kirchner T, Bakalyar SR. Advances in chromatographic simulation for method development. *J Chromatogr A*. 2010; 1217 (45): 7012–25.
40. Molnár I, Rieger HJ. Software-assisted method development under QbD principles. *J Pharm Biomed Anal*. 2012; 69: 173–80.
41. Kormany R, Molnár I, Rieger HJ. Robustness testing of monoclonal antibody analysis using DryLab®. *J Chromatogr B*. 2014; 969: 242–51.
42. Pasch H. Application of DryLab® for impurity profiling of APIs. *J Pharm Biomed Anal*. 2015; 102: 236–42.
43. Molnár I. Regulatory acceptance of DryLab® resolution maps in QbD submissions. *Pharm Tech Eur*. 2016; 28 (7): 20–7.
44. Niessen WMA. ChromSword® for automated method development in HPLC. *LCGC Eur*. 2007; 20 (3): 142–8.
45. Zhou Y, Wang L, Zhang H. Automated HPLC optimization of steroids using ChromSword®. *J Sep Sci*. 2008; 31 (5): 890–6.
46. Wilson ID, Simpson AJ. ChromSword® in impurity profiling. *J Chromatogr A*. 2009; 1216 (45): 7947–54.
47. Li X, Sun D, He J. ChromSword®-assisted optimization of pesticide residue analysis. *Food Chem*. 2012; 134 (4): 2464–70.
48. Lewis DFV. Knowledge-based retention prediction using ChromGenius™. *Chemom Intell Lab Syst*. 2006; 82 (1-2): 50–8.
49. Marchand DH, Rogers WJ. Retention prediction of new drug candidates using ChromGenius™. *J Pharm Sci*. 2007; 96 (9): 2587–94.
50. Vu T, Harrison A. Predictive impurity profiling using ChromGenius™. *J Chromatogr A*. 2009; 1216 (12): 2285–91.
51. ACD/Labs Technical Note. ChromGenius™ for QbD-based method development. Toronto: ACD/Labs; 2011.
52. Poole CF. ACD/LC Simulator in analytical method optimization. *TrAC Trends Anal Chem*. 2011; 30 (6): 899–907.
53. Rao RN, Nagaraju V. LC Simulator application in antibiotic stability studies. *J Pharm Biomed Anal*. 2012; 70: 317–23.
54. Zhao J, Wang Y. Impurity profiling of alkaloids using LC Simulator. *J Sep Sci*. 2014; 37 (14): 1893–901.
55. Lu Y, Zhang X. Virtual optimization of LC-MS bioanalytical assays using LC Simulator. *Anal Chim Acta*. 2015; 889: 123–31.
56. Zhang Z, Chen L. ChromX: AI-driven chromatographic simulation platform. *Anal Chem*. 2020; 92 (24): 16027–35.
57. Müller F, Hofmann J. ChromX optimization of peptide separations. *J Proteome Res*. 2021; 20(7): 3256–65.
58. Li H, Zhao Q. ChromX-assisted metabolomics workflow optimization. *Metabolomics*. 2022; 18 (3): 23–34.
59. McCalley DV. Virtual Column® for stationary phase selection in HPLC. *LCGC Eur*. 2011; 24 (9): 478–85.
60. Greco G, Gritti F. Column selection for basic drugs using Virtual Column®. *J Chromatogr A*. 2013; 1310: 45–52.
61. Ivanova B, Kawakami K. Virtual Column® in impurity isomer separation. *J Sep Sci*. 2015; 38 (21): 3797–804.
62. Cramer CJ. Essentials of computational chemistry. 2nd ed. Chichester: Wiley; 2013.
63. Korany MA, Mahgoub H, Haggag RS. Role of molecular modeling in QbD-based analytical method development. *J Pharm Biomed Anal*. 2019; 164: 582–92.
64. Schwaighofer A, et al. Integration of molecular modeling and AI in analytical chemistry. *TrAC Trends Anal Chem*. 2021; 142: 116328.
65. Wainer IW, et al. Docking studies in predicting enantioselectivity on cyclodextrin phases. *J Chromatogr A*. 2008; 1184 (1-2): 282–90.

66. Farkas T, et al. Docking-assisted derivatization agent selection in LC-MS. *Anal Chim Acta.* 2010; 674 (2): 158–65.
67. Xu Z, et al. Docking study of flavonoid–sorbent interactions in SPE. *J Sep Sci.* 2014; 37 (22): 3227–35.
68. Chen X, et al. Molecular docking in biosensor surface design. *Biosens Bioelectron.* 2016; 85: 797–804.
69. Markov PV, et al. MD simulation of caffeine retention in HPLC. *J Phys Chem B.* 2011; 115 (22): 6896–904.
70. Vallejo JP, et al. Peptide interactions in HILIC studied by MD. *J Chromatogr A.* 2017; 1499: 115–24.
71. Yuan H, et al. MD-based pH optimization of basic drug separations. *Anal Chim Acta.* 2015; 864: 32–41.
72. Poole CF. Simulation of volatile interactions in GC using MD. *J Chromatogr A.* 2013; 1317: 39–52.
73. Jovanovic T, et al. DFT prediction of flavonoid UV absorption. *Spectrochim Acta A.* 2012; 95: 65–72.
74. Ahmed S, et al. Ab initio vibrational analysis of drug polymorphs. *J Mol Struct.* 2015; 1081: 453–61.
75. Singh R, et al. Semi-empirical NMR predictions for alkaloids. *Magn Reson Chem.* 2016; 54 (3): 214–22.
76. Mishra P, et al. Quantum chemistry for pKa prediction in LC. *J Comput Chem.* 2017; 38 (5): 322–9.
77. Kaliszan R, et al. QSRR of pharmaceuticals in LC. *J Chromatogr A.* 2005; 1080: 1–9.
78. Zhang H, et al. QSAR models for steroid selectivity in LC. *Anal Chim Acta.* 2011; 703 (2): 192–200.
79. Vlahović F, et al. QSRR-based prediction of pesticide retention. *J Sep Sci.* 2018; 41 (15): 2990–3000.
80. Creek DJ, et al. Retention time prediction in metabolomics. *Anal Chem.* 2011; 83 (22): 8703–10.
81. Rathore AS, et al. Monte Carlo simulation of adsorption isotherms. *J Chromatogr A.* 2003; 1016: 59–72.
82. Kim J, et al. Monte Carlo prediction of SPE distribution coefficients. *J Sep Sci.* 2010; 33 (9): 1276–84.
83. Vander Heyden Y, et al. Monte Carlo robustness testing of HPLC. *J Chromatogr A.* 2007; 1158: 138–47.
84. Qiao J, et al. MEP mapping of β-blockers in ion-pair LC. *J Chromatogr B.* 2012; 887–888: 74–81.
85. Kawakami K, et al. MEP-assisted chiral amino acid separation. *J Chromatogr A.* 2014; 1363: 135–42.
86. Oliveira G, et al. Surface mapping for derivatization reagent design. *Anal Bioanal Chem.* 2018; 410: 2253–64.
87. Roy S, et al. MEP-guided ion-pair reagent selection in LC. *Anal Methods.* 2019; 11: 3082–91.
88. Ghosh S, et al. Hybrid modeling in chromatography. *J Chromatogr A.* 2020; 1624: 461220.
89. Zhang Y, et al. Hybrid QSRR–machine learning in peptide separations. *Anal Chem.* 2021; 93 (6): 2932–41.
90. Rahman S, et al. Chemoinformatics and DFT integration for LC optimization. *J Comput Aided Mol Des.* 2022; 36: 423–36.
91. Misra BB, et al. Hybrid metabolomics prediction platforms. *Metabolomics.* 2023; 19 (1): 12–25.
92. Sarkar A, Mohanty S, Das P. Multivariate statistical approach to predict biodegradability of pharmaceuticals. *Chemosphere.* 2019; 229: 517–26.
93. Chemat F, Vian MA, Cravotto G. Green extraction of natural products: concept and principles. *Int J Mol Sci.* 2012; 13 (7): 8615–27.

94. Bevilacqua M, Bucci R, Magrì AL, Magrì AD. Chemometrics in food chemistry. *J Food Chem.* 2013; 141 (2): 2031–9.
95. Kümmerer K, Dionysiou DD, Olsson O, Fatta-Kassinos D. A path to clean water. *Science.* 2018; 361 (6399): 222–4.
96. Wicker J, Lorsbach T, Gütlein M, Schmid E, Latino D, Kramer S, et al. enviPath—The environmental contaminant biotransformation pathway resource. *Nucleic Acids Res.* 2016; 44 (D1): D502–8.
97. Płotka-Wasylka J. A new tool for the evaluation of the analytical procedure: Green Analytical Procedure Index. *Talanta.* 2018; 181: 204–9.
98. Gao H, Struble TJ, Coley CW, Wang Y, Green WH, Jensen KF. Using machine learning to predict suitable conditions for organic reactions. *ACS Cent Sci.* 2018; 4 (11): 1465–76.
99. Sheridan RP, Feuston BP, Maiorov VN, Kearsley SK. Similarity to molecules in the training set is a good discriminator for prediction accuracy in QSAR. *J Chem Inf Comput Sci.* 2004; 44 (6): 1912–28.
100. Wenzel J, Matter H, Schmidt F. Predictive multitask deep learning of chemical reactivity. *J Chem Inf Model.* 2019; 59 (2): 968–78.
101. Kumar A, Kim J, Abbasi T, Abbasi SA. Artificial neural networks in green chemistry: predictive tools for sustainable design. *Clean Technol Environ Policy.* 2018; 20 (2): 257–69.
102. Coley CW, Barzilay R, Jaakkola TS, Green WH, Jensen KF. Prediction of organic reaction outcomes using machine learning. *ACS Cent Sci.* 2017; 3 (5): 434–43.
103. Schwaller P, Gaudin T, Lanyi D, Bekas C, Laino T. "Found in translation": predicting outcomes of complex organic chemistry reactions using neural sequence-to-sequence models. *Chem Sci.* 2018; 9 (28): 6091–8.
104. Lundberg SM, Lee S-I. A unified approach to interpreting model predictions. *Adv Neural Inf Process Syst.* 2017; 30: 4765–74.
105. Khosravi P, Kazemi E, Imielinski M, Elemento O, Hajirasouliha I. Deep convolutional neural networks enable discrimination of heterogeneous digital pathology images. *Bioinformatics.* 2018; 34 (13): i264–71.
106. Breiman L. Random forests. *Mach Learn.* 2001; 45:5–32.
107. Weston J, Watkins C. Multi-class support vector machines. Csd. Trinity. Cam. Ac. UK. 1998;171–84.
108. Roy K, Kar S, Das RN. A primer on QSAR/QSPR modeling: Fundamental concepts. Springer; 2015.
109. Jessop PG. Searching for green solvents. *Green Chem.* 2011; 13 (6): 1391–8.
110. Lesellier E, West C. The many faces of packed column supercritical fluid chromatography–A critical review. *J Chromatogr A.* 2015; 1382: 2–46.
111. Pan SJ, Yang Q. A survey on transfer learning. *IEEE Trans Knowl Data Eng.* 2010; 22 (10): 1345–59.
112. Kim S, Chen J, Cheng T, Gindulyte A, He J, He S, et al. PubChem 2023 update: improved data, user interfaces and chemical structure search. *Nucleic Acids Res.* 2023; 51(D1): D1373–82.
113. Van Aken K, Strekowski L, Patiny L. EcoScale, a semi-quantitative tool to select an organic preparation based on economical and ecological parameters. *Beilstein J Org Chem.* 2006; 2:3.
114. Berthold MR, Cebron N, Dill F, Gabriel TR, Kötter T, Meinl T, et al. KNIME: The Konstanz Information Miner. In: Data analysis, machine learning and applications. *Springer*; 2008; 319–26.
115. Abadi M, Agarwal A, Barham P, Brevdo E, Chen Z, Citro C, et al. Tensor Flow: Large-scale machine learning on heterogeneous distributed systems. arXiv preprint arXiv:160304467. 2016.
116. Kumar NN, Archana M. A comprehensive review of green metric tools for sustainable analytical method development. In: *Advances in Biological Sciences Research*, ICBEST 2024; Atlantis Press; 2025; 132–38.
117. Gałuszka A, Migaszewski ZM, Konieczka P, Namieśnik J. Analytical Eco-Scale for assessing the greenness of analytical procedures. *TrAC Trends Anal Chem.* 2012; 37: 61–72.

118. Varmuza K, Filzmoser P. Introduction to multivariate statistical analysis in chemometrics. Boca Raton: CRC Press; 2016.
119. Chen T, Guestrin C. XGBoost: A scalable tree boosting system. In: *Proceedings of the 22nd ACM SIGKDD Conference on Knowledge Discovery and Data Mining*; 2016; 785–94.
120. Times of India. CIMAP researchers develop AI-based solution to detect adulteration in medicinal plants. Lucknow; 2025.
121. Iqbal J, Gupta V, Bansal A, Sharma R. QbD-assisted HPTLC method with greener solvent strategies. Green Analytical Chemistry. Elsevier, 2025; 13.
122. Karelson M. Molecular descriptors in QSAR/QSPR. New York: Wiley; 2000.
123. Polishchuk PG, Madzhidov TI, Varnek A. Estimation of the size of drug-like chemical space based on GDB-17 data. *J Comput Aided Mol Des.* 2013; 27 (8):675–79.
124. Kulah E, Özdemir D. Digital twins in analytical chemistry: opportunities and challenges. *Anal Chim Acta.* 2023; 1250: 340–59.
125. Capello C, Fischer U, Hungerbühler K. What is a green solvent? A comprehensive framework for the environmental assessment of solvents. *Green Chem.* 2007; 9 (9): 927–934.
126. Pham TPT, Cho CW, Yun YS. Environmental fate and toxicity of ionic liquids: A review. *Water Res.* 2010; 44 (2): 352–372.
127. Aksenov AA, et al. Application of computational methods for predicting solvent properties in green analytical chemistry. *J Chem Inf Model.* 2016; 56 (7): 1234–1245.
128. OECD. *Guidance on Grouping of Chemicals, Second Edition.* OECD Series on Testing and Assessment No. 194. Paris: OECD Publishing; 2014.
129. Katritzky AR, Kuanar M, Slavov S, Hall CD, Karelson M, Kahn I, Dobchev DA. Quantitative correlation of physical and chemical properties with chemical structure: utility for prediction. *Chem Rev.* 2005; 105 (10): 3371–3387.
130. Ruiz-Mercado GJ, Gonzalez MA. Green engineering and sustainable process design for the chemical industry. *Curr Opin Chem Eng.* 2012; 1 (4): 467–478.
131. Cronin MTD, Madden JC. *In Silico Toxicology: Principles and Applications.* Cambridge University Press; 2010.
132. Raies AB, Bajic VB. In silico toxicology: computational methods for the prediction of chemical toxicity. *Wiley Interdiscip Rev Comput Mol Sci.* 2016; 6 (2): 147–172.
133. Cherkasov A, Muratov EN, Fourches D, et al. QSAR modeling: where have you been? where are you going to? *J Med Chem.* 2014; 57 (12): 4977–5010.
134. Benfenati E, editor. *In Silico Methods for Predicting Drug Toxicity.* Springer; 2016.
135. Płotka-Wasylka J, et al. Green analytical chemistry as an integral part of sustainable development. *TrAC Trends Anal Chem.* 2017; 91:23–36.
136. Byrne FP, Jin S, Paggiola G, et al. Tools and techniques for solvent selection: green solvent selection guides. *Sustain Chem Processes.* 2016; 4:7.
137. Worth AP, Cronin MTD. The use of QSARs in international decision-making frameworks to predict health effects of chemical substances. *Environ Health Perspect.* 2001; 109 (10): 1011–1016.
138. Voutchkova AM, Gorman DJ, Green EM, et al. Virtual screening of solvents for green chemical processes. *Green Chem.* 2010; 12: 996–1006.
139. Tropsha A. Best practices for QSAR model development, validation, and exploitation. *Mol Inform.* 2010; 29 (6-7): 476-88.
140. Groom CR, Bruno IJ, Lightfoot MP, Ward SC. The Cambridge Structural Database. *Acta Crystallogr B Struct Sci Cryst Eng Mater.* 2016; 72 (2): 171-9.
141. Oprea TI, Matter H. Integrating virtual and high-throughput screening. *Nat Rev Drug Discov.* 2004; 3 (11): 935-48.
142. Schneider G, Fechner U. Computer-based de novo design of drug-like molecules. *Nat Rev Drug Discov.* 2005; 4 (8): 649-63.
143. Paluch AS, Song LF, Shen JW. Predicting the effect of solvent on crystal morphology. *Cryst Growth Des.* 2019; 19 (2): 1302-12.
144. Neumann MA, Leusen FJJ, Kendrick J. A major advance in crystal structure prediction. *Angew Chem Int Ed Engl.* 2008; 47 (13): 2427-30.

145. Childs SL, Stahly GP, Park A. The salt–cocrystal continuum: the influence of crystal structure on ionization state. *Mol Pharm.* 2007; 4 (3): 323-38.
146. Jimenez-Gonzalez C, Constable CS, Constable DJC. *Green chemistry and engineering: a practical design approach.* Hoboken (NJ): John Wiley & Sons; 2011.
147. Murray PM, Tyler SNG, Moseley JD. Beyond the typical: a critical review of solvent selection and use in industry. *Org Process Res Dev.* 2013; 17 (1): 40-6.
148. Lo YC, Rensi SE, Torng W, Altman RB. Machine learning in chemoinformatics and drug discovery. *Drug Discov Today.* 2018; 23 (8): 1538-46.
149. Wu Z, Ramsundar B, Feinberg EN, Gomes J, Geniesse C, Pappu AS, Leswing K, Pande V. MoleculeNet: a benchmark for molecular machine learning. *Chem Sci.* 2018; 9: 513-30.
150. DeJaco RF, Ellefsen NJ, Glaser MS, Yusof NSM, Smith MD, Carreon MA, Siepmann JI. Computational screening of biomass-derived solvents for bio-oil recovery. *ACS Sustain Chem Eng.* 2020; 8 (33): 12668-78.
151. Segler MHS, Kogej T, Tyrchan C, Waller MP. Generating focused molecule libraries for drug discovery with recurrent neural networks. *ACS Cent Sci.* 2018; 4 (1): 120-31.
152. A. Klamt, "COSMO-RS: From Quantum Chemistry to Fluid Phase Thermodynamics and Drug Design," Elsevier, 2005.
153. N. J. Babu and A. Nangia, "Solubility Advantage of Amorphous Drugs and Pharmaceutical Cocrystals," *Crystal Growth & Design*, vol. 11, no. 7, pp. 2662-2679, 2011.
154. D. J. Good and N. Rodríguez-Hornedo, "Solubility Advantage of Pharmaceutical Cocrystals," *Crystal Growth & Design*, vol. 9, no. 5, pp. 2252-2264, 2009.
155. Katritzky AR, Fara DC, Kuanar M, et al. QSPR analysis of boiling points of organic compounds. *J Chem Inf Model.* 2005;45(3):499–507.
156. Reichardt C, Welton T. *Solvents and Solvent Effects in Organic Chemistry.* 4th ed. Wiley-VCH; 2011.
157. Płotka-Wasylka J, Rutkowska M, Owczarek K, Tobiszewski M. Green analytical chemistry in sample preparation for determination of trace organic pollutants. *TrAC Trends Anal Chem.* 2016;77:69–85.
158. Ruiz-Mercado GJ, Gonzalez MA. *Green Engineering: Environmentally Conscious Design of Chemical Processes.* Wiley; 2012.
159. Katritzky, A. R., Lobanov, V. S., & Karelson, M. (1995). QSPR: The correlation and quantitative prediction of chemical and physical properties from structure. *Chemical Society Reviews*, 24(4), 279–287.
160. Varnek, A., & Tropsha, A. (2008). Chemoinformatics approaches to virtual screening. *RSC Publishing.*
161. Devillers, J. (2006). *Ecotoxicological QSARs.* CRC Press.
162. Toropov, A. A., & Toropova, A. P. (2015). QSAR/QSPR modeling of ecotoxicity: A review of recent advances. *Current Drug Safety*, 10(2), 142–148.
163. Jain, N., Yalkowsky, S. H. (2001). Estimation of aqueous solubility I: Application to organic nonelectrolytes. *Journal of Pharmaceutical Sciences*, 90(2), 234–252.
164. Acree, W. E., & Abraham, M. H. (2006). Solubility predictions for crystalline nonelectrolyte solutes dissolved in organic solvents based upon the Abraham general solvation model. *Canadian Journal of Chemistry*, 84(3), 610–616.
165. Constantinou, L., & Gani, R. (1994). New group contribution method for estimating properties of pure compounds. *AIChE Journal*, 40(10), 1697–1710.
166. Gani, R., & Jiménez-Gonzâlez, C. (2011). Chemical product design for sustainability: A case study of solvents. *Clean Technologies and Environmental Policy*, 13(3), 493–507.
167. Jessop, P. G.; Izgorodina, E.; Altıntop, E. *Greener Solvents for Extraction and Chromatography.* Green Chemistry, 2016, 18, 2925–2930.
168. Marrucho, I. M.; Branco, L. C.; Rebelo, L. P. N. *Deep Eutectic Solvents: Properties and Applications.* Annual Review of Chemical and Biomolecular Engineering, 2014, 5, 527–546.
169. Fredenslund, A.; Jones, R. L.; Prausnitz, J. M. *Group Contribution Estimation of Activity Coefficients in Nonideal Liquid Mixtures.* AIChE Journal, 1975, 21, 1086–1099.

170. US EPA. *Estimation Programs Interface Suite™ (EPI Suite).* United States Environmental Protection Agency, Washington, DC, 2023. Available at: https://www.epa.gov/tsca-screening-tools/epi-suitetm-estimation-program-interface
171. Katritzky, A. R.; Karelson, M.; Sild, S.; Tamm, T.; Tammekivi, T. *Quantitative Structure–Property Relationship (QSPR) Correlations of Solvent Properties.* Journal of Chemical Information and Computer Sciences, 1996, 36, 1182–1190.

Solvent-Free Extraction Techniques

Payal Rahul Dande, Mayank Sharma, Jonnagadda Susanth, and Shital Patil

Abstract Solvent-free extraction techniques have become essential pillars of Green Analytical Chemistry, offering sustainable alternatives to conventional solvent-intensive methods. This chapter provides a comprehensive overview of principles, mechanisms, and applications of solvent-free extraction (SFE) approaches designed to minimize environmental impact while maintaining optimization and efficiency. The introductory section delineates the concept of Green Analytical Chemistry and articulates the limitations inherent in conventional solvent-based methodologies. Subsequent sections elucidate the fundamental mechanisms governing analyte transfer and isolation in solvent-free systems, emphasizing their intrinsic advantages in terms of operational safety, energy efficiency, and ecological sustainability. Various categories of solvent-free techniques are detailed, including sorbent- and phase-based methods such as solid-phase microextraction (SPME) and stir-bar sorptive extraction (SBSE), as well as fluid-based methods like supercritical fluid extraction (SFE) and phytonic extraction. Energy-assisted techniques—ultrasound-, microwave-, and thermally-driven processes—are examined for their enhanced extraction efficiency and eco-friendly attributes. Emerging strategies employing biomaterials, enzymes, and nanomaterials are presented as potential substitutes for synthetic sorbents, while headspace extraction is examined as a solvent-free approach for the isolation of volatile analytes. Comparative assessments address the efficiency, selectivity, cost-effectiveness, and sustainability metrics of the discussed techniques. The chapter culminates with insights into multidisciplinary applications across environmental monitoring, food safety, pharmaceutical, and industrial sectors, and explores future prospects emphasizing integration with green analytical methodologies, material innovations, miniaturization, automation, and alignment with circular economy principles.

P. R. Dande (✉) · S. Patil
School of Pharmaceutical Sciences, Faculty of Health Sciences, JSPM University, Pune, India
e-mail: drpayaldande@gmail.com

M. Sharma · J. Susanth
School of Pharmacy & Technology Management, SVKM's Narsee Monjee Institute of Management Studies (NMIMS) Deemed-to-be-University, Shirpur, Maharashtra, India

V. Bhusari and M. Ghante (eds.), *Green Analytical Chemistry in India*,
https://doi.org/10.1007/978-3-032-24564-9_9

Keywords Green chemistry · Solvent-free extraction · Supercritical fluid extraction · Solid-phase microextraction · Phytonic · Energy assisted extraction

1 Introduction

The science of pharmacognosy has always been deeply interlinked with the extraction and characterization of bioactive compounds from medicinal plants. For centuries, conventional solvent-based approaches such as maceration, Soxhlet extraction, and percolation have served as the mainstay for isolating phytochemicals. However, these methods require extensive use of organic solvents, which are not only costly but also hazardous to both humans and the environment. The global movement toward sustainable practices has highlighted the necessity of developing safer and greener extraction alternatives.

In this context, the concept of solvent-free extraction has emerged as a pioneering development. Such methods make use of physical adsorption, phase transfer, and innovative sorbent technologies that often eliminate or drastically reduce the need for organic solvents. They are closely aligned with the broader discipline of *Green Analytical Chemistry* (GAC), which provides guiding principles for cleaner, safer, and more environmentally responsible analytical practices [1]. Importantly, solvent-free techniques enable extraction of delicate bioactive compounds without the risk of solvent-induced degradation while simultaneously reducing environmental burdens caused by solvent production and waste disposal. Thus, they represent a paradigm shift in natural-product research and analysis.

1.1 Concept of Green Analytical Chemistry

Green Analytical Chemistry is built on the same foundation as *Green Chemistry*, but specifically applies its philosophy to analytical methods. GAC proposes twelve major principles, including the reduction or elimination of solvents, design of energy-efficient methods, minimization of sample size, and reduction of hazardous waste [1]. Its central goal is to ensure that high-quality analytical data are obtained in a manner that poses minimum risk to both humans and the environment.

In pharmacognosy, these principles are particularly critical as plants often contain thermolabile compounds such as volatile oils, flavonoids, tannins and anthocyanins that deteriorate under harsh extraction conditions. Green methods, particularly solvent-free ones, offer a means of obtaining high-quality phytochemical extracts that are chemically intact and more representative of the in vivo plant composition. Moreover, adherence to GAC in natural-product extraction directly enhances the social acceptability of herbal therapeutics in international markets where eco-certifications and “green labels” are increasingly demanded.

1.2 *Importance of Solvent-Free Extraction*

Solvent-free methods are important for multiple interrelated reasons. First, they eliminate toxic solvent residues, which is vital for herbal medicines intended for oral or topical human use. Second, they protect laboratory workers from long-term exposure to harmful organic vapors such as benzene, chloroform, or methanol, which are classified as either toxic or carcinogenic. Third, from an environmental standpoint, reducing the consumption of solvents leads to reduced emissions of volatile organic compounds (VOCs), which are known contributors to air pollution and climate change [2].

In addition to health and environmental concerns, solvent-free extraction carries significant economic benefits. Solvents comprise a substantial component of laboratory expenses, and their safe disposal demands additional infrastructure. By eliminating them, one not only simplifies workflows but also reduces costs. Beyond economics and environment, solvent-free extraction often enhances extract quality. Since no additional chemical medium is introduced, the obtained extract more faithfully mirrors the native phytochemical profile of medicinal plants, which is important for pharmacological activity studies and therapeutic applications.

1.3 *Limitations of Conventional Solvent-Based Methods*

Although traditional solvent-based procedures remain widespread, their disadvantages are significant. From a scientific perspective, they can result in loss of bioactivity due to decomposition of sensitive molecules under prolonged heating (such as in Soxhlet extraction). Many natural compounds including terpenoids, phenolics, and certain glycosides are highly sensitive to heat and oxidizing conditions. Operationally, solvent-based extraction often requires large extraction volumes, lengthy durations, and labor-intensive protocols. For example, maceration or percolation sometimes extends to several days, while still requiring multistep concentration and solvent removal processes.

Environmentally, common solvents such as chloroform, hexane, and dichloromethane are associated with serious ecological toxicity and generate persistent waste. Regulations such as those imposed by the *US Environmental Protection Agency (EPA)* and *European Medicines Agency (EMA)* increasingly restrict their use, creating a strong regulatory drive toward innovation [3, 4]. Collectively, these limitations highlight why there is a pressing need for safer, more cost-effective, and environmentally compliant technologies—hence the rise of solvent-free methodologies.

2 Principles of Solvent-Free Extraction Techniques

2.1 Fundamentals of Extraction Without Solvents

Solvent-free extraction techniques are founded on replacing harmful liquid solvents with adsorptive phases, coated fibers, or thermal/physical energy inputs. Instead of dissolving analytes in an organic medium, these methods allow volatile and semi-volatile compounds to partition between a sample matrix (such as plant tissue, essential oil glands, or powders) and an extraction phase.

For example, methods like solid-phase microextraction (SPME) use chemically selective coatings to capture molecules directly, while stir-bar sorptive extraction (SBSE) utilizes PDMS-coated stir bars to absorb analytes during stirring. The fundamental principle is that analytes will migrate across phases as governed by their partition coefficient (K) until equilibrium is reached. This equilibrium is then shifted to enrich the analyte load on the sorbent for later analysis [5].

2.2 Mechanisms of Analyte Transfer and Isolation

At the core of solvent-free extraction are diffusion, adsorption, absorption, and partitioning processes.

- In adsorption-based methods, analyte molecules adhere to the sorbent surface through van der Waals interactions, hydrogen bonding, or polarity-based interactions.
- In absorption-based methods (such as SBSE), analytes physically dissolve into a polymeric phase.
- In headspace-based SPME, compounds are volatilized into the sample headspace and subsequently trapped onto the fiber's surface.

Temperature, pH, analyte volatility, and coating polarity directly influence the extraction efficiency [6, 7]. As these methods do not rely on chemical solvents for analyte recovery, they preserve the intrinsic structural properties of sensitive phytochemicals and prevent contamination with extraneous solvent peaks during chromatographic analysis.

2.3 Advantages in Terms of Sustainability and Safety

The advantages of solvent-free techniques are multifold:

- Sustainability: Reduced chemical input lowers environmental impact, ensuring compliance with green certification standards.

- Safety: Eliminating toxic solvents prevents occupational hazards, such as exposure to carcinogenic vapors or combustion risks.
- Efficiency: Shorter sample preparation and reduced consumables lead to faster workflow and improved throughput.
- Quality: Solvent-independent extracts retain the natural bioactive composition, which is crucial when evaluating pharmacological efficacy.

These combined benefits make solvent-free extraction not just an environmentally friendly choice, but also scientifically superior for pharmacognostical research focused on natural-product analysis for therapeutic applications.

3 Sorbent- and Phase-Based Extraction Methods

3.1 Solid-Phase Microextraction (SPME)

Principle and Working Mechanism

SPME was introduced by Janusz Pawliszyn in the early 1990s and has since revolutionized green sample preparation. It involves a fused-silica fiber coated with a sorbent material, which is directly exposed to the sample (or its headspace) for analyte capture. After equilibrium is established, the fiber is inserted into a gas chromatograph (GC) injector for thermal desorption or into liquid chromatographic systems (LC) after appropriate elution.

Types of Coatings and Fibers

SPME fibers are available in diverse chemistries tailored for specific analytes. Non-polar PDMS (polydimethylsiloxane) coatings are suitable for volatile hydrocarbons and terpenes, while polar coatings such as Carbowax target alcohols and carbonyl compounds. Composite fibers like DVB-CAR-PDMS combine different sorptive properties to widen analyte coverage [8].

Applications and Limitations

SPME is highly effective for profiling essential oils, aroma constituents in herbal teas, pharmacognostic volatile analysis, and even biomarker detection in plant-derived drug formulations. However, fibers suffer from fragility, limited longevity, and selectivity issues when handling highly polar or non-volatile constituents. Despite these constraints, SPME remains a frontrunner in solvent-free pharmacognostic extraction [6, 7].

3.2 Stir-Bar Sorptive Extraction (SBSE)

Principle and Applications

SBSE is an advancement over SPME, using a magnetic stir-bar coated with polydimethylsiloxane (PDMS) to absorb analytes during stirring. The significantly greater sorbent volume provides improved extraction sensitivity. After sampling, the stir bar is thermally desorbed and transferred to GC–MS or LC systems for analysis.

In pharmacognosy, SBSE has been applied in the analytical profiling of volatile oils in aromatic plants, alkaloids in medicinal herbs, and flavonoids in dietary supplements [9]. Its high sensitivity makes it especially suitable for trace analysis of minor phytoconstituents, which is critical for detecting rare bioactive molecules present in complex mixtures.

Comparative Advantages

Compared to SPME, SBSE offers much higher analyte capacity due to its larger phase volume, leading to better reproducibility and lower limits of detection. Furthermore, PDMS coatings are more robust and longer-lasting than SPME fibers. However, limitations include restricted sorbent chemistries (mostly PDMS) and reliance on specialized desorption equipment, making it more demanding in terms of instrumentation availability.

4 Supercritical and Alternative Fluid-Based Methods

4.1 Supercritical Fluid Extraction

Supercritical fluid extraction (SFE) is a green, eco-friendly, and non-toxic extraction technique extensively used in the pharmaceuticals, food, cosmeceuticals, nutraceutical, and other industries. It utilizes the solvents that behaves like supercritical liquid at a critical temperature and pressure. Such supercritical fluid can be used as the extracting solvent that has high density, low viscosity, high solvation, diffusion and penetration power, and can be compressed as gas. The efficiency of SCFE is based on the capacity of solvent to attain phase equilibrium, required density, and its interactions with solute, that leads to better selectivity and yield [10].

The principle of SFE technique is based on the ability of a solvent to behave as supercritical fluid above their critical temperature and pressure. At such conditions, the supercritical fluid solvent possess the property of both liquid and gas. It has density of a liquid and an ability to diffuse as a gas leading to better penetration. Thus, it leads to better diffusion of component and enhance the solvation efficacy compared to conventional methods that make use of liquid organic solvents [11].

Carbon dioxide (CO_2) is the most widely preferred solvent used in SFE as it is economic, easily available, inert, non-toxic, non-flammable, and attains the nature of supercritical fluid at a critical temperature of 31.1 °C and pressure of 7.38 MPa. The density and solvation power of the CO_2 solvent can also be tuned by varying the temperature and pressure to selectively extract different components. However, a cosolvent or modifier may be required to extract more polar compounds. Ethanol, and methanol are majorly used as cosolvent as it improves the extraction of polar compounds and also helps too [12].

After extraction, the solvent can be easily removed from the extracts by releasing the temperature and pressure. The drop in pressure will revert the solvent to attain gaseous phase facilitating easy separation and leaving no solvent residue in the extracted components [13]. This technique provides pure extracts without the need of removing the solvent by distillation, evaporation as seen in conventional extraction techniques. It helps to reduce use of harmful solvents, requires less time for extraction, provides better yield and quality of extract and is more efficient than conventional extraction techniques.

4.1.1 Instrumentation and Process Design

As shown below in the Fig. 1, a SFE system has a cylinder that contain carbon dioxide as solvent and a modifier cylinder for the cosolvent (to extract polar compounds if required). Both of these are attached with the cooling unit also known as a chiller, which is used to cool the supercritical CO_2 and maintain it in liquid form before it is pumped into the extraction vessel. A high-pressure pump is used to pump the liquid CO_2 in the extraction chamber that hold the matrix (solid or liquid). The extractor is placed in an oven which controls the temperature (30–250 °C) attached to a back-pressure regulator/restrictor that exerts the required pressure to maintain CO_2 in its supercritical state. Extraction vessels are designed to withstand pressures above 40 MPa and can be configured with flow distributors to improve contact between solvent and matrix [14]. The supercritical fluid extracts the targeted components/ analytes and moves toward the collector vessel. Due to the lowering of temperature and pressure, CO_2 reverts in the gaseous phase, loses its solvation power and precipitate the extracted analyte in the collector chamber in an offline mode. The CO_2 is collected and recycled to be utilized again. However, sometimes the collector vessel is directly attached to the detector to analyze the extracted analyte and this is known as online mode of extraction [15].

Thus, the extraction can be performed in an offline mode which results in collection of extracts or in an on-line mode where it coupled with other analytical instruments such as gas, liquid, and supercritical fluid chromatography for its direct analysis [11]. Similarly, the extraction is considered as static mode, when the pump is opened only after the complete extraction has taken place while it will be dynamic when the pump sending the materials out to the collection chamber is always open during the extraction process [16]. Modern SFE instruments are increasingly automated, equipped with precise temperature–pressure control, and adapted for coupling with analytical detectors [17]. The purity, quality, and yield of extracted components/ analyte depend majorly on type of sample source/matrix, the nature and polarity of analytes to be separated, selection of solvent, need of cosolvent and a proper control of required temperature and pressure throughout the extraction process.

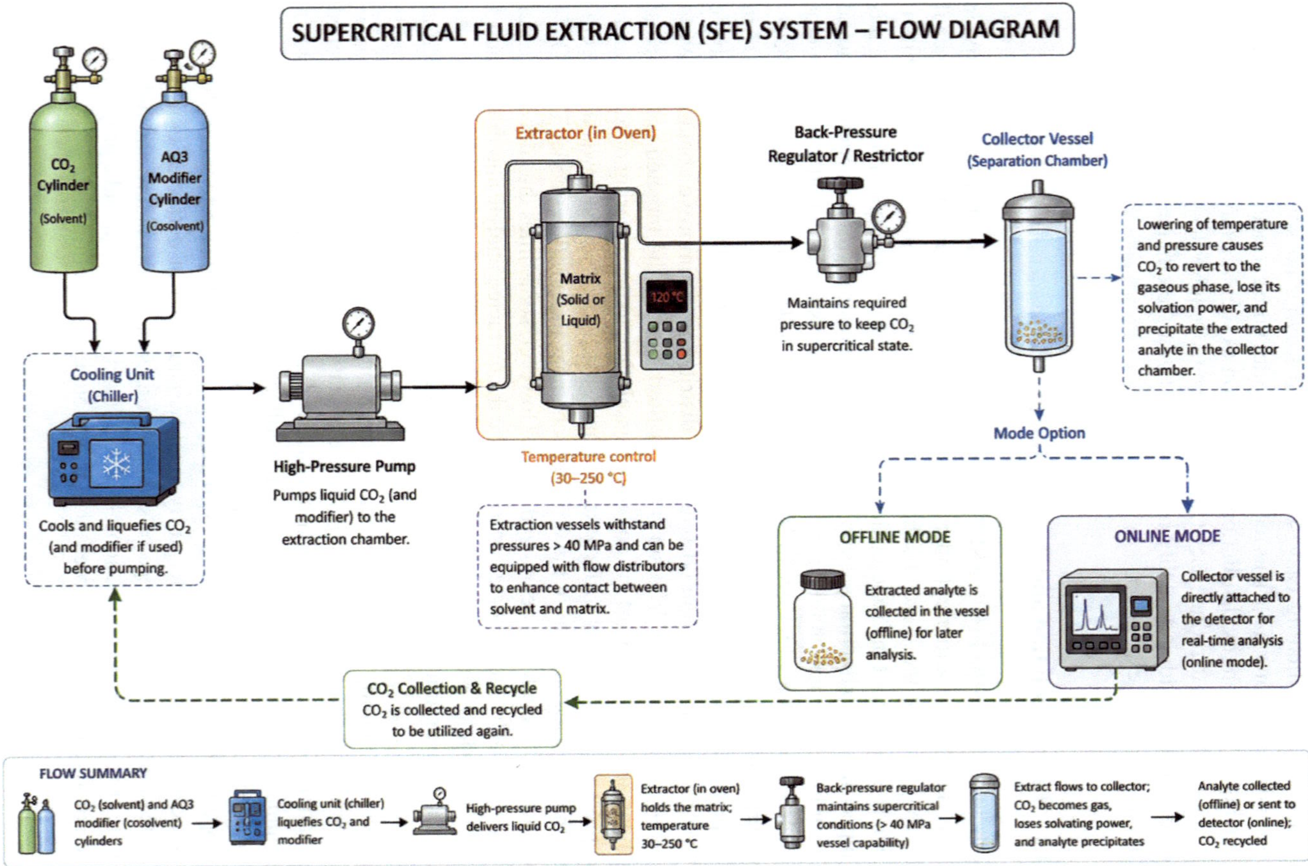

Fig. 1 Supercritical fluid extraction (SFE) system—Flow diagram

4.1.2 Applications of Supercritical Fluid Extraction

SFE has found broad application in food, pharmaceutical, cosmetic, and nutraceutical industries. In food technology, it is widely employed for decaffeination of coffee and tea, extraction of essential oils, flavors from natural products, active ingredients from spices, and recovery of valuable lipids from seeds and algae [13, 18]. In pharmaceuticals, SFE is applied to obtain bioactive compounds such as polyphenols, flavonoids, carotenoids, and cannabinoids with high selectivity and minimal solvent residues (Dey et al. 2021) [19]. The cosmetic sector benefits from SFE in producing fragrance compounds and antioxidant-rich extracts for skincare formulations [15]. It is used in nutraceuticals to extract important nutrients and fatty acids from plant and marine sources. The technique is also increasingly used in detecting the pesticides in food and in the valorization of agro-industrial by-products, enabling recovery of bioactive compounds from fruit peels, seed expeller, spent coffee grounds, and brewery that can be used in the food, pharmaceutical, and cosmetic industries [18]. This technique also offers potential environmental applications in treating solid and water waste treatments to remove hazardous contaminants [20, 21]. These applications highlight SFE as a versatile, sustainable, and selective extraction approach aligned with green chemistry principles (Table 1).

Table 1 Recent applications of supercritical fluid extraction (2020–2025)

Sector	Matrix	Target extract (s)	SFE conditions	References
Food	Spent coffee grounds	Antioxidants, polyphenols	200–300 bar, 40–60 °C, CO_2 + ethanol cosolvent	[22]
Pharma	Medicinal plants (Curcuma longa, Artemisia annua)	Curcuminoids, artemisinin	250–350 bar, 40–55 °C	[23, 24]
Cosmetics	Grape seeds, pomegranate peel	Essential oils, bioactive phenolics	200–300 bar, 40–50 °C	[25]
Cannabinoids	Cannabis sativa	CBD, THC, minor cannabinoids	250–400 bar, 40–60 °C, CO_2 with ethanol	[26]
Valorization	Agro-industrial residues (olive pomace, citrus peel)	Oils, flavonoids, carotenoids	200–350 bar, 40–55 °C	[27]
Nutraceuticals	Microalgae (Chlorella, Spirulina)	Omega-3 fatty acids, carotenoids	250–400 bar, 40–60 °C	[28]

4.1.3 Key Characteristics and Advantages of SFE

Supercritical fluid extraction (SFE) is considered an eco-friendly alternative to conventional organic solvent-based extraction techniques [29, 30]. It is a rapid, highly selective, and automated process that ensures consistency, reproducibility, and minimal environmental impact [31]. As a closed and automated system, it reduces the risk of contamination, enhances product quality and uniformity, improves process control, and increases operational efficiency [32]. Depending on specific requirements, SFE systems can be modified and integrated with chromatographic and spectroscopic techniques, making them versatile for analytical and industrial applications [31]. The technology is available in various configurations suitable for laboratory, research, and large-scale industrial use, with applications spanning food, pharmaceuticals, cosmetics, nutraceuticals, agriculture, and environmental sectors [33].

SFE commonly employs supercritical carbon dioxide (CO_2), which is non-toxic, non-flammable, odorless, cost-effective, readily available, and recyclable, making it a safe solvent for extraction and analytical purposes [30, 32]. The high diffusivity and low viscosity of supercritical CO_2 enable faster mass transfer and more efficient extraction compared to conventional solvent-based methods [34]. Additionally, the relatively low operating temperatures are particularly suitable for extracting heat-sensitive or easily oxidizable compounds [29]. CO_2 can be easily separated from the extract by depressurization, resulting in solvent-free, high-purity products [31]. The solvent properties of CO_2 can be further modified using cosolvents and fine-tuned by adjusting pressure and temperature, allowing selective extraction of target compounds [34].

The only limitation is the high cost of investment and maintenance, and it has some of the important advantages over conventional methods. It is considered as an eco-friendly alternative to organic solvent extraction methods. It is quick, highly selective, and automated technique, ensuring consistency, reproducibility, and minimal environmental impact. By being a closed automated system, it significantly reduces the risk of contamination, improves product quality and consistency, enhance process control, and increase operational efficiency. Depending on the use, it can be modified, hyphenated with other chromatographic and spectroscopic instruments based on the requirements. It is available in various models for lab, research and industrial needs. SFE has multiple uses across food, biologicals, pharmaceutical, cosmetics, nutraceutical, agriculture industry, and most importantly caters to environmental needs for a sustainable use. It makes use of supercritical fluid like CO_2 which is non-toxic, non-flammable, odorless, cost-effective, easily available, and recyclable making it safe to be used for any type of extraction and analytical needs. The high diffusivity and low viscosity of the supercritical fluid led to faster and more efficient extraction than traditional solvent methods. Low operating temperatures, particularly with supercritical CO_2, are ideal for extracting heat-sensitive or easily oxidized compounds. The CO_2 as solvent can be separated easily, resulting in pure extracts free from harmful organic solvent residue. The properties of CO2 as solvent can be modified using cosolvents and also fine-tuned by manipulating pressure and temperature, allowing for highly selective extraction of specific compounds.

4.1.4 Recent Advances and Trends

Recent developments in SFE focus on enhancing efficiency, sustainability, and integration with other green technologies. Advances include continuous SFE systems for industrial-scale processing, hybrid methods coupling SFE with microwave or ultrasound to accelerate mass transfer, and improved co-solvent strategies to broaden the extraction spectrum [35]. There has also been growing interest in applying SFE to valorise food processing residues, in line with circular bioeconomy initiatives [18]. Hyphenation of SFE with supercritical fluid chromatography (SFE–SFC) has advanced the simultaneous extraction and separation of complex mixtures, particularly in pharmaceutical and metabolomic research (Smith 2021). Furthermore, life cycle assessment studies demonstrate the reduced environmental footprint of supercritical CO_2 processes compared to conventional organic solvent extraction, positioning SFE as a key technology in the transition toward eco-friendly processing [36]. Overall, the trajectory of SFE research indicates a move toward process intensification, greener operations, and integration with modern analytical techniques.

4.2 Phytonic Extraction

The name "Phytonics" is made using two terms "phyto," which means plant, and "onics," which indicate a related science or technology and the solvent used in this process is known as phytosol. Hence this method is sometimes also called as phytosol extraction method. This is a modern, eco-friendly patented technique (by Advanced Phytonics Limited, Manchester, UK) used for extraction of highly selective and thermolabile compounds because it operates at low temperatures and pressures. It is considered as the best alternative to SFE and other conventional techniques for it offers various advantages over them [37].

This technique utilizes inert, non-toxic, non-flammable, non-chlorinated fluorohydrocarbons as a solvent for the extraction of pure extract with no solvent residue making it safe to be utilized in pharmaceuticals, cosmetics, and the food industry.

4.2.1 Principle of Phytonic Extraction

Phytonic extraction works on the same principle as that of SFE, where a solvent gas specifically 1,1,1,2-tetrafluoroethane (HFC-134a) is liquified and used to extract selective analytes from the plant matrix. After complete extraction, the phytosol is directed to the separation chamber where the solvent is evaporated using minimal heat and vacuum. The evaporated solvent is then re-liquefied using a gas compressor and a condenser, allowing it to be recycled and reused in a closed-loop system. This technique provides pure extract with no solvent residue [38].

4.2.2 Instrumentation

The instrumentation required for phytonic extraction include a solvent reservoir which contain the liquified solvent, a pump to move the solvent into an extraction vessel that holds the plant matrix, a separator which helps to evaporate the solvent and separate the extracted analyte and gas compressor that compresses the gas and direct it further to a condenser which cools it liquify and recycle back for reuse. An inline filter may be used to separate any fine particles from the solvent before it's evaporated. This setup is more capital-intensive than a simple conventional extraction apparatus but requires significantly less energy and offers superior control over the extraction process.

4.2.3 Advantages

Phyotonic techniques offers all the benefits specified in SFE with a higher efficiency at a much lower energy consumption and maintenance cost. The system is automated and closed making it environment safe. It needs minimal temperature and pressure in comparison to SFE as the solvent (THC) gets easily evaporated at a very low boiling point (– 26.1 °C), and requires just 75 psi pressure to liquify. The process can be performed at neutral pH and in an ambient operating condition making it ideal for extracting valuable and thermosensitive components without suffering any damage due to oxidation or hydrolysis. The generated waste biomass is dry and thus "eco-friendly" to handle.

4.2.4 Applications

Phytonic extraction is particularly well-suited for a variety of industrial applications, especially where high-purity, high-quality extracts are required. The low temperature used in this process makes it ideal for extracting high quality essential oil and heat-sensitive compounds without thermal degradation. It's used to extract active pharmaceutical ingredients (APIs), nutraceutical actives and other bioactive compounds from medicinal plants with high purity. As this method does not leave any undesirable solvent residues, it can safely be used to extract high quality, clean-label flavors, and aromatic oils from spices and herbs for food industry. It is also used to obtain plant-based actives for skincare and cosmetic products [39, 40].

5 Energy-Assisted Solvent-Free Techniques

5.1 Ultrasound-Assisted Extraction (UAE)

5.1.1 Principle of Acoustic Cavitation

Ultrasound-assisted extraction (UAE) is a green, non-thermal process that uses high-frequency sound waves at a high intensity to enhance the release of bioactive compounds from plants material. The principle is based on the acoustic cavitation, in which ultrasound frequencies ranging from 20 to 100 kHz is propagated through the menstruum. These waves generate alternating compression and rarefaction cycles that induce the formation, growth, and collapse of microscopic bubbles (cavities) in the solvent. The violent collapse of these bubbles releases intense localized energy that leads to disruption of cell wall, initiates mass transfer, and facilitates the diffusion of intracellular bioactive compounds into the menstruum [41, 42].

5.1.2 Advantages

Ultrasound extraction technique increases extraction efficiency, reduces extraction time, and provides higher extraction yield and better selectivity in comparison to the conventional extraction methods [42, 43]. It operates at relatively low temperatures, preventing the degradation of heat-sensitive molecules, has low solvent consumption and is more energy efficient than traditional methods, positioning it as "green" technology.

5.1.3 Process Parameters

The efficiency of UAE is strongly influenced by several process variables. The lower ultrasonic frequencies (< 40 kHz) generate more intense cavitation, while the higher frequencies (> 100 kHz) produce smaller, more uniform bubbles, leading to a mild and efficient extraction conditions [42, 44]. Besides the polarity, viscosity, and vapor pressure of the solvent also play an important role in altering the cavitation intensity and enhancing the solubilization capacity. Green solvents, such as ethanol and deep eutectic solvents, are increasingly being utilized in UAE to alter the extraction efficiency and yield [45]. Mild heating within controlled temperatures in UAE often helps to improve yields without degrading thermolabile compounds [43]. However, optimization of the process is necessary as the prolonged sonication might lead to degradation of sensitive phytochemicals [46].

5.1.4 Applications

UAE has been successfully applied across various domains. It is used in food and nutraceuticals for the extraction of polyphenols, flavonoids, carotenoids, essential oils, and proteins from fruits, vegetables, cereals, and by-products [42, 46]. UAE has shown increase in the extraction yield of phytopharmaceutical actives like alkaloids, saponins, and other secondary metabolites [43]. It is also used to extract selective antioxidants, color pigments, and aromatic oils used in cosmetic industry. This green technology is also used in the valorization of biomass to obtain biodiesel and bioactive-rich residues to create b**ioenergy** [45].

Its extraction efficiency, versatility, scalability, and eco-friendly nature make UAE an integral part of modern **green extraction technologies**, in combination with other similar techniques.

5.2 *Microwave-Assisted Extraction (MAE)*

5.2.1 Principle and Mechanism (Dielectric Heating and Mass-Transfer Effects)

Microwave-assisted extraction (MAE) is a technique that uses microwave energy to rapidly heat a solvent in contact with a sample, causing cell walls to rupture and facilitating the release of target compounds. It uses electromagnetic radiation at 915 or 2450 MHz to heat matrices volumetrically through dielectric heating [47, 48]. The radiations increase the dipolar rotation of polar molecules of the solvent and initiate ionic conduction in the oscillating field [49, 50]. Thus, causing uniform heating inside plant tissues, leading to internal pressure build-up and cell disruption, which accelerates mass transfer of solutes into the solvent leading to extraction of active components [10, 42].

The extraction depends on the major factors such as power used in microwave, irradiation time, solvent dielectric properties, matrix moisture content, water content, solvent selection, and particle size. The extraction efficiency also depends on the system operating under open reflux or closed-vessel conditions [49, 51].

5.2.2 Benefits and Limitations

Benefits: This method significantly reduces extraction time and solvent consumption in comparison to conventional techniques, and offers higher extraction efficiency, quality yield with lower costs. It is compatible with water, ethanol, and natural deep eutectic solvents, making it energy efficient, green, and sustainable alternative to conventional techniques [49, 51]. Similarly, closed-vessel system in MAE improves efficiency and gives reproducible yield of extraction [49, 50].

Limitations: There are chances of non-uniform heating and hot spots usually if the process is not optimized and includes heterogeneous sample [52]. The scale-up process for this technique is still challenging and requires technological advanced set up and larger capacity microwaves to be utilized by the large-scale industry [53].

5.2.3 Applications of MAE

MAE is being used for extracting natural compounds, antioxidant bioactive, essential oil, and aromatic products in pharmaceutical industry [50, 51]. It is widely applied for extracting pesticides, mycotoxins, and polar metabolites in food analysis [49]. It is used to extract antioxidants such as polyphenols, flavonoids and carotenoids from fruits, seeds, and also to recover bioactives from the agro-wastes using green solvents [51, 54]. Recent work is going on to combine MAE with machine learning, and other hybrid techniques to improve selectivity and efficiency to make its use at industrial scale [53, 55].

5.3 Microwave-Assisted Solvent-Free Extraction/ Solvent-Free Microwave Extraction (SFME)

5.3.1 Principle and Mechanism of Dielectric Heating

Solvent-free microwave extraction (SFME), is a green chemistry technique that combines the benefits of microwave heating along with dry distillation to extract volatile compounds from plant materials without the use of external solvents or water. It is based on the same principle as that of MAE of dielectric heating. Fresh plant material is placed directly into a microwave reactor without the addition of any solvent or water. The microwave irradiation (2450 MHz) is applied which interacts with polar molecules and ions of the moisture present in the plant matrix [47, 56]. The dipolar rotation and ionic conduction cause superheating of the water molecules present inside the plant material [49]. The sudden heating generates internal pressure, rupture glandular structures and cell walls, and facilitates the release of essential oils, or other volatile compound along with the water vapors [57]. The vapors are directed out of the microwave cavity to an external cooling system, which condenses them into a liquid distillate, the isolated and concentrated extract/volatile oil separate from water and gets collected outside the microwave reactor. This makes SFME distinct from conventional MAE, as it utilizes only the moisture content present in the sample as an extraction medium without the need of any solvent.

5.3.2 Benefits

It eliminates the need for any organic solvents, thus reduces environmental impact and makes it a greener alternative. It also offer major advantages like faster extraction, higher yields, improved quality of extracted products, lower costs, with less energy consumption, and reduced environmental impact compared to traditional methods like hydro-distillation [57, 58]. Emerging industrial designs promise the future scalability of this method to be utilized commercial extraction of volatiles and non-volatile compounds [59].

5.3.3 Applications

SFME has been successfully applied to extract essential oil from plants like rosemary, thyme, oregano, and citrus peels, producing oils of pure quality [56, 57]. It is efficiently used to valorize agro-industrial waste like citrus peels to yield natural flavors and antioxidants with reduced energy input [59]. This technique is very useful for pharmaceuticals and nutraceuticals to extract bioactive-rich extracts from aromatic herbs without solvent contamination, making them most suitable for therapeutic and nutraceutical formulations [57]. SFME is increasingly being used for rapid isolation of volatile metabolites to be used as an analytical sample for food authentication and quality control studies [49].

5.4 Thermal Desorption Methods

Thermal desorption (TD) techniques are widely used to extract, concentrate and introduce volatile and semi-volatile organic compounds (VOCs/SVOCs) from air, solid and liquid matrices to gas chromatographic (GC) systems, and allied detectors (MS, FID, IMS, etc.).

5.4.1 Principle of Thermal Desorption

At its core, thermal desorption relies on controlled heat application to an adsorbent (sorbent tube, trap, or packed bed) to overcome the adsorption energy binding analytes to the sorbent surface, releasing them into a carrier gas stream that transports them into a focusing trap and then onto the GC column. It is an analytical approach in which vapor-phase analytes sorbed or trapped onto a solid phase (adsorbent) are released by controlled heating and carried by an inert gas stream into an analytical instrument—most commonly a gas chromatograph coupled to mass spectrometry (TD–GC–MS) or other detectors (TD–GC–FID, TD–GC–IMS) for separation and detection [60, 61].

5.4.2 Advantages, Limitations, and Practical Considerations

TD techniques provide high preconcentration efficiency, solvent-free sample preparation and compatibility with sampling strategies over large sample volumes, enabling trace detection down to parts-per-trillion (ppt) levels for many compounds [62, 63].

However, sorbent choice can cause losses of very volatile analytes, degradation of certain reactive analytes, thermal decomposition or formation of artifacts can occur at high desorption temperatures [62]. Management of moisture is most critical part in such analysis as water can displace analytes or damage traps/columns.

5.4.3 Applications

TD–GC–MS with sorbent tubes is used for ambient VOC monitoring in environmental and ambient air monitoring in urban air studies [60, 62]. It helps in measurement of VOCs emitted from building materials, furnishings, and consumer products [61]. TD analysis is used in routine analysis of atmospheres at industrial workplace to evaluate occupational hygiene, compliance, and exposure assessment [60]. TD–GC analysis is applied to monitor soil vapor, sub-slab gases and vapor intrusion into buildings. It helps to sensitively quantify chlorinated solvents, aromatics and fuel components (Markes/industry application notes [64]. Direct TD coupled to GC–MS has been used to study thermal decomposition products of polymers, adhesives and other toxicant screening, fire investigations, and material emissions studies [65]. Dynamic headspace sampling using TD is used for aroma profiling, volatile fingerprinting and food authenticity studies Food, flavor and natural-product analysis where lossless preconcentration is required [63].

5.4.4 Conclusion

Thermal desorption methods are widely adopted analytical techniques essential for analysis of trace volatile and semi-volatile organic compound across environmental, occupational, industrial, forensic and food science domains. Careful method validation and adherence to standards remain central to reliable TD analysis [62, 66].

6 Biomaterial- and Enzyme-Based Green Extraction

Green extraction technologies that utilize biomaterials and enzymes are effective in conserving bioactive compounds, improving operational efficiency, and safeguarding the environment by substituting synthetic solvents with renewable alternatives.

The process of green extraction utilizes the inherent chemical and physical characteristics of biomaterials, along with the catalytic functions of enzymes, to selectively obtain bioactive compounds while minimizing environmental consequences. Biomaterials such as chitosan, alginate, starch, lignocellulosic waste, and charcoal engage with bioactives through mechanisms including hydrogen bonding, π-π stacking, electrostatics, and inclusion complexation [67, 68]. Enzyme-assisted extraction facilitates the breakdown of structural barriers in plants or microorganisms, enabling the release of bioactive compounds without the requirement of harsh solvents or extreme conditions [69].

These porous biomaterials containing functional groups (–OH, –COOH, $–NH_2$) facilitate the absorption of bioactives through mechanisms such as hydrogen bonding, ionic interactions, and the formation of hydrophobic inclusion complexes. Chitosan interacts with phenolic acids through its amino groups, while charcoal engages in the adsorption of polyphenols via π-π interactions. The selectivity can be altered through the manipulation of pH levels and ionic charges [70].

6.1 Natural Sorbents and Biopolymers

These polymers originate from renewable resources such as chitosan, alginate, starch, and lignocellulosic byproducts. These compounds possess functional groups that engage with bioactive substances, leading to selective adsorption, and enhanced stability. Biochar is produced through the pyrolysis of biomass, characterized by its extensive surface area and porosity, which render it highly effective in the adsorption and removal of contaminants [71].

6.2 Agro-Waste and Eco-friendly Adsorbents

Agricultural byproducts, including fruit peels, seed husks, and shells, consist of lignocellulosic matrices that are abundant in cellulose, hemicellulose, lignin, and silica. The processes of chemical activation and pyrolysis represent distinct methodologies for modifying surfaces, enhancing adsorption characteristics, and facilitating the extraction of bioactive compounds while effectively decolorizing wastewater [72].

6.3 Enzyme-Assisted Extraction

The application of enzymes facilitates the breakdown of glycosidic, ester, and peptide bonds within plant matrices, resulting in the weakening of cell walls and the subsequent release of targeted compounds, including flavonoids and essential oils. Enzymatic reactions take place under mild aqueous conditions, thereby safeguarding sensitive compounds like carotenoids. The synergistic action of enzymes, including cellulase and pectinase, enhances the efficiency of extraction processes [69, 73].

6.4 *Green Nanomaterials in Solvent-Free Extraction*

Metal oxide nanoparticles (ZnO, CuO, TiO_2), graphene composites, and biogenic nanoparticles serve as catalysts and nanosorbents in solvent-free extraction processes, facilitating bond breakdown and selective adsorption with minimal energy input and without the use of solvents [74]. In the process of solvent-free extraction, two key mechanisms occur: (i) the agents serve as catalysts, facilitating bond cleavage and enhancing the yield of phytochemicals, and (ii) they function as nanosorbents, effectively concentrating and selectively adsorbing specific biomolecules [75].

Applications

- Employing biopolymers and agricultural waste materials as adsorbents for the extraction of pectin from citrus peels and polyphenols from grape juice. This enhances the stability of the product and optimizes the utilization of waste materials.
- Utilization of enzymatic methods for the extraction of essential oils and antioxidants (carotenoids and anthocyanins) from medicinal plants and agricultural byproducts, ensuring the integrity of the components is maintained.
- Employing biochar and lignocellulosic bioadsorbents for the removal of contaminants and alteration of effluent color, thereby facilitating environmental remediation.
- The integration of enzyme extraction methodologies with ultrasonic or microwave technologies aims to enhance efficiency and reduce processing duration.
- A method of extraction that eliminates solvents, utilizing environmentally friendly nanomaterials to selectively recover heterocyclic compounds and phytochemicals, ensuring minimal impact on the environment.

The implementation of these green extraction methods plays a significant role in promoting sustainable development through the transformation of low-cost renewable resources and waste into valuable bioactive compounds. They additionally diminish the necessity for detrimental chemicals and energy-consuming processes.

7 Headspace-Based Extraction Techniques

Headspace extraction techniques represent established methodologies for introducing samples into gas chromatography (GC). Researchers assess volatile compounds by collecting samples from the "headspace," which is the vapor phase situated above a solid or liquid sample in a sealed environment. The methodology is predicated on the principle that volatile analytes move between the sample matrix and the headspace in accordance with their vapor pressures and equilibrium constants. This facilitates the preparation of samples and minimizes analytical interference by enabling the extraction of only volatile compounds, thereby excluding non-volatile matrix components. Two main categories of headspace extraction exist: static headspace extraction, where the vapor phase reaches equilibrium with the sample prior to sampling, and dynamic headspace extraction, which involves the continuous removal of volatile molecules from the sample to a trap, thereby enhancing the sensitivity of the process. The methodologies employed in headspace analysis hold significant importance due to their user-friendly nature, adaptability to various requirements, and contribution to maintaining equipment cleanliness.

7.1 *Extraction of Static and Dynamic Headspace*

In static headspace extraction, a sample, regardless of its state as a liquid or solid, is placed in a sealed vial and subjected to a specific temperature increase. This technique facilitates the release of volatile compounds into the vapor phase or headspace above the sample. The system can reach a state of balance, characterized by the analytes' vapor pressures and partition coefficients, facilitating the differentiation between the sample matrix and the vapor phase. Upon achieving equilibrium within the system, a small volume of the headspace vapor is extracted using a syringe or an automated valve system and subsequently introduced into the gas chromatograph for analytical assessment.

This represents an equilibrium methodology, as the principles of Henry's law illustrate the relationship between the concentration of analytes in the headspace and their concentration within the sample. The method is commonly used for routine analysis in situations where analyte concentrations are relatively high. It is easy to use, produces clean extracts, and requires little upkeep because only volatile components reach the GC apparatus.

Dynamic headspace analysis extraction, also known as purge and trap, differs from static extraction in that it continuously purges the sample vial using an inert gas, such as helium or nitrogen. This process eliminates volatile and semi-volatile analytes from the sample headspace and captures them on a solid adsorbent (trap). Subsequently, the trap undergoes heating to facilitate the thermal desorption of the analytes into the carrier gas stream of the gas chromatograph for subsequent analysis.

DHE represents a non-equilibrium exhaustive methodology based on Le Chatelier's principle, effectively extracting nearly all analytes from the sample. This enables the detection of trace or ultra-trace chemicals. The continuous flow and entrapment enhance the device's sensitivity and lower the detection thresholds.

7.2 *Utilization of Volatile and Semi-volatile Compounds*

Headspace-based extraction techniques are extensively employed for both qualitative and quantitative analyses of volatile organic compounds (VOCs) and semi-volatile organic compounds (SVOCs) within intricate matrices, including liquids, solids, and biological materials. Applications are notably prevalent in the domains of environmental analysis, food and flavor chemistry, testing for pharmaceutical residual solvents, forensic toxicology, and industrial quality assurance.

Applications

- Environmental Testing: Conducting an analysis of volatile organic compounds present in air, soil, and wastewater samples. Dynamic headspace facilitates the detection of contaminants at ultra-trace levels through comprehensive purge and trap methodologies.
- Food and Flavors Sector: Examining the volatile compounds present in food, beverages, and fragrances. The static headspace methodology enhances the extraction of aroma volatiles while ensuring the integrity of chromatography columns.
- Pharmaceuticals: Employing static headspace gas chromatography to assess the concentrations of residual solvents in pharmaceutical formulations in accordance with regulatory guidelines.
- Clinical and Biomedical Analysis: In clinical research, headspace methodologies are employed to identify volatile biomarkers present in biological fluids such as breath, blood, and urine, facilitating the diagnosis of diseases or the screening for toxic substances. These techniques enable the quantification of pharmaceutical residues and metabolic volatiles present in tissues and biological materials.
- Industrial Process Monitoring: Procedures utilizing headspace techniques for rapid quality assurance facilitate the monitoring of emissions associated with chemical manufacturing, petrochemical processing, and product off-gassing.
- Vacuum-Assisted Headspace SPME: Vacuum-induced headspace SPME (Vac-HS-SPME) represents an innovative approach that employs a vacuum to enhance the rate of analyte equilibrium attainment and to augment the recovery yield of the analyte. This facilitates the extraction of semi-volatile compounds. This enables application to samples containing thermally unstable analytes, environmental pollutants, and complex matrices where traditional HS-SPME methods may be ineffective.

The primary benefit of headspace extraction for volatiles and semi-volatiles lies in its ability to selectively concentrate and isolate analytes from intricate matrices, eliminating the need for direct infusion of the material. This minimizes interference from molecules that remain stable under heat, enhances the longevity of the instrument, and typically facilitates the identification of objects [76].

8 Comparative Assessment of Solvent-Free Extraction Techniques

Solvent-free extraction is primarily governed by temperature, pressure, sorbent type, agitation, and fiber or stir-bar geometry. Thermal control ensures maximal release of analytes without degradation, while applied pressure (particularly in SFE) tunes solvent density for selective extraction. Sorbent chemistry and surface area (e.g., PDMS, CAR/PDMS, DVB fibers, thin films, or stir bars) define analyte partitioning and affinity. Agitation or ultrasonic/microwave energy accelerates mass transfer, reducing extraction time. Fiber or stir-bar length and coating thickness modulate capacity, enabling trace-level detection or bulk extraction depending on the application. Representative metrics illustrate this: MAE extracted neem oil at 31.65% w/w in 24 min versus 10 h for Soxhlet; SFE recovered 547.27 mg/g ergosterol at 300 bar; UAE yielded 27.1 mg/g polyphenols from carob fruit in under 10 min.

8.1 *Efficiency and Selectivity*

Solvent-free extraction techniques, particularly microwave-assisted extraction (MAE), ultrasound-assisted extraction (UAE), supercritical fluid extraction (SFE), and solid-phase microextraction (SPME), have revolutionized the extraction of bioactive components from natural sources. For example, MAE has achieved dramatic reductions in extraction time and increases in yield for neem oil, where optimized conditions yielded 31.65% w/w oil in just 24 min, while conventional Soxhlet required 10 h for comparable extraction. UAE, evidenced in carob fruit polyphenolic extraction, achieved yields up to 27.1 mg/g in under 10 min, compared to much longer durations for solvent-based processes. SFE, using supercritical CO_2 to extract ergosterol from Agaricus bisporus mushrooms, produced a yield of 547.27 mg/g at 300 bar pressure with complete solvent removal and tunable selectivity. SPME is routinely applied in environmental monitoring to selectively concentrate organic pollutants from water samples without using bulk solvents, providing highly selective results with minimal waste. These examples demonstrate both high efficiency and superior selectivity, leveraging physical and chemical control parameters like temperature, pressure, and fiber characteristics to target desired compounds and minimize degradation or contamination [77, 78].

8.2 *Environmental Impact and Sustainability Metrics*

The environmental benefits of solvent-free extraction are exemplified by quantitative reductions in chemical waste, solvent toxicity, and energy consumption. For instance, SFE applied to the extraction of silymarin from Silybum marianum thistle operates with recyclability of supercritical CO_2, generating negligible by-products and almost zero solvent waste. MAE, as used for extracting gallic acid from moringa, runs at lower temperatures (30–50 °C) and microwave energy, which significantly cuts down energy, avoids volatile organic solvents, and preserves bioactive integrity. UAE in green tea leaf extraction uses water or ethanol in tiny amounts, drastically lowering chemical hazards and effluent. Solvent-free methods drastically reduce hazardous waste and lessen carbon footprints compared to conventional extraction approaches. Metrics to assess environmental impact include energy consumption, greenhouse gas emissions, and total process hazard output. SPME and stir-bar sorptive extraction (SBSE) exemplify green alternatives, as they function with minimal to no solvent, aligning with green chemistry principles. Adoption of these techniques improves safety and supports sustainability in laboratory and industrial settings [79].

Applications

- Green Analytical Chemistry for regulatory testing and routine analyses.
- Industrial extractions where large volumes of solvents previously posed storage and disposal challenges.

Examples

- Gel electromembrane extraction (G-EME) uses biodegradable gels instead of organic solvents, drastically decreasing waste production.
- SFE and subcritical water extraction use less energy and produce less hazardous waste compared to conventional solvent-based processes.

8.3 Cost-Effectiveness and Practicality

Despite higher initial investment for equipment in some cases (e.g., SFE units), overall lifecycle costs of solvent-free extraction are strongly favorable due to lower operational energy, elimination of solvent handling, and faster throughput. MAE extraction of Antioxidants from black beans, for example, achieves up to 33.82 mg/g anthocyanins with yields achieved at a solvent-to-solid ratio of 50 mL/g under microwave energy, minimizing labor and utility costs while maximizing productivity. UAE's extraction of pumpkin seed oil resulted in 447.4 mg/100 g yield within 34 min at a fraction of the electricity and none of the solvent expenses typical of traditional methods. SPME's practicality extends to field or point-of-care diagnostics with portable devices requiring minimal user training or infrastructure. SFE is now economically viable for production of pharmaceuticals (flavonolignans, triglycerides), flavors and fragrances, where fast turnover and purity offset setup costs, especially at scale. These examples underscore both the cost and practical advantages, particularly in pharma, food, and environmental monitoring sectors. Solvent-free extraction often leads to cost savings by removing the need for chemicals, reducing waste management expenses, and streamlining sample handling.

Applications

- food sector for byproduct valorization and rapid screening;
- pharmacology for cost-effective extraction of compounds from limited or precious sources
- environmental monitoring using portable, on-site extraction devices.

Examples

- Fish oil extraction from catfish heads using solvent-free technology improved cost-effectiveness and sustainability in byproduct management.
- Microwave-assisted extraction allows batch and continuous operation, minimizing labor and maximizing throughput.

Technique	Target analytes	Typical yield/ recovery	Extraction time	Key parameters	Solvent / energy benefit	Representative example
MAE (microwave-assisted extraction)	Lipophilic bioactives, essential oils	Neem oil: 31.65% w/w	24 min	Temp:60–80 °C; microwave power; solid-to-solvent ratio	Low energy, minimal solvent	Neem oil extraction; 10 × faster than Soxhlet
UAE (Ultrasound-assisted extraction)	Polyphenols, antioxidants	Carob fruit: 27.1 mg/g	< 10 min	Temp: 30–50 °C; ultrasound frequency/ intensity; agitation	Low energy, small solvent volumes	Carob polyphenols; rapid extraction
SFE (supercritical fluid extraction)	Sterols, lipids, aroma compounds	Agaricus bisporus ergosterol: 547.27 mg/g	30–60 min	Pressure: 300 bar; temp: 40–60 °C; CO_2 flow; modifiers	Complete solvent-free, tunable selectivity	Ergosterol extraction with supercritical CO_2
SPME (solid-phase microextraction)	VOCs, semi-volatiles	ng–µg scale	5–30 min	Fiber type: PDMS, CAR/PDMS, DVB; immersion/headspace; agitation; fiber length/geometry	Solvent-free, minimal sample prep	VOC monitoring in water; field-deployable
SBSE/ HSSE (stir-bar sorptive extraction)	Trace organics, volatiles	ng–µg scale	30–60 min	PDMS-coated stir bar; temperature-controlled; agitation	High preconcentration, no solvent	Benzene-series detection in groundwater
SFME (solvent-free microwave extraction)	Essential oils, hydrophobic compounds	Thymus mastichina oil: 1.8% w/w	10–20 min	Microwave power, biomass moisture, agitation	No solvent, energy efficient	Essential oil recovery from biomass

9 Applications Across Fields

Solvent-free extraction techniques are now integral in diverse scientific and industrial sectors, delivering greener, safer, and often more effective sample preparation and analysis.

9.1 Environmental Monitoring

Solvent-free techniques like solid-phase microextraction (SPME), micro solid-phase extraction (μSPE), stir-bar sorptive extraction (SBSE), and in-needle microextraction (INME) are leading innovations in environmental analysis. SPME, notably, is utilized for the detection of volatile organic compounds (VOCs), polycyclic aromatic hydrocarbons, and pesticides in air, water, and soil—with the fiber directly exposed to the sample or its headspace, extracting analytes with minimal intervention. For instance, SPME has enabled solvent-free VOC extraction from water samples for monitoring industrial pollution and assessing site remediation, while vacuum-assisted HS-MonoTrap systems detect hazardous emissions from asphalt and other sources. Integration of advanced adsorbents (graphene, metal–organic frameworks) further improves selectivity and sensitivity, empowering real-time, on-site environmental surveillance.

9.2 Food Quality and Safety

Solvent-free extractions have transformed food safety analysis by expediting detection of contaminants, pesticides, additives, and flavor compounds. Microwave-assisted extraction (MAE) and ultrasound-assisted extraction (UAE) are employed in rapid extraction of pigments and antioxidants from fruits and vegetables with higher yields and restricted solvent use. Supercritical fluid extraction (SFE) utilizes CO_2 to separate pesticides and organic pollutants from foods and food packaging without residual solvents, a crucial advantage in food authentication. The QuEChERS method (quick, easy, cheap, effective, rugged, and safe) has revolutionized multi-residue pesticide screening in food matrices by reducing solvent waste, labor, and cost while achieving high recoveries of target analytes; now, it is a standard protocol in the global food industry. These methods ensure compliance with regulatory limits and safeguard consumer health.

9.3 Pharmaceutical and Biomedical Analysis

The pharmaceutical sector leverages solvent-free microextraction to analyze drugs, metabolites, and biomarkers in complex biological samples, significantly reducing chemical exposure and increasing analytical throughput. Techniques such as SPME, SBSE, and microextraction by packed sorbent (MEPS) extract pharmaceuticals, illicit drugs, and clinical analytes from biological matrices (saliva, blood, plasma) with minimal sample volumes, often achieving lower detection limits than conventional methods. For example, SPME combined with mass spectrometry is routinely used for monitoring immunosuppressants in whole blood and therapeutic drugs like voriconazole in plasma. These techniques are critical in clinical diagnostics, therapeutic drug monitoring, forensic toxicology, and anti-doping analyses, enhancing automation and green chemistry compliance.

9.4 Industrial and Agricultural Application

In industry and agriculture, solvent-free methods streamline extraction and quality control of essential oils, bioactive compounds, and agricultural chemicals. SFE is extensively used for concentrating bioactives from medicinal plants and extracting oils from seeds or herbs such as lavender, mint, and cannabis, which then serve the food, cosmetic, and wellness industries. UAE efficiently extracts high-value oils from pumpkin and flax seeds in minutes, optimizing energy use for scalable production. In agriculture, SPME is adopted for real-time pesticide residue analysis in crops and soil, aiding compliance and risk assessment without the logistical burden of large solvent volumes. Environmental rehabilitation projects also utilize solvent-free extractions for removing pollutants from soil and water, contributing to sustainable and regulatory-driven land management [80].

10 Future Perspectives and Innovations

10.1 Integration with Green Analytical Methodologies

Solvent-free extraction is increasingly integrated with the principles of Green Analytical Chemistry, prioritizing reduction of hazardous chemicals, waste minimization, and energy efficiency. Methods such as intermittent solvent-free microwave extraction (ISFME) combine microwave heating with solventless processes for extracting essential oils from citrus peels, offering highly selective, low-impact extraction. Implementation of closed-loop systems, renewable energy sources, and solvent recycling further aligns extraction platforms with global sustainability standards by reducing emissions and lifecycle negative impacts.

10.2 Advances in Materials for Extraction Devices

Recent years have seen the creation of advanced sorbents and coatings such as functionalized graphene, metal–organic frameworks, and nano-engineered polymers dramatically enhancing the efficiency, selectivity, and reusability of devices like SPME fibers and SBSE stir bars. Modular supercritical CO_2 systems now feature corrosion-resistant alloys and smart sensors, increasing process stability and enabling real-time monitoring of extraction parameters for precise targeting and optimization. Development of biodegradable and reusable materials fosters further reduction in environmental impact, with researchers focusing on tunable surface chemistries for targeted compound isolation.

10.3 Miniaturization and Automation

AI, IoT, and robotics are revolutionizing extraction with miniaturized and fully automated platforms. Laboratories are adopting compact solventless systems that utilize microfluidics and AI-driven sensors to control parameters like temperature, pressure, and extraction time, optimizing yields and analytics in real time. Predictive maintenance and remote monitoring allow manufacturers to anticipate issues, streamline processes, and reduce downtime. Continuous-flow and batch automation enabled by real-time data analysis and programmable extraction cycles bridge the gap between research trials and full-scale industrial deployment. Portable and on-site extraction devices further democratize high-quality analytical capability for agriculture, environmental monitoring, and personalized nutrition.

10.4 Potential Role in Circular Economy and Sustainable Development

Solvent-free extraction is a key enabler in the circular economy, maximizing resource recovery, reducing waste, and creating new value chains from byproducts (e.g., recovery of antioxidants from fruit peels, transformation of crop residues into functional ingredients). Adoption across sectors allows lifecycle management of resources with carbon footprint reduction, efficient bio-based feedstock refinement, and integration with regenerative agriculture and recycling-driven industries. Companies are empowered to minimize environmental impact, meet carbon–neutral targets, and create new "green" jobs while contributing to broader sustainable development goals [81].

Challenges

- Efficiency and selectivity vary, as methods like SPME recover VOCs well but struggle with highly polar analytes.
- Scaling from lab to industry is difficult, e.g., CO_2 extraction works in small batches but gives inconsistent yields at pilot scale.
- Specialized instruments (high-pressure CO_2 units, microwave reactors) and costly sorbents limit routine use.
- Complex matrices such as fatty foods or polluted soils reduce sensitivity, similar to metabolite loss in plasma analysis.
- Regulatory gaps, limited device recycling, and high setup costs restrict broader industrial adoption compared to solvent-based methods.

References

1. GAŁUSZKA, A., MIGASZEWSKI, Z. & NAMIEŚNIK, J. 2013. The 12 principles of green analytical chemistry and the SIGNIFICANCE mnemonic of green analytical practices. *TrAC Trends in Analytical Chemistry,* 50**,** 78-84.
2. PŁOTKA-WASYLKA, J., GAŁUSZKA, A. & NAMIEŚNIK, J. 2019. Green analytical chemistry: summary of existing knowledge and future trends. *Green Analytical Chemistry: Past, Present and Perspectives.* Springer.
3. AZMIR, J., ZAIDUL, I. S. M., RAHMAN, M. M., SHARIF, K. M., MOHAMED, A., SAHENA, F., JAHURUL, M. H. A., GHAFOOR, K., NORULAINI, N. A. N. & OMAR, A. K. M. 2013b. Techniques for extraction of bioactive compounds from plant materials: A review. *Journal of Food Engineering,* 117**,** 426–436.
4. LANJWANI, M. F., KHUHAWAR, M. Y., TUZEN, M., FESLIYAN, S. & ALTUNAY, N. 2025. Chapter 7 - Advancements and innovations in solvent-based extraction techniques. *In:* LOCATELLI, M. & KAYA, S. (eds.) *Green Analytical Chemistry.* Elsevier.
5. PAWLISZYN, J. 1997. *Solid phase microextraction: theory and practice*, John Wiley & Sons.
6. SAITO, Y. & NAKAGAMI, K. 2020. Chapter 1 - Sample preparation for the analysis of drugs in biological fluids. *In:* HEMPEL, G. (ed.) *Handbook of Analytical Separations.* Elsevier Science B.V.
7. SNOW, N. H. 2000. Solid-phase micro-extraction of drugs from biological matrices. *J Chromatogr A,* 885**,** 445–55.
8. PAWLISZYN, J. 2012. 1 - Solid-Phase Microextraction in Perspective. *In:* PAWLISZYN, J. (ed.) *Handbook of Solid Phase Microextraction.* Oxford: Elsevier.
9. BALTUSSEN, E., SANDRA, P., DAVID, F. & CRAMERS, C. J. J. O. M. S. 1999. Stir bar sorptive extraction (SBSE), a novel extraction technique for aqueous samples: theory and principles. 11**,** 737–747.
10. REVERCHON, E. & DE MARCO, I. J. P. T. 2006. Supercritical antisolvent precipitation of cephalosporins. 164**,** 139–146.
11. POURMORTAZAVI, S. M. & HAJIMIRSADEGHI, S. S. J. J. O. C. A. 2007. Supercritical fluid extraction in plant essential and volatile oil analysis. 1163**,** 2–24.
12. BRUNNER, G. 2005. Brunner, G.: Supercritical fluids: technology and application to food processing. Journal of Food Engineering 67, 21–33. *Journal of Food Engineering - J FOOD ENG,* 67.
13. DE MELO, M., SILVESTRE, A. & SILVA, C. J. T. J. O. S. F. 2014. Supercritical fluid extraction of vegetable matrices: Applications, trends and future perspectives of a convincing green technology. 92**,** 115–176.
14. MUKHOPADHYAY, M. 2000. *Natural extracts using supercritical carbon dioxide*, CRC press.

15. HERRERO, M., MENDIOLA, J. A., CIFUENTES, A. & IBÁÑEZ, E. J. J. O. C. A. 2010. Supercritical fluid extraction: Recent advances and applications. 1217, 2495-2511.
16. Donato P, Dugo P, Mondello L. Chapter 9 - Separation of Lipids. In: Fanali S, Haddad PR, Poole CF, Schoenmakers P, Lloyd D, editors. *Liquid Chromatography*, https://doi.org/10.1016/B978-0-12-415806-1.00009-7. Amsterdam: Elsevier; 2013. p. 203–48.
17. Herzyk F, Piłakowska-Pietras D, Korzeniowska M. Supercritical Extraction Techniques for Obtaining Biologically Active Substances from a Variety of Plant Byproducts. 2024; 13(11): 1713, https://www.mdpi.com/2304-8158/13/11/1713
18. ZHOU, J., GULLON, B., WANG, M., GULLON, P., LORENZO, J. M. & BARBA, F. J. J. P. 2021. The application of supercritical fluids technology to recover healthy valuable compounds from marine and agricultural food processing by-products: A review. 9, 357.
19. TYŚKIEWICZ, K., KONKOL, M. & RÓJ, E. J. M. 2018. The application of supercritical fluid extraction in phenolic compounds isolation from natural plant materials. 23, 2625.
20. LESMANA, S. O., FEBRIANA, N., SOETAREDJO, F. E., SUNARSO, J. & ISMADJI, S. J. B. E. J. 2009. Studies on potential applications of biomass for the separation of heavy metals from water and wastewater. 44, 19-41.
21. SUNARSO, J. & ISMADJI, S. J. J. O. H. M. 2009. Decontamination of hazardous substances from solid matrices and liquids using supercritical fluids extraction: a review. 161, 1-20.
22. RODRIGUEZ GARCIA, S. L., RAGHAVAN, V. J. C. R. I. F. S. & NUTRITION 2022. Green extraction techniques from fruit and vegetable waste to obtain bioactive compounds—A review. 62, 6446–6466.
23. Widmann AK, Wahl MA, Kammerer DR, Daniels R. Supercritical Fluid Extraction with CO(2) of Curcuma longa L. in Comparison to Conventional Solvent Extraction. Pharmaceutics. 2022;14(9). https://doi.org/10.3390/pharmaceutics14091943.
24. Oladipo B, Ojumu TV. Supercritical CO(2) extraction of artemisinin from Artemisia annua plant and the biotechnological production of artemisinin precursors: A dual-focus review. Biotechnology notes (Amsterdam, Netherlands). 2025; 6:143–56. https://doi.org/10.1016/j.biotno.2025.05.003.
25. MORO, K. I. B., BENDER, A. B. B., DA SILVA, L. P., PENNA, N. G. J. F. & TECHNOLOGY, B. 2021. Green extraction methods and microencapsulation technologies of phenolic compounds from grape pomace: A review. 14, 1407–1431.
26. SAN NICOLAS, M., VILLATE, A., OLIVARES, M., ETXEBARRIA, N., ZULOAGA, O., AIZPURUA-OLAIZOLA, O. & USOBIAGA, A. J. A. C. A. 2023. Exploratory optimisation of a LC-HRMS based analytical method for untargeted metabolomic screening of Cannabis Sativa L. through Data Mining. 1279, 341848.
27. DA SILVA, J., TORRES, T. M. S., RIBEIRO, P. R. V., DE BRITO, E. S., FERREIRA, S. R. S. J. F. & TECHNOLOGY, B. 2025. An eco-friendly high-pressure biorefinery approach for the recovery of valuable compounds from cashew nut testa shell (Anacardium occidentale L.). 18, 1997–2016.
28. PATERSON, S., GÓMEZ-CORTÉS, P., VILLANUEVA-BERMEJO, D., HERNÁNDEZ-LEDESMA, B. & DE LA FUENTE, M. A. J. L. 2025. Effect of supercritical CO2 extraction on the recovery of bioactive lipids and pigments from pretreated Nannochloropsis gaditana biomass. 223, 117748.
29. ANUSHA, V. AND SIVAKUMAR, P. (2023) *A review on supercritical fluid extraction.* Current Journal of Applied Science and Technology, 42(41), pp. 1–12.
30. ARGUN, M.E. AND ARGUN, M.Ş. (2025) *Recovery of valuable compounds using supercritical CO_2 as a green separation method.* Journal of Food Measurement and Characterization.
31. FRAGUELA-MEISSIMILLY, H. et al. (2023) *New trends in supercritical fluid technology for extraction of bioactive compounds.* Molecules, 28(11), 4421.
32. TZIMA, S. et al. (2023) *Recent advances in supercritical CO_2 extraction of bioactive compounds.* Molecules, 28(3), 1410.
33. RODRÍGUEZ-GONZÁLEZ, I. et al. (2025) *Supercritical CO_2 extraction of bioactives from biomass.* Journal of Applied Phycology, 37, pp. 2779–2791.

34. SHI, J. et al. (2025) *Exploring selectivity of supercritical CO_2 extraction.* Food and Bioprocess Technology, 18, pp. 2754–2764.
35. Jahnavi P, Indirabanu S, Chellappan RD, Mubeen MIBk, Bodapati A, Dharmamoorthy G, et al. Supercritical Fluid Extraction: A Green and Sustainable Approach for the Isolation of High-Value Compounds from Natural Sources %J Asian Journal of Green Chemistry. 2025; 9(5):605–29. https://doi.org/10.48309/ajgc.2025.516853.1728.
36. HERRERO, M. & IBAÑEZ, E. J. T. J. O. S. F. 2018. Green extraction processes, biorefineries and sustainability: Recovery of high added-value products from natural sources. 134, 252–259.
37. COSTELLO, R. C. 2016. Advanced Phytonic Extraction. 08/16/2016 ed.: COSTELLO.
38. ABUBAKAR, A. R., HAQUE, M. J. J. O. P. & SCIENCES, B. 2020. Preparation of medicinal plants: Basic extraction and fractionation procedures for experimental purposes. 12, 1–10.
39. FENG, W., HAO, Z., LI, M. & ZHANG, J. 2019. Analytical Methods of Isolation and Identification. *In:* RAO, A. V., MANS, D. R. A. & RAO, L. (eds.) *Phytochemicals in Human Health.* London: IntechOpen.
40. HANDA, S. S., KHANUJA, S., LONGO, G. & RAKESH, D. D. 2008. Extraction technologies for medicinal and aromatic plants. *International centre for science and high technology*, 21–25.
41. AMEER, K., SHAHBAZ, H. M., KWON, J. H. J. C. R. I. F. S. & SAFETY, F. 2017. Green extraction methods for polyphenols from plant matrices and their byproducts: A review. 16, 295–315.
42. CHEMAT, F. & KHAN, M. K. J. U. S. 2011. Applications of ultrasound in food technology: Processing, preservation and extraction. 18, 813–835.
43. TIWARI, B. K. J. T. T. I. A. C. 2015. Ultrasound: A clean, green extraction technology. 71, 100–109.
44. VILKHU, K., MAWSON, R., SIMONS, L., BATES, D. J. I. F. S. & TECHNOLOGIES, E. 2008. Applications and opportunities for ultrasound assisted extraction in the food industry—A review. 9, 161–169.
45. ZAINAL-ABIDIN, M. H., HAYYAN, M., HAYYAN, A. & JAYAKUMAR, N. S. J. A. C. A. 2017. New horizons in the extraction of bioactive compounds using deep eutectic solvents: A review. 979, 1–23.
46. ISLAM, M., MALAKAR, S., RAO, M. V., KUMAR, N. & SAHU, J. K. 2023. Recent advancement in ultrasound-assisted novel technologies for the extraction of bioactive compounds from herbal plants: a review. *Food Sci Biotechnol,* 32, 1763-1782.
47. CHEMAT, F., ROMBAUT, N., MEULLEMIESTRE, A., TURK, M., PERINO, S., FABIANO-TIXIER, A.-S., ABERT-VIAN, M. J. I. F. S. & TECHNOLOGIES, E. 2017a. Review of green food processing techniques. Preservation, transformation, and extraction. 41, 357–377.
48. LEONELLI, C., VERONESI, P. & CRAVOTTO, G. 2012. Microwave-assisted extraction: An introduction to dielectric heating. *Microwave-assisted extraction for bioactive compounds: theory and practice.* Springer.
49. BODEA, I. M., CĂTUNESCU, G. M., DAVID, A. P., VIDICAN, R., POP, C. R., STĂNILĂ, A., ROTAR, A. M. & PALOP GÓMEZ, A. 2024. Optimization of microwave assisted extraction of bioactive compounds in oregano and lovage ethanolic extracts. *LWT,* 212, 116973.
50. NOUR, A. H., OLUWASEUN, A. R., NOUR, A. H., OMER, M. S. & AHMED, N. 2021. Microwave-assisted extraction of bioactive compounds. *Microwave Heating-Electromagnetic Fields Causing Thermal and Non-Thermal Effects.* IntechOpen.
51. TAPIA-QUIRÓS, P., GRANADOS, M., SENTELLAS, S. & SAURINA, J. 2023. Microwave-assisted extraction with natural deep eutectic solvents for polyphenol recovery from agrifood waste: Mature for scaling-up? *Science of The Total Environment,* 912, 168716.
52. DESTANDAU, E., MICHEL, T. & ELFAKIR, C. 2013. Microwave-Assisted Extraction.
53. MELIKOGLU, M. 2025. Microwave-assisted extraction: Recent advances in optimization, synergistic approaches, and applications for green chemistry. *Sustainable Chemistry for Climate Action,* 7, 100122.
54. ŞEN ARSLAN, H. 2024. Eco-friendly microwave-assisted extraction of fruit and vegetable peels demonstrates great biofunctional properties. 12, 8930–8938.

55. MOBASHERI, F., KHAJEH, M., GHAFFARI-MOGHADDAM, M., PIRI, J. & BOHLOOLI, M. 2025. Machine learning optimization of microwave-assisted extraction of phenolics and tannins from pomegranate peel. *Sci Rep,* 15, 19439.
56. LUCCHESI, M. E., CHEMAT, F., SMADJA, J. J. F. & JOURNAL, F. 2004. An original solvent free microwave extraction of essential oils from spices. 19, 134–138.
57. ARAUJO, A. R. T. S., PÉRINO, S., FERNANDEZ, X., CUNHA, C., RODRIGUES, M., RIBEIRO, M. P., JORDAO, L., SILVA, L. A., RODILLA, J., COUTINHO, P. & CHEMAT, F. 2021. Solvent-Free Microwave Extraction of Thymus mastichina Essential Oil: Influence on Their Chemical Composition and on the Antioxidant and Antimicrobial Activities. 14, 709.
58. CHEMAT, F., ROMBAUT, N., SICAIRE, A.-G., MEULLEMIESTRE, A., FABIANO-TIXIER, A.-S. & ABERT-VIAN, M. 2017b. Ultrasound assisted extraction of food and natural products. Mechanisms, techniques, combinations, protocols and applications. A review. *Ultrasonics Sonochemistry,* 34, 540–560.
59. XIE, Y., ZHANG, L., WU, W., XIE, J., GAO, B., XIAO, Y., ZHU, D. J. B. & BIOPROCESSING 2025b. Sustainable and green extraction of citrus peel essential oil using intermittent solvent-free microwave technology. 12, 48.
60. EPA, U. 1999. Determination of Volatile Organic Compounds in Ambient Air Using Active Sampling onto Sorbent Tubes. Compendium of Methods for the Determination of Toxic Organic Compounds in Ambient Air-Compendium Method TO-17. US Environmental Protection Agency (US EPA) Washington, DC, USA.
61. STANDARDIZATION, I. O. F. 2021. Indoor air-Determination of organic compounds (VVOC, VOC, SVOC) in indoor and test chamber air by active sampling on sorbent tubes, thermal desorption and gas chro-matography using MS or MS FID. 08/2021 ed. Geneva: ISO 16000-6:2021.
62. HELIN, A., HAKOLA, H. & HELLÉN, H. J. A. M. T. 2020. Optimisation of a thermal desorption–gas chromatography–mass spectrometry method for the analysis of monoterpenes, sesquiterpenes and diterpenes. 13, 3543–3560.
63. RODRÍGUEZ-NAVAS, C., FORTEZA, R. & CERDÀ, V. J. C. 2012. Use of thermal desorption–gas chromatography–mass spectrometry (TD–GC–MS) on identification of odorant emission focus by volatile organic compounds characterisation. 89, 1426–1436.
64. CHANG, C.-P., LIN, T.-C., LIN, Y.-W., HUA, Y.-C., CHU, W.-M., LIN, T.-Y., LIN, Y.-W. & WU, J.-D. J. A. O. O. H. 2016. Comparison between thermal desorption tubes and stainless steel canisters used for measuring volatile organic compounds in petrochemical factories. 60, 348–360.
65. BECKER, R., SCHOLZ, P., JUNG, C. & WEIDNER, S. J. A. M. 2023. Thermal desorption gas chromatography-mass spectrometry for investigating the thermal degradation of polyurethanes. 15, 1979–1984.
66. WINGFORS, H., MÖRÉN, L., WIKTELIUS, D. & MAGNUSSON, R. 2022. The potential of thermal desorption-GC/MS-based analytical methods for the unambiguous identification and quantification of perfluoroisobutene and carbonyl fluoride in air samples. *J Sep Sci,* 45, 2968-2976.
67. DA SILVA, R. P., ROCHA-SANTOS, T. A. & DUARTE, A. C. J. T. T. I. A. C. 2016. Supercritical fluid extraction of bioactive compounds. 76, 40–51.
68. MARTÍNEZ-ABAD, A., GIUMMARELLA, N., LAWOKO, M. & VILAPLANA, F. J. G. C. 2018. Differences in extractability under subcritical water reveal interconnected hemicellulose and lignin recalcitrance in birch hardwoods. 20, 2534–2546.
69. PURI, M., SHARMA, D. & BARROW, C. J. J. T. I. B. 2012. Enzyme-assisted extraction of bioactives from plants. 30, 37–44.
70. MUSSATTO, S. I., BALLESTEROS, L. F., MARTINS, S., TEIXEIRA, J. A. J. S. & TECHNOLOGY, P. 2011. Extraction of antioxidant phenolic compounds from spent coffee grounds. 83, 173–179.
71. AZMIR, J., ZAIDUL, I. S. M., RAHMAN, M. M., SHARIF, K. M., MOHAMED, A., SAHENA, F., JAHURUL, M. H. A., GHAFOOR, K., NORULAINI, N. A. & OMAR, A. K. J. J. O. F. E. 2013a. Techniques for extraction of bioactive compounds from plant materials: A review. 117, 426–436.

72. MARAN, J. P., SWATHI, K., JEEVITHA, P., JAYALAKSHMI, J. & ASHVINI, G. J. C. P. 2015. Microwave-assisted extraction of pectic polysaccharide from waste mango peel. 123, 67–71.
73. HERNÁNDEZ BECERRA, E., DE JESÚS PÉREZ LÓPEZ, E. & ZARTHA SOSSA, J. W. J. F. A. M. 2021. Recovery of biomolecules from agroindustry by solid-liquid enzyme-assisted extraction: A review. 14, 1744–1777.
74. BOROWSKA, M., PSZCZOŁA, J., PAWLAK, K., RUZIK, L., OMBUGADU, J. N., TRZASKOWSKI, M., JANKOWSKI, K. J. C., PHYSICOCHEMICAL, S. A. & ASPECTS, E. 2025. Microwave-Assisted Green Synthesis of Selenium Nanoparticles Using Citrus Extracts: Insights into Size-Controlled Formation and Surface Characteristics. 137516.
75. PACHAR, P. K. & SACHDEVA, H. J. I. C. C. 2025. Synthesis, characterization and catalytic activity of GO/ZnO nanocomposite for green synthesis of dihydropyrimidinones/thiones as potential antibacterial agents. 115026.
76. BICCHI, C., CORDERO, C. E. I., LIBERTO, E., SGORBINI, B. & RUBIOLO, P. 2012. Headspace sampling in flavor and fragrance field. *Comprehensive sampling and sample preparation.* Elsevier, Academic Press.
77. BHADANGE, Y. A., CARPENTER, J. & SAHARAN, V. K. 2024. A Comprehensive Review on Advanced Extraction Techniques for Retrieving Bioactive Components from Natural Sources. *ACS Omega,* 9, 31274–31297.
78. SAHA, M., MAKHIJA, R. & ASATI, V. 2025. Chapter 8 - Solvent free extraction procedures. *In:* LOCATELLI, M. & KAYA, S. (eds.) *Green Analytical Chemistry.* Elsevier.
79. XIE, Y., ZHANG, L., WU, W., XIE, J., GAO, B., XIAO, Y. & ZHU, D. 2025a. Sustainable and green extraction of citrus peel essential oil using intermittent solvent-free microwave technology. *Bioresour Bioprocess,* 12, 48.
80. CHERIYAN, B. V., KARUNAKAR, K. K., ANANDAKUMAR, R., MURUGATHIRUMAL, A. & KUMAR, A. S. 2025. Eco-friendly extraction technologies: A comprehensive review of modern green analytical methods. *Sustainable Chemistry for Climate Action,* 6, 100054.
81. BISHT, A., SAHU, S. C., KUMAR, A., MAQSOOD, S., BARWANT, M. M. & JAISWAL, S. G. 2025. Recent advances in conventional and innovative extraction techniques for recovery of high-added value compounds for food additives and nutraceuticals. *Food Physics,* 2, 100047.

Waste Management and Recycling in Analytical Laboratories

Vidhya K. Bhusari, Shital D. Godse, Sanjay R. Chaudhari, and Supriya G. Jagtap

Abstract The analytical laboratories produce a variety of waste products—organic solvents, effluents of various aqueous sources, solid polluted products, chemical substances residues—which can be harmful to nature and to humans, when not adequately managed. The chapter is a practical, all-encompassing overview of the modern waste management and recycling techniques applicable to the environment of an analytical laboratory manager, researcher, or student. Topics covered include regulatory frameworks, waste segregation, storage and transport, solvent recovery and recycling, biological and hazardous waste handling, on-site treatment options, monitoring and metrics, environmental management systems, economic considerations, and future directions such as green analytical chemistry and circular economy approaches. Practical tools include sample SOPs, checklists, audit templates, and a decision flowchart for waste handling.

Keywords Analytical laboratories · Waste management · Solvent recycling · Hazardous waste · Green analytical chemistry · ISO 14001 · Wastewater treatment · Waste minimization · Waste audits · Sustainability

V. K. Bhusari (✉)
Department of Pharmaceutical Chemistry, KJ's Educational Institute, Trinity College of Pharmacy, Pune, Maharashtra, India

S. D. Godse
School of Pharmacy, Vishwakarma University, Kondhwa, Pune, Maharashtra, India

S. R. Chaudhari
Department of Pharmacognosy, KJ's Educational Institute, Trinity College of Pharmacy, Pune, Maharashtra, India

S. G. Jagtap
Department of Pharmacognosy, Sinhgad Technical Education Society's, Smt. Kashibai Navale College of Pharmacy, Kondhwa (Bk.), Pune, Maharashtra, India

V. Bhusari and M. Ghante (eds.), *Green Analytical Chemistry in India*,
https://doi.org/10.1007/978-3-032-24564-9_10

1 Introduction

The analytical laboratories are necessary in the pharmaceutical, environmental monitoring, food safety, and clinical diagnostics. They generate high-value data that underpin quality assurance, regulatory compliance, and innovation across industries. However, the processes that produce this data also generate diverse categories of waste, ranging from hazardous solvents to infectious biological residues. With growing emphasis on sustainability and environmental stewardship, laboratory waste management has emerged as both a scientific and ethical imperative [1, 2].

Laboratory waste is distinct from general municipal waste because it often contains toxic, flammable, reactive, or infectious materials. For instance, routine use of solvents such as acetonitrile, methanol, and dichloromethane in chromatographic methods contributes significantly to hazardous waste volumes. Similarly, heavy metals from spectroscopy, acids and alkalis from titrations, and pharmaceutical actives from stability studies all create streams that require careful handling. The mismanagement of such waste not only endangers laboratory staff but also poses risks to health and the environment.

Globally, environmental agencies and professional societies have raised concerns about the growing footprint of research and testing facilities. The pharmaceutical industry, in particular, is under scrutiny for solvent usage, energy consumption, and hazardous by-products. In this context, analytical laboratories, though smaller in scale compared to manufacturing units, become important focal points for implementing sustainable waste minimization and recycling strategies [3–5].

Waste management in laboratories is based on the general hierarchy of waste: prevention, reduction, reuse, recycling, energy recovery, and safe disposal. This is a point of hierarchy that the best way to deal with waste is not to deal with it when it is in its end life but avoid it at its origin. This requires redesigning of methods, miniaturizing assays, and green analytical chemistry (GAC) practices, which are required of analytical scientists. As an illustration, the substitution of the traditional HPLC protocols which consume liters of solvent in a day to ultra-high performance liquid chromatography (UHPLC) with reduced solvent usage is a practical sustainability enhancement [6, 7].

Recycling technologies also play a vital role in reducing laboratory waste burdens. Solvent recovery through distillation, re-purification of acids, and regeneration of chromatographic columns extends the usable life of resources while cutting disposal costs. Advances in closed-loop recycling systems now allow laboratories to achieve near-zero waste discharge for certain operations. Such practices not only comply with regulatory expectations but also improve economic efficiency, offering laboratories dual benefits of compliance and cost savings.

Waste segregation is another foundational element of laboratory sustainability. Proper separation of hazardous, non-hazardous, recyclable, and biomedical waste ensures that each stream is treated with the appropriate method, reducing contamination and maximizing recovery. The use of color-coded bins, clear labeling, and training of laboratory personnel are low-cost yet high-impact strategies to embed sustainability into daily operations [8].

The regulatory landscape governing laboratory waste is expanding globally. Agencies such as the U.S. Environmental Protection Agency (EPA), the World Health Organization (WHO), and national pollution control boards mandate strict protocols for handling chemical, biomedical, and e-waste. International standards such as ISO 14001 further emphasize environmental management systems, encouraging laboratories to integrate sustainability into their quality frameworks. Non-compliance carries significant penalties, reputational damage, and potential risks to accreditation [9, 10].

At the same time, societal and institutional expectations are shifting toward greener research environments. Universities, research centers, and industrial labs are increasingly judged on their ability to balance scientific productivity with environmental responsibility. Funding agencies and accreditation boards now reward laboratories that demonstrate tangible sustainability initiatives, from reducing solvent usage to implementing recycling programs.

Education and training are equally vital in shaping laboratory culture around waste management. Laboratory staff, students, and researchers must be sensitized to the risks of poor waste practices and trained in the use of personal protective equipment (PPE), spill management, and emergency protocols. A culture of safety and sustainability ensures that laboratory waste is not seen merely as an afterthought but as an integral part of experimental design and execution [11–13].

This chapter explores the full spectrum of waste management and recycling strategies relevant to analytical laboratories. It highlights the types of waste generated, principles of waste minimization, best practices for segregation, treatment options, and innovative recycling technologies. The discussion also incorporates regulatory frameworks, green analytical chemistry approaches, and case studies that illustrate practical applications. By synthesizing technical, regulatory, and operational perspectives, this chapter aims to serve as a comprehensive resource for laboratory managers, researchers, and policymakers seeking to align analytical practice with sustainability goals.

In doing so, it underscores a fundamental shift in laboratory science: from a focus on precision and accuracy alone to a broader vision of responsible innovation that safeguards both data integrity and planetary health.

2 Types of Laboratory Waste

Laboratory operations generate a wide range of waste types. Understanding and classifying these streams is the first step toward effective management. Waste types can broadly be grouped into chemical, biological, radioactive, electronic, and general non-hazardous waste, depicted in Fig. 1. Each category requires specific handling, segregation, and disposal measures.

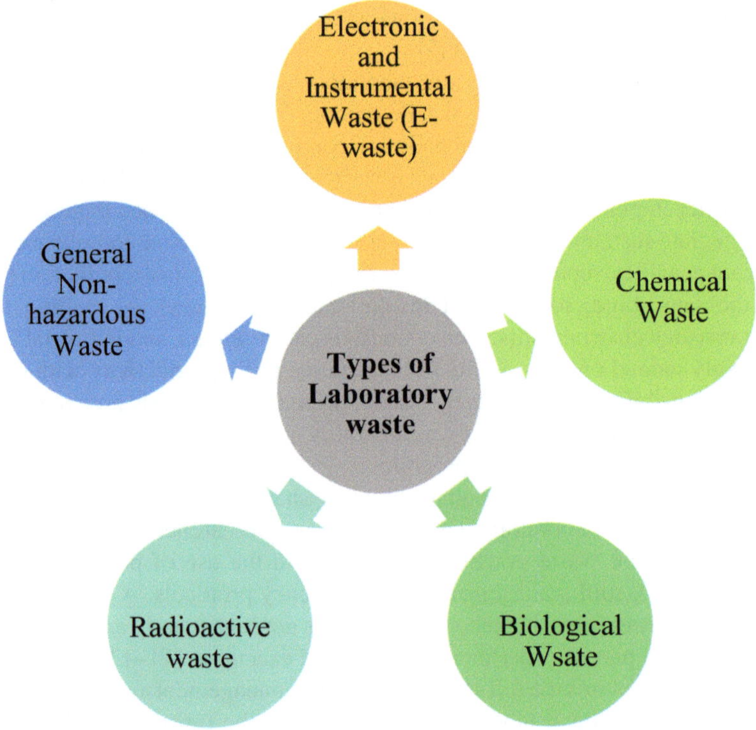

Fig. 1 Types of laboratory waste

2.1 *Chemical Waste*

Chemical waste is one of the most common types produced in analytical laboratories. It includes solvents, acids, bases, reagents, and samples containing residual chemicals. Organic solvents such as methanol, acetonitrile, hexane, and dichloromethane are widely used in chromatographic separations, extraction procedures, and synthesis reactions. These solvents are often volatile, flammable, and toxic, requiring proper containment and storage.

Inorganic acids (e.g., HCl, nitric acid, H_2SO_4) and bases (e.g., NaOH, KOH) are also commonly generated. These chemicals pose corrosive hazards and can react dangerously if mixed improperly. Residual reagents from titrations, derivatizations, or analytical standards fall under this category and must be collected separately to prevent incompatibilities.

Chemical waste may also include heavy metals, dyes, and other analytical reagents. Laboratories conducting trace metal analysis, spectrophotometry, or electrochemical studies often generate aqueous streams containing toxic metals such as lead, cadmium, and mercury. These must not enter the municipal wastewater system without proper treatment [14–16].

2.2 Biological Waste

Biological or biomedical waste is a major category of laboratory waste generated predominantly in clinical, microbiological, molecular biology, and biochemistry laboratories, where handling of infectious agents and human or animal specimens is routine. This waste stream encompasses a wide range of potentially infectious materials, including cultures, blood and tissue samples, agar plates, contaminated pipette tips, syringes, gloves, gowns, and other disposable consumables that have been exposed to pathogens. In addition, genetically modified organisms (GMOs), viral vectors, and recombinant DNA samples from research laboratories may also be classified under biomedical waste.

If improperly managed, biohazardous waste poses significant risks not only to laboratory personnel but also to waste handlers, the community, and the environment. Exposure can result in the spread of infectious diseases, needle-stick injuries, or accidental release of pathogens into the environment. Consequently, stringent biosafety protocols are mandated in most countries, guided by frameworks such as the WHO Laboratory Biosafety Manual, the U.S. Centers for Disease Control and Prevention (CDC) Biosafety Levels (BSL-1 to BSL-4), and national biomedical waste management rules.

The first and most critical step in management is segregation at the point of generation, wherein biomedical waste must be collected separately from general laboratory waste using color-coded, puncture-proof, and leak-proof containers labeled with the biohazard symbol. Common treatment methods include:

- Autoclaving (steam sterilization at 121–134 °C under pressure): The most widely accepted method for decontaminating infectious solids and liquids.
- Chemical Disinfection: Using agents such as sodium hypochlorite, phenolics, or glutaraldehyde for liquid waste and surface decontamination.
- Incineration or Plasma Pyrolysis: Applied to pathological waste and certain contaminated plastics where sterilization is not feasible.
- Shredding After Sterilization: To render sharps and plastics unrecognizable and reduce risk of reuse.

In addition to treatment, laboratories are required to maintain detailed records of waste generation, sterilization logs, and disposal certificates as part of regulatory compliance and accreditation. For instance, in India, the Biomedical Waste Management Rules (2016, amended 2018, 2019) mandate record keeping and reporting to State Pollution Control Boards. Similarly, international laboratories must follow Occupational Safety and Health Administration (OSHA) standards in the U.S. or equivalent regulatory agencies in other countries.

Failure to comply with biomedical waste management regulations can result in legal penalties, revocation of laboratory accreditation, reputational damage, and serious biosafety hazards. Therefore, implementing robust waste management SOPs, staff training, and periodic audits is essential for ensuring a culture of biosafety and sustainability in laboratory operations [17, 18].

2.3 *Radioactive Waste*

Laboratories that utilize isotopes for research, tracer studies, medical diagnostics, or analytical purposes inevitably generate radioactive waste, which is considered one of the most hazardous forms of laboratory waste. Such waste typically includes contaminated consumables such as glassware, pipettes, vials, gloves, and absorbent paper, as well as liquid solutions containing radionuclides used in tracer assays or scintillation counting. Depending on the isotope, the waste may be classified as short-lived (e.g., ^32P, ^35S) or long-lived (e.g., ^14C, ^3H), and each category requires a distinct management approach.

Radioactive waste should be monitored, shielded, and segregated to avoid radiation and pollution to the environment. Common strategies include decay-in-storage (DIS), where short-lived isotopes are stored securely until their activity falls below permissible regulatory limits, followed by disposal as regular laboratory waste. For liquid waste, containment in shielded storage vessels with appropriate labeling and secondary containment is essential to avoid leaks and spills.

It is necessary that personnel who handle radioactive materials receive specific training in radiological safety, and adhere to the principle of as low as reasonably achievable (ALARA) to keep the exposure to a minimum. These involve the use of personal protective gear (PPE), including lab coats, two pairs of gloves, face masks, and dosimeters, and also engineering solutions such as fume hoods or shielded enclosures. This includes the use of personal protective equipment (PPE) such as lab coats, double gloves, face shields, and dosimeters, as well as engineering controls like fume hoods or shielded enclosures. Continuous monitoring with Geiger-Müller counters, scintillation detectors, or dosimetry badges ensures that exposure levels remain within legally prescribed limits.

Furthermore, compliance with national regulations and licensing requirements is mandatory. In the United States, laboratories must adhere to the standards of the Nuclear Regulatory Commission (NRC) or Agreement State programs, whereas in India, oversight is provided by the Atomic Energy Regulatory Board (AERB). International agencies such as the International Atomic Energy Agency (IAEA) and the World Health Organization (WHO) also publish guidelines for the safe handling and disposal of radioactive materials. Non-compliance can result in severe legal consequences, revocation of licenses, financial penalties, and significant risks to both laboratory accreditation and public health [19, 20].

2.4 *Electronic and Instrumental Waste (E-Waste)*

E-waste from laboratories includes outdated instruments, computers, batteries, fluorescent lamps, and other electronic components. Many of these items contain heavy metals or hazardous chemicals that, if improperly discarded, can leach into the environment.

Certified e-waste recyclers and manufacturers' take-back programs are recommended for disposal. Proper segregation and collection at dedicated e-waste stations reduce contamination of other laboratory waste streams [21].

2.5 General Non-hazardous Waste

General laboratory waste comprises paper, cardboard, plastic packaging, and other materials that are not chemically or biologically contaminated. While relatively safe, these items should still be collected for recycling when possible. Segregating non-hazardous waste from hazardous streams reduces disposal costs and increases recycling efficiency [22].

3 Principles of Waste Minimization

Effective waste minimization in analytical laboratories is guided by several core principles that aim to reduce environmental impact, improve safety, and optimize costs.

3.1 Avoidance and Source Reduction

The first step in waste minimization is to prevent the generation of waste at the source. This can be achieved by designing experiments that use smaller sample sizes, selecting analytical methods that require less reagent consumption, and employing microscale techniques. For instance, using microplate assays or miniaturized extraction methods can drastically reduce solvent and reagent use [23].

3.2 Segregation of Waste Streams

Segregating waste at the point of generation prevents cross-contamination, reduces disposal hazards, and facilitates recycling. Hazardous, non-hazardous, biological, radioactive, and e-waste should all be segregated using color-coded containers and clearly labeled bins [24].

3.3 Substitution and Green Chemistry

Replacing hazardous chemicals with less toxic alternatives is an essential principle of waste minimization. Green analytical chemistry encourages the use of water-based solvents, ethanol, or ethyl lactate instead of halogenated solvents. Choosing environmentally friendly reagents and avoiding persistent chemicals reduces both immediate and long-term environmental risks [25].

3.4 Recycling and Recovery

Where feasible, laboratories should recover and recycle materials. Solvent recovery via distillation, reuse of glassware, and regeneration of chromatographic columns help to conserve resources. Recovered solvents must be analyzed for purity before reuse to ensure that analytical accuracy is maintained [26].

3.5 Efficient Inventory Management

Proper inventory control minimizes waste due to expired chemicals or reagents. Laboratories should purchase chemicals in quantities that match actual usage, maintain a first-in-first-out system, and track expiration dates. Sharing reagents within departments or research groups can also reduce waste accumulation [27].

3.6 Training and Awareness

A well-informed workforce is central to waste minimization. Safe handling, segregation, spill response, and recycling procedures should be trained to personnel. Promoting a laboratory culture focused on sustainability encourages staff to consider environmental impact when designing experiments.

3.7 Process Optimization

Optimizing laboratory processes can reduce waste generation. This includes selecting analytical methods with higher efficiency, using UHPLC instead of conventional HPLC, automating sample preparation, and reducing solvent-intensive cleaning cycles.

3.8 *Life Cycle Assessment*

Applying life cycle thinking to laboratory operations helps identify waste hotspots and informs decisions about method selection, energy use, and material substitution. Life cycle assessment (LCA) is based on the overall environmental performance of raw material purchase to disposal, and it directs more environmentally friendly lab practices.

3.9 *Energy and Water Efficiency*

Conserving energy and water reduces indirect waste. Optimizing fume hood usage, recycling cooling water, and maintaining energy-efficient instruments are key strategies. Reduced energy consumption lowers carbon footprint and can prevent waste from byproduct generation in chemical processes.

3.10 *Documentation and Monitoring*

Maintaining accurate records of waste generation, disposal methods, and recycling helps laboratories monitor working ability and to know areas for improvement. Important points may include volume of hazardous waste per experiment, percent of solvent recovered, and compliance with regulatory thresholds [1, 28–30].

4 Waste Segregation and Handling in Analytical Laboratories

Laboratory waste management begins with proper segregation and handling, which is important to ensure safety, compliance with regulations, and environmental protection. Inefficient or improper handling of laboratory waste can lead to chemical accidents, biological contamination, radioactive exposure, and environmental pollution. Effective segregation minimizes these risks and facilitates recycling and safe disposal.

4.1 *Importance of Waste Segregation*

Segregation is the practice of separating waste at the point of generation based on its type and hazard potential. The main benefits include:

1. Prevention of Chemical Reactions: Mixing incompatible chemicals, such as acids with bases or oxidizers with organics, can result in explosions or release of toxic gases^3^.
2. Enhanced Recycling: Properly separated waste, like solvents or glassware, can be reused or recycled efficiently.
3. Regulatory Compliance: Environmental authorities mandate segregation of hazardous, biological, and radioactive waste. Non-compliance can lead to legal penalties, fines, or suspension of laboratory operations.
4. Cost Efficiency: Segregating hazardous from non-hazardous waste reduces disposal costs since specialized treatment is only applied to dangerous streams [31–33].

Figure 2 illustrates color-coded containers (e.g., yellow for biological, red for sharps, black for chemical, labeled bins for recyclable glass).

Fig. 2 Laboratory waste segregation diagram

4.2 *Categories of Laboratory Waste and Segregation Guidelines*

4.2.1 Chemical Waste

Chemical waste includes solvents, acids, bases, reactive compounds, and analytical reagents. Key guidelines:

- Segregate by Chemical Class: Organic solvents, halogenated solvents, acids, and bases must be collected separately.
- Use Compatible Containers: Glass or HDPE containers with secure caps are recommended to prevent leaks or reactions. Containers should never exceed recommended volume limits.
- Label Clearly: Include the chemical name, hazard classification, and accumulation start date [34, 35].

Example: Methanol and chloroform should never be combined in the same container due to reactivity and disposal restrictions.

4.2.2 Biological Waste

Biological waste arises from clinical, microbiology, or biotechnology labs. It includes cultures, blood, tissues, contaminated PPE, and disposable lab materials.

- Red or Biohazard-Labeled Containers: Autoclavable bags or sharps containers are standard.
- Decontamination: Before disposal, biological waste should be autoclaved or chemically disinfected.
- Segregation by Hazard: Pathogenic and non-pathogenic biological waste should be separated to avoid unnecessary treatment of non-infectious materials.

Example: Contaminated pipette tips from microbial cultures should be placed in dedicated biohazard bags, while gloves from routine chemical handling go in chemical waste containers [36].

4.2.3 Radioactive Waste

Radioactive wastes consist of materials that have been contaminated with radionuclides, e.g., vials, pipettes, or liquid wastes, etc.

- Shielded Containers: Lead or specialized plastic containers labeled with radiation hazard symbols.
- Decay Storage: Short-lived isotopes can be stored until radioactivity falls below regulatory limits.

- Segregation by Isotope and Half-Life: Avoid mixing isotopes with significantly different decay rates [37].

Example: Tritium-containing liquid waste is collected separately from iodine-131 contaminated glassware due to differences in half-life and disposal requirements.

4.2.4 Sharps Waste

Needles, scalpels, broken glass, and other object capable of puncturing skin.

- Puncture-Resistant Containers: Red sharps boxes or rigid plastic containers.
- Never Recap or Bend Needles: This prevents accidental injury and cross-contamination.
- Segregate by Type: Glass vials vs. hypodermic needles may require separate disposal routes [38].

4.2.5 General Waste

Non-hazardous materials like paper, cardboard, and clean plastics can be collected in regular bins for recycling or composting. It should never be mixed with chemical, biological, or radioactive waste to prevent contamination. The classification of laboratory waste and recommended disposal methods are summarized in Table 1.

4.3 Handling Procedures

Proper handling includes careful collection, labeling, storage, and transportation:

1. Designated Storage Areas: Separate locations for chemical, biological, and radioactive waste prevent accidental mixing and facilitate monitoring.

Table 1 Types of laboratory waste and disposal methods

Sr. no	Waste type	Examples	Container/label requirement	Disposal/treatment method
1	Chemical waste	Solvents, acids, heavy metals	Labeled chemical-resistant container	Neutralization, solvent recovery, incineration
2	Biological waste	Cultures, tissues, sharps	Biohazard bags, sharps container	Autoclaving, incineration
3	Radioactive waste	Contaminated gloves, vials, solutions	Shielded and labeled container	Decay-in-storage, specialized disposal
4	General/ recyclable	Glassware, plastics, packaging	Recyclable bins (color-coded)	Recycling, reuse, landfill if non-recyclable

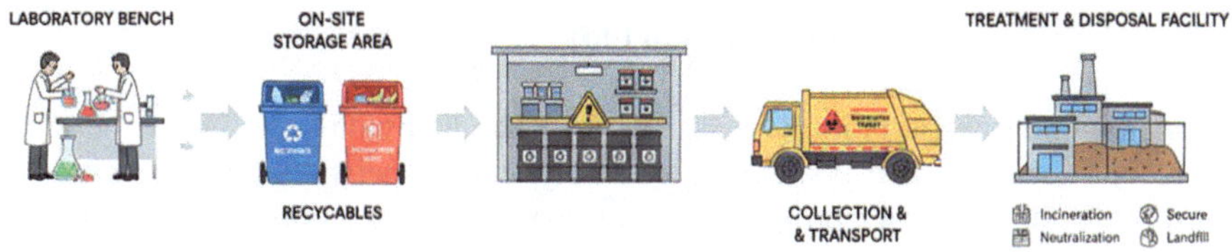

Fig. 3 Waste collection and transportation pathway

2. Secondary Containment: Use trays or bins to catch spills from primary containers.
3. Personal Protective Equipment (PPE) and other safety measures like gloves, lab coats, goggles, and masks are mandatory when handling waste.
4. Spill Response: Labs should have emergency spill kits and trained personnel to address accidental releases promptly. The stepwise process of waste collection and transportation is shown in Fig. 3.

4.3.1 Labeling and Documentation [39]

The first step in waste minimization is avoidance and source reduction, which focuses on preventing waste generation at its origin. This can be achieved by designing experiments that use smaller sample sizes, selecting analytical methods that require less reagent consumption, and employing microscale techniques. For example, microplate assays and miniaturized extraction methods can significantly reduce the amount of solvents and reagents used. Equally important is proper labeling and documentation of waste containers. Each container must clearly indicate the waste type (chemical, biological, or radioactive), appropriate hazard symbols, a detailed description of the contents, the date of first accumulation, and the initials of the responsible staff member.

Documentation ensures traceability and compliance during audits or inspections. Many labs maintain electronic logs for real-time monitoring of waste accumulation. The mandatory parameters for labeling laboratory waste containers are presented in Table 2.

Table 2 Labeling requirements for waste containers

Sr. no	Parameter	Requirement example
1	Waste type	Chemical, biological, radioactive
2	Hazard symbols	Flammable, toxic, biohazard, radioactive
3	Contents	Full description (e.g., “acetone waste”)
4	Date of first accumulation	DD/MM/YYYY
5	Responsible staff initials	e.g., “AB”

4.3.2 Collection and Transportation [40]

Collection and transportation of waste must be handled with strict safety measures to prevent accidents and contamination. Only trained personnel are permitted to transport waste, ensuring proper handling and compliance with safety protocols. Depending on the type of waste, specific requirements apply: chemical and biological waste may need to be moved using leak-proof trolleys to avoid spills, while radioactive waste must follow strict transport regulations that include appropriate shielding, clear labelling, and careful route planning to minimize risks and ensure regulatory compliance.

4.3.3 Integration with Recycling Programs [41]

Integration with recycling programs is an important constituent of sustainable waste management in laboratories and research facilities. Solvent recovery, for instance, allows organic solvents to be distilled and reused, reducing both costs and environmental impact. Similarly, clean glassware, such as containers and bottles, can be reused within the lab or sent to certified recyclers for proper processing. In addition, certain non-recyclable chemical wastes can undergo controlled incineration to enable energy recovery, ensuring that even waste materials contribute to resource efficiency and minimizing their overall environmental footprint.

5 Treatment and Recycling Strategies in Analytical Laboratories

Laboratory operations generate a variety of hazardous wastes, including acids, bases, organic solvents, heavy metals, and contaminated consumables. To reduce environmental impact and improve sustainability, laboratories employ treatment and recycling strategies that convert waste into less harmful forms or recover valuable resources. These strategies form an integral part of green analytical chemistry and modern laboratory management practices.

Chemical Neutralization: It is a primary method for treating acidic and basic waste streams. Neutralization converts reactive substances into less hazardous compounds through controlled chemical reactions.

Acidic Waste Neutralization: Strong acids such as HCl, H_2SO_4, or nitric acid are neutralized using bases like sodium NaOH, KOH, $CaCO_3$, or $NaHCO_3$.

Basic Waste Neutralization: Strong bases such as NaOH or KOH are neutralized using acids such as HCl or acetic acid [42, 43].

Procedure:

1. Perform the reaction in a ventilated fume hood to avoid inhalation of vapors.

2. Add the neutralizing agent gradually to the waste while stirring continuously to prevent violent reactions.
3. Monitor the pH of the solution until near-neutrality (pH 6–8) is achieved.
4. Neutralized liquid can often be disposed of via municipal wastewater systems if permitted by regulations.

Advantages:

- Cost-effective and simple.
- Reduces corrosivity and immediate hazard.

Limitations:

- Does not remove heavy metals or organic pollutants.
- Requires careful handling to prevent exothermic accidents.

Example: Neutralizing a solution of HCl generated from titration experiments with $NaHCO_3$ before disposal.

Solvent Recovery: Organic solvents are useful in analytical laboratories for chromatography, extraction, and sample preparation. Solvent recovery minimizes environmental impact, reduces purchase costs, and promotes sustainability [44, 45].

Methods of Solvent Recovery:

- Evaporation: Volatile solvents are evaporated under controlled conditions, and vapors are condensed for reuse.
- Distillation: The solvents are separated by the fractional or simple distillation, which is dependent on the boiling points.
- Closed-Loop Recovery Systems: Automated systems capture, purify, and recycle solvents with minimal human exposure.

Key Considerations:

- Segregate solvents by chemical class (halogenated vs. non-halogenated).
- Check purity before reuse to avoid interference in analytical procedures.
- Use drying agents or filtration to remove particulates and water.

Example: Recovered acetonitrile from HPLC runs can be distilled and reused for subsequent experiments.

Advantages:

- Reduces hazardous waste volume.
- Cost-saving through reuse.
- Environmentally friendly.

Limitations:

- Requires equipment investment (distillation units or recovery systems).
- Requires monitoring of solvent purity.

Distillation: Distillation is efficient for recovering solvents from waste streams, particularly volatile organic compounds (VOCs). It relies on differences in boiling points to separate components [46].

Types:

- Simple distillation: For single solvents or widely separated boiling points.
- Fractional distillation: For solvent mixtures with similar boiling points.
- Vacuum distillation: Reduces boiling points, minimizing thermal degradation.

Applications in Laboratories:

- Recovery of ethanol, methanol, acetone, and other common solvents.
- Purification of solvents contaminated with small amounts of other chemicals.

Advantages:

- Efficient recovery of high-purity solvents.
- Reduces hazardous waste disposal volume.

Limitations:

- Energy-intensive for large volumes.
- Not suitable for highly reactive or thermally unstable chemicals.

Example: Fractional distillation of a solvent mixture from organic synthesis reactions, allowing separation and reuse of acetone and hexane.

Catalytic Treatment: Catalysis can convert hazardous chemicals into less harmful substances or facilitate the recovery of valuable intermediates. Common catalytic approaches include:

Catalytic oxidation: Organic solvents or hazardous organics are oxidized into CO_2, water, or less toxic products using catalysts such as TiO_2, MnO_2, or transition metal salts.

Reduction reactions: Catalytic reduction converts nitro compounds, halogenated organics, or nitrates into less harmful substances.

Enzymatic catalysis: Biocatalysts can degrade specific organic pollutants in biological or chemical waste.

Advantages:

- Reduces the toxicity of organic wastes.
- Can recover intermediates or usable products.

Limitations:

- Requires proper selection of catalysts.
- May generate secondary waste streams requiring further treatment^7^.

Example: Catalytic decomposition of residual nitroaromatic waste in organic synthesis laboratories [47].

Energy Recovery: Energy recovery is the burning of the hazardous waste that cannot be recycled to produce heat or electricity. Modern incinerators operate at high temperatures (800–1200 °C) to ensure complete combustion of organic materials, including plastics, paper, and solvent-soaked wipes.

Process:

1. Waste is segregated and fed into a controlled incineration chamber.
2. Combustion converts organic materials into CO_2, water, and ash.
3. Heat energy is captured to produce steam or electricity in waste-to-energy facilities.

Advantages:

- Reduces waste volume significantly.
- Recovers energy from materials that cannot be recycled.
- Minimizes landfill disposal.

Limitations:

- Requires sophisticated emission controls for NO_n, SO_n, and VOCs.
- High capital and operational costs.

Example: Solvent-soaked laboratory wipes and contaminated plastics are incinerated in an industrial laboratory waste incinerator, with energy recovery to power laboratory heating systems [48].

Integration of Treatment Strategies.

A multi-step approach ensures maximum safety, efficiency, and resource recovery:

1. Segregate waste streams at source (chemical, biological, radioactive).
2. Neutralize acids and bases before further treatment.
3. Recover solvents via distillation or closed-loop systems.
4. Apply catalytic treatments for hazardous organics.
5. Dispose of non-recyclable residues via controlled incineration or energy recovery.

Advantages:

- Optimizes resource use.
- Minimizes environmental footprint.
- Ensures regulatory compliance and laboratory safety.

Summary: The treatment and recycling of laboratory waste are essential for implementing green laboratory practices. Integrating neutralization, solvent recovery, distillation, catalytic treatments, and energy recovery reduces environmental risks, promotes sustainability, and enhances safety for personnel and the community.

6 Green Analytical Chemistry Approaches [49–51]

GAC has emerged as a critical framework for laboratories aiming to minimize environmental impact while maintaining analytical efficiency, reliability, and accuracy. It builds on the broader principles of green chemistry but applies them specifically to analytical sciences, where the consumption of solvents, reagents, and energy can be substantial. One of the core aspects of GAC is the reduction or complete elimination of toxic and hazardous reagents, replacing them with safer alternatives that pose minimal risk to health and the environment. For instance, researchers are increasingly using water or bio-derived solvents in place of acetonitrile, methanol, or dichloromethane, which are not only costly to dispose of but also contribute to hazardous waste generation.

Another important dimension of GAC is the miniaturization of analytical procedures. Techniques such as microextraction, lab-on-a-chip devices, and microfluidics allow laboratories to perform reliable analysis with minimal quantities of reagents and samples. This significantly lowers waste volumes and reduces the environmental footprint of laboratory operations. Coupled with advancements in HPLC & UHPLC, miniaturization has led to reduced solvent consumption, faster analysis times, and increased sensitivity. Similarly, capillary electrophoresis, which requires only nanoliters of sample and electrolyte, has gained prominence as a green alternative to traditional separation techniques.

Energy efficiency also forms a vital pillar of GAC. Analytical instruments such as mass spectrometers, gas chromatographs, and spectrophotometers can consume significant amounts of electricity during continuous operation. Laboratories adopting GAC principles are moving toward instruments with lower energy requirements, incorporating automation, and designing workflows that minimize idle instrument time. In addition, coupling renewable energy systems like solar panels with laboratory operations has been recommended to decrease C footprint associated with analytical work.

Method redesign and simplification further enhance the sustainability of analytical science. This includes direct analysis of samples without extensive pretreatment, thereby avoiding the use of excessive reagents, filters, or extraction solvents. For example, solid-phase microextraction (SPME) enables analyte isolation and concentration in a single step without generating liquid waste. Direct mass spectrometry analysis of biological samples without chromatography is another emerging area, aligning with waste-free principles. By reducing the number of steps, not only is waste minimized but also time, cost, and energy consumption are optimized.

The application of renewable or recyclable materials in sample preparation also supports the GAC philosophy. Laboratories are experimenting with biodegradable sorbents, natural fibers, and polymeric supports that can be reused or safely disposed of without environmental harm. Similarly, the use of greener derivatization agents, biodegradable chelating compounds, and safer catalysts reduces the toxicological impact of analytical chemistry.

Lastly, GAC focuses on lifecycle thinking, i.e., methods used in a laboratory are not just considered based on their performance but based on their total environmental impact during the production of reagents and their disposal. The measurement of the environmental friendliness of analytical procedures has been devised as metrics like the Analytical Eco-Scale, National Environmental Methods Index (NEMI) criteria and the Green Analytical Procedure Index (GAPI). These tools help researchers compare methods objectively and select approaches that balance analytical reliability with sustainability.

In summary, Green Analytical Chemistry promotes a paradigm shift in laboratory practice by integrating environmental considerations into every stage of analytical method development and application. By adopting safer solvents, miniaturized techniques, energy-efficient instruments, simplified workflows, and sustainable materials, analytical laboratories can significantly reduce waste generation and environmental impact while continuing to produce accurate and high-quality data. This approach not only aligns with global sustainability goals but also prepares laboratories for future regulatory requirements and societal expectations in environmental stewardship.

7 Regulatory Frameworks and Compliance (EPA, WHO, ISO 14001, Local Rules) [52–55]

The effective management of laboratory waste and the promotion of recycling practices in analytical laboratories depend heavily on adherence to established regulatory frameworks and international standards. The frameworks do not only safeguard human health and environment but also present laboratories with systematic principles on how to handle, store, treat, and dispose various waste streams. Regulatory compliance ensures that laboratory practices are sustainable, legally acceptable, and aligned with global environmental protection goals.

In the US, the main regulatory body that provides control over hazardous waste management of laboratories is the Environmental Protection Agency (EPA). Resource Conservation and Recovery Act (RCRA) establishes the legal basis of classification, treatment, storage, and disposal of hazardous waste. Within this framework, laboratories should determine the presence of their wastes in the list of hazardous wastes or hazardous properties (i.e., ignitability, corrosivity, reactivity, and toxicity). The EPA also introduced the Academic Laboratories Rule (Subpart K), which offers specific provisions for universities, teaching hospitals, and non-profit research institutions, granting flexibility while ensuring compliance with hazardous waste requirements (1,2).

Globally, the World Health Organization (WHO) provides comprehensive guidance on health-care and laboratory waste management. WHO emphasizes waste minimization, segregation at the source, safe storage, and environmentally sound treatment technologies. Their framework highlights the risks of improper handling, particularly with infectious and chemical waste, which can endanger laboratory staff, patients, and communities (3). WHO guidelines are essential in developing countries where formalized waste management systems may be limited, offering adaptable models that prioritize low-cost, safe, and sustainable practices.

Simultaneously, the International Organization of Standardization (ISO) has come up with standards, including the ISO 14001, which offers an internationally recognized system of Environmental Management Systems (EMS). Laboratories adopting ISO 14001 commit to systematically identifying their environmental impacts, setting waste reduction and recycling goals, and continuously improving their performance (4). ISO certification not only demonstrates compliance with environmental legislation but also enhances institutional reputation, promotes resource efficiency, and ensures that waste management strategies are regularly audited and improved. Laboratories that integrate ISO 14001 into their operations often report reduced operating costs, improved stakeholder trust, and better preparedness for evolving regulatory requirements.

At the national and local levels, specific waste management laws and regulations provide additional layers of compliance. For instance, in India, the Central Pollution Control Board (CPCB) under the Ministry of Environment, Forest and Climate Change enforces the Biomedical Waste Management Rules (2016, amended 2018) and Hazardous Waste Management Rules (2016). These regulations mandate waste segregation, barcoding of biomedical waste bags, periodic audits, and authorization for waste disposal facilities (5). Similarly, the European Union enforces the Waste Framework Directive (2008/98/EC), which applies the waste hierarchy of prevention, reuse, recycling, recovery, and disposal across laboratories and industries (6). Local municipal authorities may also impose additional requirements, such as safe transport permits, reporting obligations, or restrictions on the use of incineration technologies.

Collectively, these regulatory frameworks stress the importance of a cradle-to-grave approach, in which laboratories remain responsible for their waste from the point of generation to final disposal. Compliance requires laboratories to implement clear standard operating procedures (SOPs), provide regular staff training, maintain accurate documentation, and establish relationships with licensed waste treatment and disposal providers. Increasingly, regulations also encourage or mandate recycling and resource recovery, shifting laboratories from a waste disposal mindset to one of circular economy integration.

Ultimately, compliance with EPA, WHO, ISO 14001, and national or local waste management regulations ensures that laboratories function responsibly while minimizing environmental risks. By embedding these frameworks into daily operations, analytical laboratories can achieve a balance between scientific innovation and environmental stewardship, while also safeguarding their legal standing and institutional credibility.

8 SOP Templates and Best Practices

Standard operating procedures (SOPs) are the cornerstone of efficient and safe waste management in analytical laboratories. Laboratories generate a wide range of wastes—including solvents, acids, alkalis, reagents, heavy metals, biological residues, and plastic consumables—which, if not managed systematically, can pose risks to laboratory personnel, the environment, and regulatory compliance. SOPs provide a stepwise, standardized, and legally defensible approach to handling such wastes, ensuring uniformity across personnel and laboratory sections.

8.1 Importance of SOPs in Waste Management

SOPs act as a bridge between regulatory requirements and practical laboratory operations. For instance, while guidelines from agencies such as the Environmental Protection Agency (EPA), World Health Organization (WHO), and ISO 14001 prescribe general waste management frameworks, laboratory-specific SOPs adapt these to real working conditions. In pharmaceutical analytical laboratories, for example, SOPs govern solvent recovery units, management of expired reagents, and disposal of HPLC mobile phases. In environmental laboratories, SOPs guide the segregation of mercury-containing samples from other chemical wastes to prevent cross-contamination [56–58].

8.2 Components of an Effective SOP Template for Waste Management and Recycling

A well-structured standard operating procedure (SOP) serves as the backbone of laboratory waste management systems. It ensures not only compliance with legal and regulatory requirements but also promotes workplace safety, sustainability, and operational consistency [57, 59–61]. The following components are typically included in an effective SOP for waste management and recycling:

8.2.1 Title and Identification

Every SOP should begin with a clear title and unique identification details. This ensures traceability, version control, and alignment with laboratory quality systems.

Content: SOP title, unique SOP number, version number, date of issue, date of next review, and approving authority (e.g., Quality Manager or Safety Officer).

Example: *"SOP-ENV-05: Collection, Segregation, and Disposal of Organic Solvent Waste in HPLC Analysis."*

8.2.2 Purpose

The purpose defines why the SOP is needed, linking it to safety, regulatory compliance, and environmental responsibility.

Example: *"To ensure safe collection, storage, recycling, and disposal of organic solvents used in chromatography while minimizing environmental hazards and complying with local waste regulations."*

8.2.3 Scope

The scope clearly defines the **boundaries of application**—which labs, sections, or processes the SOP covers.

Example: *"Applicable to all analysts handling chromatographic solvents in QC, R&D, and Microbiology laboratories."*

8.2.4 Definitions and Abbreviations

This section reduces ambiguity by explaining technical terms and abbreviations used in waste management.

Examples:

Biohazard: Biological substances that pose health risks.
Neutralization: Process of adjusting acidic or basic waste to a neutral pH before disposal.
Distillation Recovery: Solvent recycling method using fractional distillation.
Segregation Stream: Categorized waste pathways (e.g., organic solvents, halogenated solvents, sharps, recyclable plastics).

8.2.5 Responsibilities

Clear assignment of duties prevents lapses and ensures accountability.

Example:

Analysts: Segregate waste at the source and label containers.
Supervisors: Verify compliance and approve disposal forms.
Safety Officer: Oversee hazardous waste transport and ensure contractor certification.

8.2.6 Procedure (Step-by-Step Instructions)

This is the **core** of the SOP, outlining practical, sequential instructions:

Segregation: Categorize waste into organic solvents, acids, bases, heavy metals, sharps, biological waste, and general waste.
Labeling and Storage: Use of **color-coded bins**: red (biological), yellow (hazardous chemical), blue (recyclable plastics), black (general waste).
Treatment:
Acids/bases → neutralization before disposal.
Organic solvents → distillation recovery systems.
Biological waste → autoclaving or incineration.
Recycling: Recovery of solvents, reuse of uncontaminated glassware, recycling of clean plastics.
Final Disposal: Partnering with certified waste disposal agencies or authorized treatment, storage, and disposal facilities (TSDFs).

8.2.7 Safety and Precautions

This section ensures risk minimization during waste handling.

Required PPE: gloves (nitrile/latex), lab coats, safety goggles, face shields, and fume hoods for volatile solvents.
Emergency Procedures: spill kits, eyewash stations, first-aid kits, and incident reporting protocols.
Safe Handling of Sharps: Use puncture-proof containers.

8.2.8 Documentation and Records

Record-keeping ensures **traceability** and provides an audit trail for compliance.
Waste logs (daily/weekly entries).
Recycling registers.
Certificates of disposal from third-party vendors.
Neutralization and recovery records.
SOP revision history.

Example: *"Each solvent container must be accompanied by a Waste Disposal Form signed by the analyst and verified by the laboratory supervisor."*

Standardized waste segregation charts, disposal forms, flow diagrams, and sample checklists.

8.3 Best Practices in SOP Development and Implementation

In addition to structured templates, adherence to best practices ensures that SOPs are practical and sustainable.

- Clarity and Precision: Use simple, action-oriented language. For example, "Collect waste solvent in amber-colored containers" instead of "Amber containers are preferred."
- Visual Aids: Incorporate flowcharts, pictograms, and color-coded signage for easier understanding.
- Version Control: Maintain revision history to reflect updates in regulations or laboratory processes.
- Periodic Training: Conduct regular training for new employees and refresher sessions for existing staff.
- Accessibility: Store SOPs both in hard copy (at laboratory workstations) and digitally (via Laboratory Information Management Systems, LIMS).
- Integration with Green Chemistry: Promote solvent recycling, substitution of hazardous reagents with greener alternatives, and adoption of microscale techniques.
- Audit and Review: SOPs should be reviewed annually or biennially to ensure compliance with evolving regulations.

8.4 Real-World Applications and Examples

- Pharmaceutical QC Labs: SOPs often detail recovery of acetonitrile and methanol from HPLC waste using fractional distillation. Some labs have installed automated solvent recovery systems that achieve up to 80% recycling efficiency.
- Food Testing Labs: SOPs guide disposal of pesticide residues and heavy metals through licensed hazardous waste contractors, while clean glassware is reused to reduce waste.
- Academic and Research Labs: Universities implement SOPs that integrate teaching modules, ensuring students learn waste segregation alongside experiments.
- Environmental Labs: SOPs mandate separation of mercury, cyanide, and chromium waste, with secure transport to government-approved treatment facilities.

8.5 Benefits of SOP-Based Waste Management

- Ensures regulatory compliance with national and international standards.
- Enhances safety culture within laboratories.
- Promotes sustainability through recycling and waste minimization.
- Provides traceability and accountability during audits and inspections.
- Reduces operational costs by recovering solvents and reusing materials.

9 Case Studies and Examples

The practical implementation of SOPs in analytical laboratories is best understood through real-world examples where structured waste management has significantly improved compliance, safety, and sustainability. The following case studies highlight best practices adopted by different laboratories across pharmaceutical, environmental, academic, and food-testing contexts.

9.1 Case Study 1: Solvent Recovery in a Pharmaceutical Quality Control Laboratory [62]

Pharmaceutical QC laboratories, particularly those relying on HPLC, generate large volumes of organic solvent waste such as acetonitrile, methanol, tetrahydrofuran, and water mixtures. In one documented case, a medium-scale Indian pharmaceutical facility generated 500–700 L of solvent waste per month, which was initially outsourced for disposal at high cost.

9.1.1 SOP Intervention

- An SOP was developed for "Collection, Storage, Recovery, and Reuse of HPLC Solvent Waste."
- Waste was segregated by solvent compatibility: polar (e.g., methanol, acetonitrile) and non-polar (e.g., hexane, toluene).
- Distillation recovery units were installed in compliance with Good Manufacturing Practice (GMP) guidelines.
- Analysts were trained to use dedicated amber containers with hazard labels, as per OSHA Hazard Communication Standards.
- Recovered solvents were tested for purity using GC–FID before reuse in non-critical analyses.

9.1.2 Outcome

- Up to 70% of solvents were recovered with high purity.
- Disposal costs reduced by nearly 45% annually.
- The initiative aligned with ISO 14001:2015 Environmental Management Systems and demonstrated corporate environmental responsibility.

9.2 *Case Study 2: Hazardous Waste Segregation in an Environmental Testing Laboratory [63]*

Environmental testing laboratories often deal with hazardous wastes containing mercury, arsenic, cadmium, and chromium. Improper disposal of such wastes poses severe ecological and public health risks. A case report describes an accredited laboratory in East Asia where heavy metal waste was inadvertently mixed with general laboratory solvents, increasing both treatment costs and contamination risk.

9.2.1 SOP Intervention

- A dedicated SOP for "Heavy Metal Waste Management" was drafted based on EPA Resource Conservation and Recovery Act guidelines.
- Waste segregation was performed using color-coded containers (yellow: toxic metals; red: organics; blue: recyclables).
- Analysts maintained logbooks recording type, volume, and disposal route.
- Waste was transported to a Treatment, Storage, and Disposal Facility (TSDF) in compliance with local environmental laws.

9.2.2 Outcome

- Elimination of cross-contamination incidents.
- Successful accreditation renewal under ISO/IEC 17025:2017, with auditors citing waste management practices as exemplary.
- Cost savings due to reduced need for large-scale hazardous waste treatment.

9.3 *Case Study 3: Green Chemistry Integration in a University Research Laboratory [64]*

Academic institutions also face challenges with laboratory waste due to high student throughput and frequent experimentation. A chemistry department in Europe implemented an SOP incorporating Green Analytical Chemistry (GAC) principles.

9.3.1 SOP Practices

- Microscale analytical techniques reduced reagent volumes by 60–70%.
- Hazardous solvents like chloroform and dichloromethane were replaced with greener alternatives (e.g., ethyl acetate, ethanol).
- Organic solvent waste was distilled and recycled for use in teaching labs.

- Students received training in 12 Principles of Green Chemistry, ensuring integration of sustainability into curricula.

9.3.2 Outcome

- Hazardous waste volumes reduced by over 50% in the first year.
- Students demonstrated higher awareness of sustainable practices.
- The approach directly supported UN Sustainable Development Goal 12—Responsible Consumption and Production.

9.4 Case Study 4: Biological Waste Management in a Food Microbiology Laboratory [65]

Food testing laboratories working with pathogens face biosafety risks if wastes are not properly treated. A study documented a case where a food microbiology laboratory adopted strict SOPs for biological waste.

9.4.1 SOP Intervention

- All biological waste (culture plates, pipette tips, liquid media) was collected in autoclavable biohazard bags.
- Autoclaving was carried out at 121 °C for 30 min, following the WHO Laboratory Biosafety Manual.
- Autoclaved plastics were sent to authorized recyclers, while glassware was sterilized and reused.
- Checklists ensured daily verification of decontamination steps.

9.4.2 Outcome

- Zero biosafety incidents reported in three consecutive years.
- Reduced plastic disposal to landfills by 30% due to recycling initiatives.
- Laboratory successfully complied with national biosafety and food safety audit standards.

Annex: Practical Implementation Roadmap for SOP-Based Waste Management in Analytical Laboratories

Step no.	Step	Key activities	Tools/resources required	Relevant standards
1.	Initial assessment and gap analysis	Conduct a waste audit, categorize waste types (chemical, biological, recyclable, sharps), evaluate current practices, and identify gaps	Waste inventory forms, audit checklist, analytical balance for weighing waste, and hazard assessment sheets	ISO 14001:2015; EPA RCRA (2016); Local Pollution Control Board Guidelines
2.	SOP development and customization	Draft SOPs using a standardized template; define scope, responsibilities, safety, and record-keeping; integrate green chemistry practices	SOP templates, flowcharts, waste segregation guidelines	WHO Biosafety Manual (2020); ICH Q10; ISO/IEC 17025:2017
3.	Infrastructure and resource setup	Establish segregation points, color-coded bins, and a labelling system; install solvent recovery units, autoclaves, neutralization tanks; engage certified disposal agencies	Color-coded containers, autoclaves, solvent distillation units, chemical neutralization kits, storage cabinets, and TSDF contracts	OSHA Hazard Communication (2012); EPA guidelines; CPCB (India) Hazardous Waste Rules
4.	Training and capacity building	Train staff on SOPs, PPE use, waste segregation, and emergency response; conduct refresher sessions annually	Training manuals, posters, digital LMS, safety drills	ISO 45001 (Occupational Health & Safety); WHO Training Modules; OSHA Lab Safety Standards
5.	Documentation and record keeping	Maintain waste logs, disposal certificates, and solvent recovery records; track revisions in SOPs	Laboratory information management system (LIMS), waste tracking registers, disposal forms, and logbooks	ISO/IEC 17025:2017 (traceability); GMP documentation standards
6.	Monitoring and continuous improvement	Perform periodic audits, inspections, track KPIs (% waste recycled, disposal cost savings), and review SOPs annually	Audit checklists, KPI dashboards, corrective action reports	ISO 9001 (Quality Management); ISO 14001 (Environmental Management)
7.	Sustainability integration	Incorporate green chemistry, minimize waste generation, report sustainability metrics, and align with institutional CSR	Green chemistry toolkits, recycling partnerships, and sustainability reporting templates	UN SDG 12 (Responsible Consumption and Production); ISO 14064 (Carbon Footprint)

(continued)

(continued)

Step no.	Step	Key activities	Tools/resources required	Relevant standards
8.	External compliance and certification	Ensure regulatory compliance; pursue ISO 14001, NABL (India), or Green Lab Certification; maintain audit readiness	Compliance calendars, certification guidelines, and external auditor checklists	EPA RCRA; WHO Biosafety Manual; ISO 14001:2015; NABL/ISO 17025

References

1. Nnonyelu P, Dongjie N. Strategies for enhancing solid waste management practices in urban secondary schools in developing countries. Eur. J. Theor. Appl. Sci.. 2024;2:770–86.
2. Poojary H, Selvamuthukumar K, Parveen R, Illanad G, Mandal D, Koo S, Ghosh C. Current advances in solid-phase microextraction technique as a green analytical sample preparation approach. Green Chemistry Letters and Reviews. 2025 Dec 31;18(1): 2486135.
3. Srivastava RR, Rajak DK, Ilyas S, Kim H, Pathak P. Challenges, regulations, and case studies on sustainable management of industrial waste. Minerals. 2022 Dec 28;13(1):51.
4. Bułkowska K, Zielińska M, Bułkowski M. Implementation of blockchain technology in waste management. Energies. 2023 Nov 24;16(23):7742.
5. Ali M, Wang W, Chaudhry N, Geng Y. Hospital waste management in developing countries: A mini review. Waste Management & Research. 2017 Jun;35(6):581-92.
6. Rao TR, Srilaxmi B. Green Analytical Chemistry: A Comprehensive Review of Eco-Scale, Greenness Metrics, and Sustainability Approaches. Journal of Drug Delivery & Therapeutics. 2025 Apr 1;15(4).
7. Abd-Elhafeez M, Arafa MM, Amro FH, Youssef FS. Green analytical chemistry to eco-friendly HPLC techniques in pharmaceutical analysis: a review. Egyptian Journal of Veterinary Sciences. 2024 May 1;55(3):795–801.
8. Mbama CA, Otegbulu A, Beverland I, Beattie TK. Solid waste recycling within higher education in developing countries: a case study of the University of Lagos. Journal of Material Cycles and Waste Management. 2023 Mar;25(2):886–98.
9. Aare FF, Tekaron OA, Ntor GG, Ogiri TO. Strategic management of plastic pollution in Nigeria: Balancing best approaches. International Journal of Civil Law and Legal Research. 2024;4(1):5–14.
10. Farooq M, Muhammad H. Climate Governance and Sustainability Reporting. Corporate Social Responsibility and Environmental Management. 2025 May 21.
11. Javed Y, Alenezi M. A case study on sustainable quality assurance in higher education. Sustainability. 2023 May 17;15(10):8136.
12. Singh H, YT K, Mishra AK, Singh M, Mohanto S, Ghumra S, Seelan A, Mishra A, Kumar A, Pallavi J, Ahmed MG. Harnessing the foundation of biomedical waste management for fostering public health: strategies and policies for a clean and safer environment. Discover Applied Sciences. 2024 Feb 22;6(3):89.
13. Almuhaideb AM, Saeed S. Fostering sustainable quality assurance practices in outcome-based education: Lessons learned from ABET accreditation process of computing programs. Sustainability. 2020 Oct 12;12(20):8380.
14. Fatemi F, Dehdashti A, Jannati M. Implementation of chemical health, safety, and environmental risk assessment in laboratories: a case-series study. Frontiers in public health. 2022 Jun 14;10:898826.

15. Ritter JA, Bibler JP. Removal of mercury from waste water: large-scale performance of an ion exchange process. Water Science and Technology. 1992 Feb 1;25(3):165–72.
16. Munaf E, Zein R. The use of rice husk for the removal of toxic metals from wastewater. Environmental technology. 1997 Mar 1;18(3):359–62.
17. Reddy MA, Gaurav A, Ushasukhanya S, Rao VC, Bhattacharya S, Boopathi S. Bio-Medical Wastes Handling Strategies During the COVID-19 Pandemic. In Multidisciplinary Approaches to Organizational Governance During Health Crises 2023 (pp. 90–111). IGI Global.
18. Harnessing the foundation of biomedical waste management for fostering public health: strategies and policies for a clean and safer environment
19. Aloy A, Amoravian S, Burcl R. Strategy and methodology for radioactive waste characterization. IAEA, Vienna. 2007.
20. Moore R, Pole SB. A comparison of the costs of treating wastes from a radioanalytical laboratory. Lockheed Idaho Technologies Co., Idaho Falls, ID (United States); 1996 Apr 1.
21. Saha L, Kumar V, Tiwari J, Rawat S, Singh J, Bauddh K. Electronic waste and their leachates impact on human health and environment: Global ecological threat and management. Environmental Technology & Innovation. 2021 Nov 1;24:102049.
22. Bulucea CA, Mastorakis NE, Bulucea CA, Boteanu NI, Stinga AU. Systemic approach of hazardous and non-hazardous waste management. In4th WSEAS International Conference on Energy Planning, Energy Saving, Environmental Education, EPESE 2010 (Vol. 10, pp. 181–189).
23. Rau EH, Alaimo RJ, Ashbrook PC, Austin SM, Borenstein N, Evans MR, French HM, Gilpin RW, Hughes Jr J, Hummel SJ, Jacobsohn AP. Minimization and management of wastes from biomedical research. Environmental Health Perspectives. 2000 Dec;108(suppl 6):953–77.
24. Nwachukwu NC, Orji FA, Ugbogu OC. Health care waste management–public health benefits, and the need for effective environmental regulatory surveillance in Federal Republic of Nigeria. Current topics in public health. 2013 May 15;2:149–78.
25. Ilomäki M, Melanen M. Waste minimisation in small and medium-sized enterprises—do environmental management systems help?. Journal of Cleaner Production. 2001 Jun 1;9(3):209–17.
26. Zweckmair T, Hell S, Klinger KM, Rosenau T, Potthast A, Böhmdorfer S. Recycling of analytical grade solvents on a lab scale with a purpose-built temperature-controlled distillation unit. Organic Process Research & Development. 2017 Apr 21;21(4):578–84.
27. Foster BL. The chemical inventory management system in academia. Chemical Health & Safety. 2005 Sep 1;12(5):21–5.
28. Herat S. Waste management training and capacity building for local authorities in developing countries. Waste Management & Research. 2015 Jan;33(1):1–2.
29. Freese T, Elzinga N, Heinemann M, Lerch MM, Feringa BL. The relevance of sustainable laboratory practices. Rsc Sustainability. 2024;2(5):1300–36.
30. Freese T, Kat R, Lanooij SD, Böllersen TC, Roo CM, Elzinga N, Beatty M, Setz B, Weber RR, Malta I, Gandek TB. The relevance of sustainable laboratory practices and Green Labs. Integrating sustainable chemistry into research and industry for a circular economy.:317.
31. Kihila JM, Wernsted K, Kaseva M. Waste segregation and potential for recycling-A case study in Dar es Salaam City, Tanzania. Sustainable Environment. 2021 Jan 1;7(1):1935532.
32. Chepa S, Singh S, Dutt H, Sharma A, Naik S, Mahajan H. A comprehensive study of distinctive methods of waste segregation and management. In2021 third international sustainability and resilience conference: climate change 2021 Nov 15 (pp. 440–444). IEEE.
33. Ugale M, Rane S, Rajadhyaksha S, Ped-nekar P. Importance of waste segregation using a segregating robot. International Journal of Engineering Research. 2021;9(3).
34. Middha R, Srivastava N, Saxena N. Chemical Principles in Waste Segregation and Recycling. In Waste Management for Smart Cities 2024 Dec 19 (pp. 1–46). Singapore: Springer Nature Singapore.
35. Pande H, Varughese JC, Singh K, Sonecha S, Prajapati D, Joshi K, Kumar H, Bhutani S, Singh M. Segregation of biomedical waste: Methodologies and importance. Biomedical Waste Management: Bioremediation and Recycling. 2024 Sep 23;1:65–84.

36. Mehta S, Sabharwal U. Segregation of biomedical waste: methodologies and importance. Biomedical Waste Management: Bioremediation and Recycling. 2024 Sep 23:147.
37. Deng D, Zhang L, Dong M, Samuel RE, Ofori-Boadu A, Lamssali M. Radioactive waste: a review. Water environment research. 2020 Oct;92(10):1818–25.
38. Adam AA, Anyiam FE, Shube MA, Mohamed HD, Ahmed HA, Osman NH. Assessment of Medical Waste Segregation, Disposal Practices for Infectious and Sharps Waste in Healthcare Facilities in Somalia: Implications for Infection Prevention and Control. Infection and Drug Resistance. 2025 Dec 31:3605–15.
39. Albizzati PF, Cristóbal J, Antonopoulos I, Egle L, Foster G, Gaudillat P, Marschinski R, Pierri E, Tonini D. Harmonised labelling of waste receptacles matching product labels. Publications Office of the European Union, Luxembourg (Luxembourg). doi. 2023 Feb;10:09021.
40. Pires A, Martinho G, Rodrigues S, Gomes MI. Sustainable solid waste collection and management. Springer International Publishing:; 2019.
41. Rodrigues S, Martinho G, Pires A. Waste collection systems. Part A: a taxonomy. Journal of Cleaner Production. 2016 Feb 1;113:374–87.
42. González VG, Velasco FE, Candelo-Becerra JE. Strategies for use, treatment, management and final disposition of wastes in academic laboratories. TELKOMNIKA (Telecommunication Computing Electronics and Control). 2020 Dec 1;18(6):3130–41.
43. Beriault* DR, Rodriguez-Capote K, Hauff K, Kinniburgh D, Langman L, Mattman A, Konforte D, Olechuk C, Beach L, Chen Y, Elnenaei M. Reduce, Reuse, Recycle: Top 10 Choosing Wisely Canada's Recommendations for Conserving Laboratory Resources. Canadian Journal of General Internal Medicine. 2022 Nov 18;17(4):86–95.
44. Srishti S, Anilkumar A, Roy Y. A Review of Advancements in Solvent Recovery from Hazardous Waste. Journal of the Indian Institute of Science. 2025 Apr 16:1–5.
45. Chea JD, Christon A, Pierce V, Reilly JH, Russ M, Savelski M, Slater CS, Yenkie KM. Framework for solvent recovery, reuse, and recycling in industries. InComputer Aided Chemical Engineering 2019 Jan 1 (Vol. 47, pp. 199–204). Elsevier.
46. Lee A, Naquash A, Lee M, Chaniago YD, Lim H. Exploitation of distillation for energy-efficient and cost-effective environmentally benign process of waste solvents recovery from semiconductor industry. Science of the Total Environment. 2022 Oct 1;841:156743.
47. Chen C, Yu J, Yoza BA, Li QX, Wang G. A novel "wastes-treat-wastes" technology: Role and potential of spent fluid catalytic cracking catalyst assisted ozonation of petrochemical wastewater. Journal of Environmental Management. 2015 Apr 1;152:58–65.
48. Consonni S, Giugliano M, Massarutto A, Ragazzi M, Saccani C. Material and energy recovery in integrated waste management systems: Project overview and main results. Waste management. 2011 Sep 1;31(9-10):2057–65.
49. Kaya SI, Cetinkaya A, Ozkan SA. Green analytical chemistry approaches on environmental analysis. Trends in Environmental Analytical Chemistry. 2022 Mar 1;33:e00157.
50. Tobiszewski M, Mechlińska A, Namieśnik J. Green analytical chemistry—theory and practice. Chemical Society Reviews. 2010;39(8):2869–78.
51. Armenta S, Garrigues S, de la Guardia M. Green analytical chemistry. TrAC Trends in Analytical Chemistry. 2008 Jun 1;27(6):497–511.
52. Roht-Arriaza N. Developing countries, regional organizations, and the ISO 14001 environmental management standard. Geo. Int'l Envtl. L. Rev.. 1996;9:583.
53. Baer Jr HP. ISO 14000: Potential Compliance and Prevention Guidelines for EPA and DOJ. FoRDHAM ENvTL. LJ. 1995;7:927.
54. Thomas WL. Using IS0 14001 to Comply with the Management System Requirements of the US EPA's RMP Rule and the EU's Seveso II Directive. European Energy and Environmental Law Review. 1997 Dec 1;6(12).
55. Kolln K, Prakash A. EMS-based environmental regimes as club goods: Examining variations in firm-level adoption of ISO 14001 and EMAS in UK, US and Germany. Policy sciences. 2002 Mar;35(1):43–67.
56. Urban AB, Wallingford IE, Cosper SD, Rice AM, Wolf WK, Anderson HG, Wolford M, Gemeinhardt CA, Stokes DW. Standard Operating Procedures (SOP) for Open-Air Solid Waste Burning in Contingency Locations. 2022 Sep 13.

57. Okweso JA. Report on the review of the Kenya national guidelines for safe management of health care waste, injection safety and safe disposal of medical waste national communication strategy and health care waste management standard operating procedures. Nairobi: Ministry of Environment and Natural Resources. 2016:1–81.
58. Agarwal AK. Standard Operating Procedures (sop) For Hospitals In India. Atlantic Publishers & Dist; 2007 Jun 21.
59. Ahmad DM, Gáspár L, Shaheen HM, Al-Shihabi TA, Maya RA, Pinto FS. Improving Material Tracking for Sustainable Construction: A Standard Operating Procedure (SOP) Framework for Resource Efficiency. Buildings. 2025 Jun 4;15(11):1941.
60. Moh Y. Solid waste management transformation and future challenges of source separation and recycling practice in Malaysia. Resources, Conservation and Recycling. 2017 Jan 1; 116:1–4.
61. Rogoff MJ. Solid waste recycling and processing: planning of solid waste recycling facilities and programs. Elsevier; 2013 Nov 18.
62. Abejón R, Garea A, Irabien A. Organic solvent recovery and reuse in pharmaceutical purification processes by nanofiltration membrane cascades. Chemical Engineering Transactions, 43, 1057-1062 DOI: https://doi.org/10.3303/CET1543177. 2015.
63. Okrikata E, Nwosu LC. Heavy metals and pesticides as hazardous wastes and strategies for minimizing their hazards. Int J Res Publ Rev. 2023;2582:7421.
64. Lee NE, Gurney R, Soltzberg L. Using green chemistry principles as a framework to incorporate research into the organic laboratory curriculum. Journal of Chemical Education. 2014 Jul 8;91(7):1001–8.
65. Dogho MO. Sustainable bio-based approaches to food waste management in quality control laboratories. Journal of Scientific Research and Reports. 2025 May 1;31(5):261–72.

MIX
Papier aus verantwortungsvollen Quellen
Paper from responsible sources
FSC® C105338

If you have any concerns about our products, you can contact us on
ProductSafety@springernature.com

In case Publisher is established outside the EU, the EU authorized representative is:
Springer Nature Customer Service Center GmbH
Europaplatz 3, 69115 Heidelberg, Germany

Printed by Libri Plureos GmbH
in Hamburg, Germany